Ludwig Boltzmann

Ludwig Boltzmann at the age of 58, when he was a professor in Vienna.

Ludwig Boltzmann

The Man Who Trusted Atoms

CARLO CERCIGNANI

Dipartimento di Matematica
Politecnico di Milano

Oxford   New York   Melbourne
OXFORD UNIVERSITY PRESS
1998

Oxford University Press, Great Clarendon Street, Oxford OX2 6DP

Oxford New York

Athens Auckland Bangkok Bogota Bombay
Buenos Aires Calcutta Cape Town Dar es Salaam
Delhi Florence Hong Kong Istanbul Karachi
Kuala Lumpur Madras Madrid Melbourne
Mexico City Nairobi Paris Singapore
Taipei Tokyo Toronto Warsaw

and associated companies in
Berlin Ibadan

Oxford is a trade mark of Oxford University Press

Published in the United States
by Oxford University Press Inc., New York

A catalogue record of this book is available from the British Library

Library of Congress Cataloging in Publication Data
Cercignani, Carlo.
Ludwig Boltzmann : the man who trusted atoms / Carlo Cercignani.
Includes bibliographic references and index.
1. Boltzmann, Ludwig, 1844–1906.   2. Atomic structure–History.
3. Physicists–Austria–Biography.   I. Title.
QC16.B64C47   1998   530'.092–dc21   98-17743

ISBN 0 19 850154 4

Typeset by Digital by Design, Cheltenham
Printed in Great Britain by Bookcraft (Bath) Ltd, Midsomer Norton, Avon

*To my daughter Anna,
who helped me through Boltzmann's dense German*

FOREWORD

Our now standard picture of matter, as presented by the physics of today, tells us that ordinary macroscopic materials are made up of *atoms*. Although the essentials of this picture go back to early Greek times, its general acceptance is remarkably recent. From about the middle of the nineteenth century, a gradually increasing number of physicists were indeed coming to accept the reality of atoms, but there were still a great many who regarded this "atomic hypothesis" as merely a convenient fiction which did not reflect any genuine reality at a submicroscopic level.

This "hypothesis" did, nevertheless, enable (non-obvious) macroscopic qualities of substances to be deduced. In principle, knowing the laws governing the individual atoms provides a means for deducing the overall properties of materials. Yet, there is no remotely practical procedure for calculating the behaviour of a macroscopic body from a detailed calculation of the motions of all of its constituent atoms. The number of atoms making up any ordinary macroscopic system is far too enormous. A cubic centimetre of air, for example, contains some 10^{19} atoms. Thus, to deduce how a macroscopic material must behave, according to our standard picture, it is necessary to employ *statistical* arguments. The laws governing macroscopic behaviour are obtained from the forming of appropriate statistical averages of physical parameters of individual atoms. Such procedures were beginning to be worked out by a number of physicists in the second half of the nineteenth century, but the outstanding figure among these was *Ludwig Boltzmann*.

Boltzmann stands as a link between two other great theoretical physicists: James Clerk Maxwell in the nineteenth century and Albert Einstein in the twentieth. Maxwell, who is best known for his discovery of the laws governing electric and magnetic fields and light, first found the formula for the probability distribution of velocities of particles in a gas in equilibrium, but it was Boltzmann who derived the equation governing the *dynamical evolution* of the probability distribution, according to which the state of a gas, not necessarily in equilibrium, will actually change. Boltzmann's ideas were central to Max Planck's later analysis of black body radiation at the turn of the century, in which he introduced the *quantum of action*, thereby firing the opening shot of the quantum revolution. In 1905, Einstein not only picked up on this idea and developed it further (in effect showing that the "atomic hypothesis" applied even to light itself!) but was also influenced by Boltzmann's concepts in two of his other famous papers of 1905, one in which he provided a method of determining molecular dimensions and the other in which he explained the nature of Brownian motion, whereby small particles in suspension in a fluid jiggle around owing to the impact of the molecules that constitute the fluid. Both of these papers gave enormous support to the "atomic hypothesis", leading to the confidence that we now feel in this picture of reality.

The Boltzmann equation was also *mathematically* important, in being the first equation describing the time-evolution of a probability. But it was also fundamental in other ways. It opened up profound issues of physical significance—and even of deep philosophical interest—some of which are only partially resolved even today. For unlike the underlying dynamical equations governing the constituent particles themselves, the Boltzmann equation does not remain unchanged when the direction of time is reversed. The time-asymmetry of the Boltzmann equation arises as an aspect of the *second law of thermodynamics*, according to which the *entropy* of a system out of equilibrium increases with time. The crude meaning of the term "entropy" is "disorder"; so the second law tells us, roughly speaking, that the order in a system is continually being reduced. But it was another of Boltzmann's fundamental contributions to give *precision* to the very notion of entropy, by identifying it with a specific multiple of the logarithm of the volume in phase space defined by the macroscopic parameters specifying the state of the system. Accordingly, Boltzmann showed how the second law could become amenable to precise mathematical treatment.

How can it be that a time-asymmetrical dynamical equation can arise, describing the overall behaviour of some macroscopic system, when its constituent particles all satisfy time-symmetrical laws? Boltzmann thought deeply about these issues, in the face of much contemporary criticism, and realized (correctly) that the origin of the asymmetry must be traced back to a highly special state in the remote past, and must ultimately have its roots in cosmological considerations. However, virtually nothing was known about cosmology in Boltzmann's day, so he was in no position to move that part of the argument much further than this, except for the introduction of some intriguing but speculative ideas. These issues are very much alive today, when a good deal is now known about the overall spatial and temporal nature of the actual universe; for they tell us something very significant about the nature of the universe's "big bang" origin, and about the as yet unknown physical principles which came importantly into play at that crucial moment.

What is the present status of the "atomic hypothesis" to which Boltzmann was so strongly committed? Although this is indeed the presently accepted picture of the submicroscopic nature of ordinary matter, there is now an additional twist, which is provided by the very quantum theory for which Boltzmann's ideas acted as an unwitting midwife. Quantum particles are not like classical ones, in several respects. The statistics that they satisfy is subtly different from that of Boltzmann, which applies to the classical situation where each particle is allowed to have its own identity. Moreover, quantum particles do not have uniquely specified locations or velocities, and collections of them must, strictly speaking, be treated as a holistic ("entangled") whole, rather than as a collection of individuals each of which has a state on its own. Moreover, the distinction between a continuous field and a collection of individual particles is by no means as clear as it was in the classical picture (as Planck's and Einstein's treatment of black body radiation was beginning to reveal). In view of these features, it is a remarkable fact that Boltzmann's "classical" atomic picture works so extraordinarily well under normal conditions. There are, it seems to me, still important unresolved issues in relation to this.

Yet Boltzmann himself was no dogmatist in holding to only one picture of things at the expense of all others. It is hard to imagine that he could have been, in view of the uncertainties and restlessness in his character, as manifested particularly in his later years, and of the depth of that sensitivity which drove him to his final tragic end.

Carlo Cercignani provides us, here, with a most fascinating and authoritative account of Boltzmann's life and scientific influence, and of the effects that Boltzmann's contemporaries had on him and he on them. This is a very valuable account of an important era in scientific history.

Roger Penrose

PREFACE

In this book I have tried to present the life and personality, the scientific and philosophical work of Ludwig Boltzmann, as well as the milieu in which he lived and the contacts he had with the other great scientists of his time. This enterprise, which appears to be beyond one person's abilities, especially if he is not professionally devoted to the history of science, has been made possible by the fact that the subject has attracted many scientists and historians of science. I have consulted their studies, which will be quoted when need be for those who intend to fathom the subject further and make due comparisons. Responsibility for the form in which the material is presented is of course mine.

Boltzmann appears as a singular figure in the history of science. He was recognized by his contemporaries as a great scientist, but he had to fight for his ideas, which were seriously criticized. Frequently this was due to his lack of accuracy in declaring his assumptions and even to the fact that a really innovative scientist does not realize how much he is departing from the accepted theories. Revolutions in science are frequently carried out in a rather conservative way. They are recognized as such by posterity.

It is remarkable that, with a few exceptions, Boltzmann's scientific papers have not been translated into English, whereas this task has been accomplished for other scientists of equal or lesser importance. Because of this, much of Boltzmann's work is known through somebody else's presentation, not always faithful. Yet he was the man who did most to establish the fact that there is a microscopic, atomic structure underlying macroscopic bodies. His work influenced modern physics, especially through the work of Planck on light quanta and of Einstein on Brownian motion. Thus it does not seem an exaggeration to think of Boltzmann as the link between the physics of the nineteenth and twentieth centuries.

This circumstance is nowadays recognized. Boltzmann was the centre of a scientific revolution and he was right on many crucial issues. Yet when Boltzmann's name is mentioned, it is frequently accompanied by an echo of the criticism of his contemporaries. His answers to these criticisms were crystal clear from the viewpoint of the physics of his time and remain essentially accurate. His basic results, when properly understood, can also be stated as mathematical theorems. Some of these have been proved; others are still at the level of likely but unproven conjectures.

As time passes, of course, the cultural world-view changes. What one generation sees as a problem or a solution is not interpreted in the same way by another generation. This may be true for Boltzmann's ideas as well; certainly Boltzmann did not foresee quantum mechanics or general relativity, but he was acutely aware of the fact that there were problems to be better understood in atomic theory and classical mechanics. Some passages in his treatises look like prophecies of future developments.

Boltzmann's discoveries are of both conceptual and practical importance. In spite of this, sometimes people talk about them as problems in philosophy. Now it is true that Boltzmann was also a philosopher, as we shall discuss in Chapter 10 of this book, but he did not philosophize about his own scientific work. He preferred rather to discuss the basic problems of the general theory of knowledge and philosophy of science. In particular, he anticipated Kuhn's theory of scientific revolutions and proposed a theory of knowledge based on Darwin's theory of evolution.

It seems however that today, Boltzmann's basic teaching is forgotten. Most people (even scientists) know about his scientific work through secondary sources only and even write books in which the clearest of his results are surrounded by an aura of mystery. According to a popular view, for some philosophers nothing is better than some obscure but mysterious ideas that nobody really cares about and certainly cannot test, because then they have plenty of room for clever argument. However, this should not be true of scientists and philosophers seriously interested in scientific concepts and theories.

In style, this book has a twofold character: the main text is practically devoid of equations and is written for those who are not specialists in statistical mechanics. Some chapters however have appendices which go deeper into certain technical aspects of the subject, in particular those related to the Boltzmann equation. This material, placed at the end of the volume, may make the book useful as a textbook for a course (or part of a course) on statistical mechanics from a historical but rigorous standpoint (obviously some chapters might then be omitted).

Everywhere I have tried not to be ambiguous, as is unfortunately sometimes the case with many presentations addressed to a wide public. In other words, I have tried to avoid those sentences which seem very profound but, on more serious analysis, turn out to have a double meaning: one of them is correct but trivial, the other seems deep, but unfortunately is not true. This choice of mine may make the reading harder, but hopefully the reader's efforts will be rewarded by the possibility of understanding the views of one of the great scientists who shaped our understanding of nature.

I wish to thank the granddaughter of Ludwig Boltzmann, Mrs Ilse Fasol-Boltzmann, and his grandson, Professor Dieter Flamm, for their permission to reproduce photographs and cartoons from the books they edited (quoted as refs 1 and 2 of Chapter 1, respectively).

I also wish to thank the staff of the Cologne office of Oxford University Press and in particular Sönke Adlung, for their constant help and encouragement during the preparation of this volume.

Milan   C.C.
September 1997

CONTENTS

Contents

FIGURE ACKNOWLEDGEMENTS

FIG. 1.1. Ludwig Boltzmann at the age of 24, when he was a lecturer in Vienna. (Courtesy of Professor Dieter Flamm.)

FIG. 1.2. Ludwig Boltzmann at the age of 31, when he was a Professor in Vienna and was engaged in the explanation of Loschmidt's paradox. (Courtesy of Professor Dieter Flamm.)

FIG. 1.3. Ludwig Boltzmann and his family in 1886. The children are (from left to right) Henriette, Ida, Ludwig, and Arthur. (Courtesy of Dr Ilse Fasol-Boltzmann.)

FIG. 1.4. The farm of Oberkroisbach where Boltzmann and his family lived while he was Professor in Graz. (Courtesy of Dr Ilse Fasol-Boltzmann.)

FIG. 1.5. Ludwig Boltzmann's tombstone with the formula relating entropy and probability, called *Boltzmann's principle* by Einstein. (Courtesy of Professor Dieter Flamm.)

FIG. 1.6. Ludwig Boltzmann at the age of 40, when he was a Professor in Graz and was studying the thermodynamics of radiation as well as what later became known as the theory of ensembles, after Gibbs. (Courtesy of Dr Ilse Fasol-Boltzmann.)

Frontispiece and FIG. 1.7. Ludwig Boltzmann at the age of 58, when he was a Professor in Vienna. (Courtesy of the *Zentralbibliothek für Physik* in Vienna.)

FIG. 1.8. Ludwig Boltzmann lecturing in California, as imagined in a cartoon by K. Przibram. (Courtesy of Professor Dieter Flamm.)

FIG. 1.9. A view of Duino with the New and Old Castle. (From: August Selb, *Memorie di un viaggio pittorico nel littorale austriaco*, A.A. Tischbein, Trieste, 1842.)

FIG. 1.10. A cartoon by K. Przibram showing Ludwig Boltzmann during a lecture. (Courtesy of Professor Dieter Flamm.)

FIG. 9.1. Boltzmann applying the principles and basic equations of mechanics, as pictured in a cartoon by K. Przibram. (Courtesy of Professor Dieter Flamm.)

FIG. 10.1. Boltzmann pondering on the principles of philosophy, as pictured in a cartoon by K. Przibram. (Courtesy of Professor Dieter Flamm.)

Introduction

The existence of irreversible processes is well known from everyday life: one cannot go backwards in time. This holds not only for living beings but also for those objects of macroscopic dimensions which we usually deal with. Everybody laughs when a film is run backwards, and not only because people then walk backwards. Just imagine a film showing a cup of coffee falling from a table, breaking into pieces, and dispersing its contents on the floor, and think of what you would see when the film ran backward. I shall not dwell on this example, since a similar one which goes back to 1874 and is due to Sir William Thomson (Lord Kelvin) is quoted in detail in Chapter 5.

The strange fact is that all the fundamental laws of physics are symmetric with respect to the inversion of the time arrow; the cup which reunites itself from the broken pieces, recovers the coffee from the floor, and jumps back to the table does not violate any law of mechanics.

The man who first gave a convincing explanation of this paradox was an Austrian physicist, Ludwig Boltzmann. Boltzmann was born in Vienna in 1844 and committed suicide in Duino in 1906. He was one of the main figures in the development of the atomic theory of matter. His fame will be forever related to two basic contributions to science: the interpretation of the concept of entropy as a mathematically well-defined measure of what one can call the "disorder" of atoms, and the equation aptly known as the Boltzmann equation. This equation describes the statistical properties of a gas made up of molecules and is, from a historical standpoint, the first equation ever written to govern the time evolution of a probability.

From this equation Boltzmann was able to derive a proof of the irreversibility of macroscopic phenomena. It is the difference in scale between the objects that we observe in everyday life on the one hand, and molecules on the other, that explains this irreversibility through the laws of probability. As a matter of fact, an enormous collection of molecules has an incredibly large number of interactions (collisions) in a dynamics that takes place over extremely small distances (say, one millionth of a millimetre). There are a vast number of possible sequences of interactions (with imperceptible differences between them) that describe the fall and breaking into pieces of the coffee cup, whereas there is essentially only one which describes the inverse process (in which every imperceptible difference would change the entire unfolding

1

of the phenomenon). We never observe certain strange events, not because they are impossible (i.e. forbidden by some physical law), but only because they are extremely improbable.

In the physics that was known when Boltzmann started his career, the fact that we do not witness certain phenomena was ascribed to their impossibility, sanctioned by the famous Second Law of Thermodynamics. Today, following Boltzmann, we hold that this principle states only the extreme improbability of these events.

The thermodynamic measure of the level of probability of a macroscopic state is described by the variable mentioned above, called entropy, related to the probability of the microscopic state by a relationship found by Boltzmann. This relationship, written on his tombstone in Vienna, is not to be confused with the Boltzmann equation, hinted at above. Instead of referring to probability, one can speak of the measure of the disorder of the atoms, because the equivalent disordered states (for a given macroscopic state) are very many and the probability that one of them may occur is extremely high.

Boltzmann's theory tells us that the entropy, i.e. the disorder at atomic level, of the Universe always tends to increase.

We remark that a good understanding of the Second Law of Thermodynamics is related to understanding how life is possible. Our metabolism (from the Greek word μεταβάλλειν), our exchange with the external world, is frequently understood as (and, grossly, it is!) an exchange of matter. Subsequently one thinks of energy (the famous calories). As a matter of fact, for a child who is growing or an adult who is becoming fatter, the exchange of matter is important; and the energy exchange is important to enable us to walk and perform other physical activities (such as eating itself and digestion). But what do we exchange to keep ourselves alive? Not energy, which is consumed in work or sweat (frequently at the same time), but entropy. More exactly, negative entropy; in other words, we get rid of entropy to keep our state well ordered (i.e. in good health). Where does this negative entropy come from? From food, that is, from animals or plants (we exclude, for simplicity, drugs and food produced artificially, where order is obviously introduced by the production process). And whence does the negative entropy of food come? From other food, if it is meat or fish. But in the end we always arrive at plants. Where does the negative entropy of plants come from? The answer is: from the sun, through photosynthesis (which is the way plants "eat").

The sun, this high-temperature source of concentrated energy, radiates low-entropy light which is exploited by plants. Boltzmann, in an address to a formal meeting of the Imperial Academy of Sciences on 29 May 1886, described this important process in the following way:

Between sun and earth [...] energy is thus not at all distributed according to the laws of probability [...]. The general struggle for existence of animate beings is therefore not a struggle for raw materials—these, for organisms, are air, water and soil, all abundantly available—nor for energy which exists in plenty in any body in the form of heat (albeit unfortunately not transformable), but a struggle for entropy, which becomes available through the transition of energy from the hot sun to the cold earth. In order to exploit this transition as much as possible, plants spread their immense surface of leaves and force the sun's energy, before it falls to the earth's temperature, to perform...certain chemical syntheses...The products of this chemical kitchen constitute the object of struggle of the animal world.

But, one might say, if entropy continuously increases in the Universe, it must have been very low at the instant of the big bang that originated the Universe itself. It seems that things are exactly that way. When the Universe was a small ball of primeval fire, the order must have been very high, the entropy very low. Roger Penrose (see *The emperor's new mind*, Oxford University Press, 1989) estimated the probability of such an ordered state to be $10^{-10^{123}}$, i.e. the inverse of a number whose *number of figures* is given by 1 followed 123 zeros!

While Boltzmann's ideas constitute the very foundations of our understanding of the Universe whenever we relate the microscopic description to whatever we see with the naked eye and experiment with every day, there is still a surprising confusion, even among scientists, about the degree of rigour of these ideas. This confusion, which is undoubtedly due to the originality of Boltzmann's vision (and also to some obscure statements in his early papers), today perpetuates the objections raised by his contemporaries. These objections have in fact proved to be not well-founded and are based on misunderstandings of what was actually stated. Nowadays we possess rigorous mathematical theorems that show the meaning and the accuracy of Boltzmann's vision.

When we study questions as deep as those fathomed by Boltzmann, we realize that an understanding of the basic laws of physics, which rule the microscopic aspects of our conception of the world, is not equivalent to understanding the significant aspects of everyday experience. The extremely small size of the basic constituents of matter is such that we cannot immediately obtain from it an image of the world at a macroscopic level. There are hierarchies of structures, and new concepts arise at each level. Even if the real world is made up of atoms (or even smaller particles), it is too difficult to describe what occurs in that world in terms of those basic constituents. What we can do is to establish a bridge between the various levels in order to form a coherent picture; the whole of Boltzmann's work is a masterpiece of this procedure, i.e. how to construct, starting from atoms, a description that explains everyday life. It is thus not surprising if nowadays the Boltzmann equation is used for practical purposes. When an aerospace engineer studies the re-entry of a shuttle, he must take into account that the description of air as a continuous medium, usually adopted in the design of airplanes, is no longer valid in the higher, rarefied atmosphere and he must use the atomistic description provided by the Boltzmann equation. If we want to study the motion of the very minute particles that pollute our atmosphere, we must again, because of the tiny size of these aerosols, abandon the traditional model of air as a continuum and use the Boltzmann equation.

Suitable modifications of the same equation are used to study important phenomena in other fields of modern technology, from the motion of neutrons in nuclear fission reactors and that of charged particles in the fusion reactors of the future, from the radiation produced in a combustion chamber to the motion of the charge carriers of the submicron-size chips nowadays used in electronic equipment, particularly computers. The idea is also spreading into other fields with not so small particles, such as granular materials and road traffic.

Boltzmann would have been pleased by these technological applications. He was very much interested in technology (in particular he predicted the superiority of the aeroplane over the dirigible airship) and more than once paid a handsome tribute to the role of technology in the development of science, as one may see from the two quotations in Chapter 12.

Boltzmann is also known for his ability to popularize science. His book *Populäre Schriften* presents, together with sketches of life in his days, his own conception of the nature of science in general and theoretical physics in particular, in a plain style and many flights of his sense of humour. Memorable from this viewpoint is his long description of a trip to California in 1905 entitled *Reise eines deutschen Professors ins Eldorado*. Here, among other things, he tells of his difficulties in finding wine: to ask where you can find it, he says, is something that produces the same embarrassment as if in a European country you had asked where to find certain girls "whose motto is 'Give me money, I give you honey'". More succinctly, in a letter to his assistant Stefan Mayer he says: "The wine one almost hides like a schoolboy his cigar. This is what they call freedom."

The less humorous and more profound of these lectures were the aspect of his thought which, alongside his own contributions to theoretical physics, attracted the attention of members of the Vienna Circle and of related thinkers such as Wittgenstein. As we shall see, he opposed "idealistic philosophy"; it is interesting to remark that Lenin quotes with approval the views of Boltzmann, who thus became a hero of scientific materialism in the former Soviet Union. Chapter 10 of this volume is devoted to Boltzmann's philosophical views.

Boltzmann met with many difficulties in making his theory understood by his contemporaries; this theme will appear repeatedly in this book. His viewpoint however is increasingly being confirmed and has exerted an essential influence on the development of twentieth-century physics, especially on the work of Planck and Einstein.

In many respects, Boltzmann appears to us as a pioneer of modern physics and of a modern vision of knowledge in which everything appears to be unrelated and independent, autonomous and different, but is in fact subtly tied up with strings that are difficult to take hold of. These are the strings that make reductionism true and false at the same time, and underline the importance of all the forms of knowledge, from physics to literature, from chemistry to psychoanalysis, even if constrained within the frame of a single vision based on a unique culture that many great men, among whom Boltzmann is to be numbered, have contributed to create. It is the culture that is needed to understand the complex world in which we live, in order to avoid fighting against phantoms and windmills and if we are to construct a better life for our children and grandchildren.

1

A short biography of Ludwig Boltzmann

1.1 Youth and happy years

We do not know why Gottfried Ludwig Boltzmann, born in Berlin in 1770, moved to Vienna in his early years to become a maker of musical boxes. There he married and had a son, Ludwig Georg, who became a taxation officer and married Maria Pauernfeind, the daughter of a Salzburg merchant, in 1837. Ludwig Georg's and Maria's eldest son, Ludwig Eduard Boltzmann,was born in Vienna on 20 February 1844. He was to become a well-known physicist. The night of his birth marked the passage from Shrove Tuesday to Ash Wednesday and Boltzmann used to say that his birth date explained why his temper could suddenly change from great happiness to deep depression. Two years later a second child, Albert, was born; he died of pneumonia when he was at secondary school. Ludwig also had a sister, Hedwig. The three children were baptized and grew up in the Roman Catholic religion, professed by their mother, while their ancestry on the father's side was Protestant.

Boltzmann's elementary education took place under a private teacher in his parents' home. His father's salary was not large but was compensated by his mother's fortune; she came in fact from a rather rich family (in Salzburg there is still a Pauernfeindgasse and even a Pauernfeindstrasse) [1]. His father had to move to Wels and subsequently to Linz, where Boltzmann began his studies in the local gymnasium. He was almost always the most proficient in his class and showed great enthusiasm for mathematics and science. Boltzmann later ascribed the deterioration of his sight, from which he suffered in the last years of his life, to the long evenings spent in study by candlelight. In Linz he also took piano lessons from Anton Bruckner. The lessons came to a sudden end when the mother of the future scientist made an unfavourable remark on the fact that the master had put his wet raincoat on a bed. In spite of this, Boltzmann continued to play the piano throughout his life, improving his ability, and subsequently used to accompany his son Arthur Ludwig, who played the violin.

When Boltzmann was fifteen, his father died of tuberculosis. This tragic event left an indelible mark on the boy. At the age of nineteen the future physicist enrolled in the University of Vienna as a student of mathematics and physics. The Institute of Physics had been founded just 14 years before by Christian Doppler (1803–54), the well-known discoverer of the Doppler effect. Because of this effect the frequency of sound changes with the relative motion of the source with respect to the person hearing the sound, as commonly experienced in the change in pitch of the sound from a moving vehicle, as it approaches and then goes past. An analogous effect, also enunciated by Doppler, occurs in optics, but is more complex because of relativistic effects. The institute enjoyed considerable autonomy within the University of Vienna. Andreas von Ettingshausen had just left the position of director to Josef Stefan (1835–93), who was still rather young at that time and was later to become famous for his experimental discovery of the relation between radiant heat and temperature. Stefan was one of the few non-British physicists who were receptive to the idea of local action mediated by a field, the new approach to electromagnetism developed by Maxwell on the basis of the insights and experiments of Faraday (see next chapter). Boltzmann especially appreciated the close contact that Stefan had with his students. On this, many years later, he wrote in Stefan's obituary a sentence frequently quoted: "When I deepened my contacts with Stefan, and I was still a university student at the time, the first thing he did was to hand me a copy of Maxwell's papers and since at that time I did not understand a word of English, he also gave me an English grammar; I had received a dictionary from my father."

Three years after enrolling, Boltzmann got his PhD (he had already published two papers). One might wonder what was the subject of his thesis. The answer is very simple: there was no thesis in the curriculum of studies in philosophy at the University of Vienna before 1872–73.

During the subsequent year (1867) Boltzmann became Assistant Professor. At that time he also became a friend of Josef Loschmidt (1821–95), who was already working at the Institute of Physics. This institute had only a small laboratory in a house at number 15 Erdbergerstrasse, but its members were full of ideas. About this small group, Boltzmann subsequently wrote:

Erdberg has remained for all my life a symbol of honest and inspired experimental work. When I succeeded in injecting a bit of life into the Institute of Graz, I used to call it, jokingly, Little Erdberg. By this I did not mean that the available space was scarce, because it was quite ample, probably twice as much as in Stefan's Institute; but I had not succeeded in equalling the spirit of Erdberg as yet. Even in Munich, when the young PhDs came to tell me that they did not know what to work on, I used to think: "How different we were in Erdberg! Today there is beautiful experimental apparatus and people are looking for ideas on how to use it. We always had plenty of ideas and were only preoccupied with the lack of apparatus."

In 1868 Boltzmann was awarded the *venia legendi* (entitlement to lecture) and, when he was just twenty-five years old, in 1869, he obtained the Chair of Mathematical Physics in Graz, the main city of Styria and today the second in Austria, called "the city of greenery" because of its splendid position on the Mur river, with plenty of gardens, in a landscape of pleasant hills. The University of Graz was then undergoing such rapid development as to reach the level of the most important European universities of that time.

In Graz Boltzmann had become a colleague of August Toepler, the Director of the Institute, who had arrived there just before him. Toepler, a very cordial person, was also very active in the administration. He soon became a great friend and adviser of Boltzmann. After his arrival in Graz, Toepler planned, worked and struggled to establish a new building (soon to become famous), new apparatus and much larger research funds. Thus for the young scientist his arrival in Graz marked the start of a period of intense

FIG. 1.1. Ludwig Boltzmann at the age of 24, when he was a lecturer in Vienna.

scientific activity, which culminated in 1872 in the publication, in the Proceedings of the Imperial Academy of Sciences of Vienna, of the paper with the hardly informative title of "Further researches on the thermal equilibrium of gas molecules". It is in this paper that the celebrated equation called after Boltzmann was introduced. Although this subject will be treated in detail in some of the subsequent chapters, in particular in Chapters 4 and 5, it seems appropriate to devote a few preliminary words to this basic contribution, with which Boltzmann's fame will be forever associated. This equation describes the statistical properties of a gas made up of molecules and is, from a historical standpoint, the first equation ever written to govern the temporal evolution of a probability.

In the same paper Boltzmann was able to derive a proof of the irreversibility of macroscopic phenomena. It is the difference of scale between the objects that we observe in everyday life on the one hand, and molecules on the other hand, which explains this irreversibility through the laws of probability. It turns out that the vast number of molecules contained in a volume of macroscopic size undergo an incredibly large number of collisions in a dynamics that takes place over incredibly small distances, of the order of one millionth of a millimetre. There is an exceptionally large number of possible sequences of interactions which describe the natural processes that occur in nature. They are imperceptibly different from each other; a small modification in the details of the initial state, such as the position or velocity of a few molecules, would not show up as a major difference in our perception of the phenomenon, provided that these changes do not affect our gross image. In contrast, there exists essentially only one sequence which describes the inverse process, formed by our series of sensations backwards in time; each imperceptible difference in some details of the initial state (the previous final state with *inverted velocities*) would drastically change the entire series of our sensations and hence our perception of the phenomenon. We never observe certain strange events, not because they are impossible (i.e. forbidden by some physical law), but only because they are extremely improbable. One can use the following (highly inadequate) example to illustrate this circumstance: if we put a large number of black and white powder grains in a box, accurately placing the black ones in the right-hand half of the box and the white ones in the left-hand half, then, provided they are not tightly packed, if we shake the box in all possible ways, after a while we shall obtain a grey mixture in which the two powders are finely intermingled. On the other hand it would be impossible, starting from the grey mixture and shaking for days or years, to obtain separation of the two powders. There is no mechanical impossibility, it is merely the fact that there are so many more possible positions of the various powder grains that will give a grey appearance, as compared to the much smaller number of configurations in which the grains are well ordered. So it is a matter of probability. The elementary constituents of a gas have no colours but have positions and velocities; the act of shaking the powder grains is replaced by the collisions of the molecules which bring them into more probable states. Boltzmann wrote an equation that allows us to compute the evolution in time of the probability that a molecule will be in a certain position with a certain velocity in the most various situations, from the air in a room to that around a shuttle during re-entry. At first he did not realize what he had accomplished; he thought that he had remained within the boundaries of mechanics, that he was computing actual numbers of molecules, without realizing how much probability was involved. True, he says in the introduction of his paper that he is going to use statistics and probability, but he also says that it would be "erroneous to believe that the mechanical theory of heat is therefore afflicted with some uncertainty because the principles of probability theory are used…It is only doubly imperative to handle the conclusions with the greatest strictness." Now, any use of probability theory certainly involves some uncertainty, as the prototype of any probabilistic law, the so-called "law of large numbers" shows: one can only compute averages and there will be a randomly distributed deviation about these averages. One may surmise that Boltzmann was well aware of this from the beginning. But he also seemed to think that he had obtained a result which, except for these fluctuations, followed from the equations of mechanics without exception. He had actually obtained a result which holds with an unthinkably large probability, but many subtleties (mainly concerning initial data), of which he did not seem to be aware in 1872, are involved in his derivation. As we shall see, the objections of friends and adversaries forced him to reshape his ideas and create another view, which is uncompromisingly new and opened a novel era in physics. A more detailed discussion of this basic point will be given in Chapter 5.

In the physics of that time, the fact that we do not witness certain phenomena was ascribed to their impossibility, sanctioned by the famous Second Law of Thermodynamics. Today, following Boltzmann, we hold that this principle states only the extreme improbability of these events.

The thermodynamic measure of the level of probability of a macroscopic state is described by a variable called entropy; it had emerged, as we shall see in the next chapter, from intricate thermodynamic calculations. The connection between thermodynamics and mechanics was established through the kinetic theory of gases, as we shall see in the third and subsequent chapters. The need for this connection was felt particularly when the use of steam engines had become common and thus heat had to be related to the mechanical conception of Nature erected by many scientists on the foundations established by Galileo and Newton.

In those years, in addition to his theoretical ideas, Boltzmann was also carrying on an experimental study of the relation between dielectric permittivity and refractive index, which he published in 1873.

In his obituary notice for Josef Loschmidt (written in 1895), Boltzmann tells the following anecdote which refers to his early researches:

> At that time I had in mind to do some experiments with spheres made of sulphur crystals. Since there was nobody who could grind these spheres, Loschmidt proposed to do it together while we were queuing to buy tickets at the Burgtheater. He also hoped that the carbon disulphide that we planned to use would succeed in forcing the queuing people to leave.

Thanks to the funds obtained by Toepler, Boltzmann used to take short spells of leave in order to work with Robert Wilhelm von Bunsen (1811–99), the famous chemist, and the mathematician Leo Königsberger (1837–1921) in Heidelberg and with Gustav Kirchhoff (1824–87) and Hermann Ludwig Ferdinand von Helmholtz (1821–94) in Berlin.

His colleagues in Berlin were among the most famous physicists of that time. Kirchhoff is known especially for the extension of Ohm's law relating electric current and potential difference to three-dimensional materials (as opposed to one-dimensional wires), for the identification of Ohm's electroscopic force with the electrostatic potential difference, for celebrated investigations of the electric propagation along a telegraph wire and three-dimensional conductors, for having developed mathematical analogies between perfect fluids and electric conductors, and for his laws on electric networks. He also studied radiant heat, where he introduced the concepts of emissive power and absorption coefficient, showing how their ratio for any material equals the emissive power of a so-called black body at the same temperature. He also worked on the foundations of mechanics; his viewpoint was that physics only describes facts, whereas forces have no deep meaning: they are just a tool capable of giving a simpler description of the laws of motion, as discovered by Newton.

Helmholtz had very wide interests ranging from mathematics and physics to aesthetics, from physiology to psychology. He was born in Potsdam and was the son of a Gymnasium professor. He became a military surgeon until the end of 1848, when he was appointed Assistant of the Anatomical Museum in Berlin, and Teacher of Anatomy at the Academy of Arts. In the following year he went to Königsberg (today Kaliningrad) as Professor of Physiology. In 1856 he became Professor of Anatomy and Physiology at the University of Bonn, in 1859 Professor of Physiology at the University of Heidelberg, in 1871 Professor of Natural Philosophy at the University of Berlin.

It was during his career as a military surgeon that he published his most celebrated essay on "The conservation of force", where "force" had the meaning, then common, of "energy". There is no doubt that this essay imparted a very great impulse to the problem of understanding the role and meaning of this basic principle of physics. In particular, Helmholtz clarified the assumptions that have to be made about a mechanical system in order to ensure that energy is conserved. As a result of his work the principle of conservation of energy became the unfailing guide to organizing physical facts and theories in a clear scheme.

Then he turned to physiological optics and invented the ophthalmoscope. Later he studied the sensation of tone in physiological acoustics, which later he applied to a theory of musical instruments.

He also wrote very important papers on vortex motion in fluid dynamics and made a great effort to provide a comprehensive presentation of the contributions of different authors to electromagnetic theory, so that they could be easily compared.

In January 1872 Boltzmann wrote to his mother:

Yesterday I spoke at the Berlin Physical Society. You can imagine how hard I tried to do my best not to put our homeland in a bad light. Thus, in the previous days, my head was full of integrals...Incidentally there was no need for such an effort, because most of the listeners would have not understood my talk anyway. However, Helmholtz was also present and an interesting discussion developed between the two of us. Since you know how much I like scientific discussions, you can imagine my happiness. Especially because Helmholtz is not so accessible otherwise. Although he has always worked in the laboratory nearby, I had not had a chance of talking much to him before.

From a scientific point of view, he had the greatest possible opinion of Helmholtz, although the behaviour of the Prussian Secret Councillor chilled him. Once Boltzmann said: "On some problems I can talk to just one person, and that is Helmholtz, but he is too distant."

In 1873 Boltzmann could not resist the temptation to accept a chair in Vienna as a professor of mathematics. To be a professor in Vienna was considered in Austria—and still seems to be—the highest possible step in one's academic career, though often it turned out to amount to loss of peace and concentration on one's research. Of course, Boltzmann might have had another motivation, such as the wish to conduct experimental work, and in this respect he was not deceived.

In order to appoint Boltzmann, who was known to be a physicist, to the new position in mathematics, the Vienna faculty argued that although his researches originated in physics, they were also "excellent as mathematical works, containing solutions of very difficult problems of analytical mechanics and especially of probability calculus". The mark of a "decided mathematical talent" was recognized in his use of higher analysis in the theory of heat.

In 1875 Boltzmann's basic equation encountered an objection from his friend Loschmidt. This objection (the "reversibility paradox") and Boltzmann's answer will be discussed at length in Chapter 5.

FIG. 1.2. Ludwig Boltzmann at the age of 31, when he was a Professor in Vienna and was engaged in the explanation of Loschmidt's paradox.

Before leaving Graz, Boltzmann had met his future wife, Henriette von Aigentler, a girl with long blonde hair and blue eyes, ten years his junior, who was living as an orphan in the house of the parents of the composer Wilhelm Kienzl in Stainz, south of Graz. She was a teacher and, after meeting Ludwig Boltzmann, she decided to study mathematics. If Professor Hirzel, the Dean of the Faculty of Philosophy of the Graz University, found the girl's desire to study mathematics and physics a bit strange, it should occasion little surprise. In his eyes a woman's destiny was to cook food and clean rooms, the only basis, he thought, of a solid family life. In the first semester she was allowed just to listen to the lectures, because there were no laws as yet to banish women from universities. When the second semester started, the Faculty had already approved a rule to exclude female students. Then the girl presented a petition to the Minister of Public Education, a former colleague of her father (who had recently died) at the Graz tribunal. The minister exempted her from the rule voted by the Faculty, but in the subsequent semester the problem arose again. Finally, after her engagement to

Boltzmann, the young lady decided to follow Professor Hirzel's advice and learned how to cook in the home of the Lord Mayor of Graz, who had also been a great friend of her father.

Recently, the letters that Boltzmann and his fiancée exchanged during the period of their engagement have been published in a book entitled *Hoch geehrter Herr Professor! Innig geliebter Louis! Ludwig Boltzmann, Henriette von Aigentler, Briefwechsel* [2] ("Illustrious Professor: Dearly beloved Louis: L.B., H. v. A. Correspondence") and edited by their grandson, Professor Dieter Flamm. Some interesting aspects of Boltzmann's life and personality emerge from these letters. Already from the title we see that when their intimacy grew, Henriette used "Louis" in place of "Ludwig"; in turn, Henriette, first addressed to as *Hochgeehrtes Fräulein*, became just "Jetty." The change occurred in the week between 30 November and 6 December: evidence is provided by two letters with these dates.

Boltzmann sent a written offer of marriage to Henriette von Aigentler on 27 September 1875; he thought that this kind of thing was best dealt with in writing. From this letter we learn that even in those days inflation posed problems. After the apologetic sentence "As a mathematician, You will not find that numbers, which dominate the world, are very poetical", Boltzmann describes his financial problems as follows: "Last year my yearly income was 5400 florins. This will be sufficient to support our existence, but if we take into account the huge increase of prices in Vienna, it is not sufficient to offer You many distractions and amusements."*

Boltzmann continues with a description of his views on marriage: "Although rigorous frugality and care for his family are essential for a husband whose only capital is his own work, it seems to me that permanent love cannot exist if [a wife] has no understanding and enthusiasm for her husband's efforts, and is just his maid and not the companion who struggles alongside him."

From another letter written by Henriette on 25 November, we learn that Boltzmann had been offered a position in Freiburg. After claiming that the matter is one of indifference for her ("since, after all, I just want you"), Henriette examines the advantages and disadvantages of the move, since Boltzmann had asked her advice. Among the advantages she mentions less professional engagement (compared with Vienna), with the consequence that he would have more time for his studies and his family. Then there is the matter of becoming director of a physics institute, upon which Boltzmann had not been so explicit; Henriette thinks that this position would please him. The main disadvantage is the salary (about one half of his income in Vienna, which was made up of the salary plus three supplementary allowances). She wonders whether these allowances would be available in Freiburg as well. In any case the lower income would be compensated by a free apartment and the lower prices of foodstuffs. Another advantage would be the attractive situation of Freiburg, near the Black Forest and Lake Constance.

Boltzmann was short and stout. He had curly hair and blue eyes. His fiancée used to call him, according to their grandson Dieter Flamm [3], to whom many of the intimate

---

*Here and in the following chapters of the book we have written You with a capital Y, to underline the fact that in the original German the respectful form *Sie* is used instead of the intimate form *Du*.

details of this biography are due, "sweet fat darling" (*Mein liebes dickes Schatzerl*). Whenever he had to leave her, Boltzmann, who was tender-hearted, was not able to hold back his tears. Whenever a favour was requested from him, he was not able to say no. If a student in a poor financial situation failed in an examination, he felt strongly responsible. In the last years of his life, no student failed an examination with him. He was very conscientious and hence the administrative work, much heavier in Vienna than in Graz, became a big load and almost an obsession for him.

In Graz, the Institute of Physics had just been completed and transformed into an ideal centre for high-quality research in the most advanced physics of the time. August Toepler, who, upon his arrival in Graz, had complained that the physics cabinet contained "for the most part antiquated junk", had already got the state to commit 100 000 florins for a new institute building. Between 1873 and 1876, Toepler bought 28 000 florins worth of apparatus.

However, in a letter written to Boltzmann in 1876, Toepler, who held the chair of general and experimental physics and was the creator of this wonderful place, laments the lack of money. Having expended his energy on the new institute (and his health too, having descended from the second floor to the basement of the institute), Toepler believed that Graz needed a new physicist and, in the summer of the same year, left Graz for Dresden. The chair left available by Toepler in Graz had a strong appeal for Boltzmann, for various reasons. He would have inherited the laboratory of Toepler and taught physics rather than mathematics. The administrative load in Graz was less than in Vienna. We should not forget that Boltzmann wanted to marry and he had difficulties in finding an apartment in Vienna. Last but not least, his future wife was from Graz.

It was not easy for even a Boltzmann to be appointed to the chair available in Graz. Among his rivals too was Ernst Mach (1838–1916), who had already had a chair of mathematics there from 1864 to 1867.

Mach is very famous for his celebrated treatise *Die Mechanik in ihrer Entwickelung historisch-kritisch dargestellt*, which had a great influence on an entire generation of scientists, the most remarkable and famous case being that of Einstein, as well as on the development of the philosophy of science. The main thesis of this treatise, published in 1883, was that the traditional view prevailing in his days, which considered mechanics as the cornerstone of all the branches of physics and wanted to find a mechanical explanation for every physical phenomenon, is essentially a prejudice. In order to prove his thesis, he made use of all the available tools, from psychology to evolutionist biology, together with his detailed knowledge of the old texts and remarkably refined ability to analyse conceptual structures. In addition to a basic analysis of the concepts of mass and force, akin to those of Kirchhoff and Helmholtz, he stressed that the main feature of science is economy of thought; science strives to formulate its principles in such a way as to condense in a few concepts and propositions a wealth of knowledge that accumulates as a result of experimental observations. Thus he was against experimental pictures or models that could go beyond observed facts. For this reason he denied the existence of atoms. As we shall see in Chapter 10, Boltzmann's philosophy was different in many details and was very much in favour of pictures, as models of reality helping us in making discoveries. The economic formulation was, in his view, the knell tolling for some dead branch of science.

FIG. 1.3.   Ludwig Boltzmann and his family in 1886.   The children are (from left to right) Henriette, Ida, Ludwig , and Arthur.

In 1876 Mach was in Prague and wanted to go back to Graz. Hence there followed days of anxiety for the young couple, waiting for their marriage, already fixed for July 17, without knowing whether they should try to find a house in Vienna or in Graz. To increase their worries, there was the danger that their honeymoon would be spoiled by the need to discuss the appointment to the chair in Graz with the Ministry. About this unpleasant situation Boltzmann wrote: "I hate this continuous secret battle; I know much better how to integrate than how to intrigue." Five days before the marriage, the Ministry had yet not decided, but the couple decided to marry anyway. The much longed-for decision also came.

The Boltzmanns spent 14 years in Graz. Except for the last two years, as we shall see, their life was happy there. They had two sons, Ludwig Hugo (1878–89) and Arthur

(1881–1952), and two daughters, Henriette (1880–1945) and Ida (1884–1910). A third daughter, Elsa (1891–1966), was born after the family had left Graz. The only sad event was the death of Boltzmann's mother in 1885. These happy years are to be contrasted with the subsequent events in Boltzmann's life, which were marked by restlessness and a desire to move and change.

In those years Boltzmann, honoured and respected by the academic community and the government, developed his ideas on the statistical conception of nature. In order to help him, a position of *Extraordinarius* (equivalent to an Associate Professor of today) was created and offered to Arthur von Ettingshausen (1850–1932) (a nephew of Andreas, whom we met before), who took care of practically all the administrative matters. He is the discoverer of the thermomagnetic effect that carries his name and co-discoverer of the Nernst–Ettingshausen effect (a galvano-thermomagnetic effect).

In 1878 Boltzmann became Dean of the Faculty, in 1881 Government Councillor, in 1885 a member of the Imperial Academy of Sciences, in 1887 Rector of Graz University, and in 1889 Court Councillor, not to speak of the long list of academic recognitions that were bestowed on him by foreign countries.

In addition to his chair, which had been Toepler's, there was another, of theoretical physics, that had been his own during his first stay in Graz. This position of full professor was held by Heinrich Streintz, a very respectable and distinguished person, who remained in the shadow of his famous colleague. He was not very active; he published some ten articles on different topics and a monograph on the foundations of mechanics, which was fairly widely known for a while until it was outshone by Mach's treatise, which we have already mentioned. Boltzmann never had problems with Streintz, to whom he gave all his personal and financial support when needed.

Boltzmann loved nature very much. He used to take long walks, during which he was always in good humour and gave explanations of botany to his children; these walks, together with ice-skating in winter, were also meant to make up for his infrequent physical exercise in his youth. For this reason he also installed gymnastic apparatus in his house and insisted that his children used it. Although, like other scientists of the Institute, he had a flat for his family in the *Physikalisches Institut*, Boltzmann had also bought a farm near Oberkroisbach, with a commanding view over a large part of Styria, and lived in the country with his family. He knew the plants well, had a herbarium and possessed a collection of butterflies. This lover of nature could look at the wonderful scenery near his farmhouse from his working room. His neighbours were farmers. He had an Alsatian dog, who each day came down from the farm at noon, waited for him outside the Institute and accompanied his master to a nearby pub, where he lay at Boltzmann's feet during his lunch. It is also said that Boltzmann himself drove a cow, which he had just bought, through the roads of Graz, after consulting his colleagues in zoology about the best way to milk her.

During this time Boltzmann was invited to court on several occasions. He was however a slow eater, perhaps because of his myopia. On the official occasions, the Emperor Franz-Joseph barely touched his food and the court etiquette did not allow his guests to continue their meal when the Emperor had finished his own. Our scientist was much disappointed when the waiter carried away his dish so quickly that he had barely

FIG. 1.4.   The farm of Oberkroisbach where Boltzmann and his family lived while he was
Professor in Graz.

succeeded in sampling the delicious food. But this was certainly not the reason why he
refused the title of nobility that was offered to him. He used to say: "Our middle-class
name was good enough for my ancestors and it will be for my children and grandchildren
as well."

Boltzmann was very fond of his children. Here we can report an anecdote that refers to
a later period of his life. His youngest daughter wished to have a pet monkey, but his wife
did not like to have animals at home. Boltzmann hit on the compromise of buying a few
rabbits for his daughter, and installed a cage for them in his own library. The daughter's
bedroom was near his study and in the evening Boltzmann used to knock lightly at her
door as a sign of affection.

We have already mentioned that Boltzmann liked ice-skating and walking. His
hobbies also included swimming. He also liked to spend evenings with friends, and
the parties lasted till late hours. These receptions at his house were quite frequent and
among the guests were his PhD students as well. He liked to socialize and he was a
welcome guest at parties because he entertained everybody with his exceptional sense
of humour. On these occasions he used to write jocular poems, among which was one
entitled "Beethoven in Heaven" (*Beethoven im Himmel*), which we shall say more about
later (Section 1.9).

FIG. 1.5. Ludwig Boltzmann's tombstone with the formula relating entropy and probability, called *Boltzmann's principle* by Einstein.

Boltzmann was very knowledgeable about classic German literature, which he liked to quote. He dedicated his book *Populäre Schriften* ("Writings addressed to the Public") to Friedrich Schiller (1759–1805), whose poetry is permeated by the concept of personal freedom and never separates the ethical from the artistic message. Schiller was the poet he preferred, whereas among composers Boltzmann liked Ludwig van Beethoven most. He liked to play the Beethoven symphonies on the piano in the arrangements by Liszt. Together with friends and his son Arthur he frequently played chamber music. He also liked to attend concerts and was a subscriber to the Opera House in Vienna.

At the university he had plenty of funds, considerable space, the help of Ettingshausen, a couple of PhD students, and visitors, and every opportunity to pursue his own experimental interests, which arose from either his theoretical problems or his own choice. Even the duties as a Dean were not heavy, since whenever possible they were left to the Vice-Dean. These were idyllic circumstances. Boltzmann was in contact with the most famous physicists of those days: Hendrik Antoon Lorentz (1853–1928) (who discovered an error in the proof of irreversibility based on the Boltzmann equation in the case of polyatomic gases; see Chapters 5 and 8), Helmholtz (whom we have already met), Wilhelm Ostwald (1853–1932) (see Chapter 11), and the successor of Maxwell to the Cavendish professorship in Cambridge, John William Strutt (Lord Rayleigh by inheritance, 1842–1919). In spite of this, the few letters written in that period indicate that somehow he felt he was far from the centres of modern science, where he might have enjoyed more contacts and discussions. However, he confessed to Toepler, with a certain amount of satisfaction, that marriage makes one lazy, more so than he had expected.

It is perhaps for this reason that his successor Leopold Pfaundler, with all his respect for Boltzmann, declared that he had found a pigsty in the building of the Institute. It seems that the famous scientist, kept busy by his researches, did the least possible amount of work as a university professor.

In 1877 he published his paper "Probabilistic foundations of heat theory", in which he formulated what Einstein later called the *Boltzmann principle*; the interpretation of the concept of entropy as a mathematically well-defined measure of what one can call the "disorder" of atoms, which had already appeared in his work of 1872, is here extended and becomes a general statement. As we have already mentioned, in physics prior to Boltzmann, the effects of the disordered motions of atoms were studied without even mentioning the atomic structure of matter, in a discipline called thermodynamics; there was a rather mysterious quantity, the aforementioned entropy, that played an important role whenever certain processes were not allowed. Thanks to Boltzmann's work it turned out that entropy is none other than the measure of the level of probability of a macroscopic state, which can be related to the probability of the microscopic state, describing the world of molecules, by a relation found by Boltzmann (not to be confused with the Boltzmann equation, which we hinted at before) and written on his tombstone in Vienna. Rather than probability, one can speak of a measure of the disorder of the atoms, because the equivalent disordered states (for a given macroscopic state) are very many and the probability that one of them occurs is extremely high. We shall discuss this paper in detail in Chapter 6.

In 1884 Boltzmann, who had heard about the work of the Italian physicist Adolfo Bartoli on radiation pressure, was stimulated by it and gave a brilliant theoretical deduction of Stefan's law of radiant heat, according to which the energy irradiated by a source is proportional to the fourth power of its absolute temperature (see Chapter 9). In the same year he also wrote a fundamental paper, generally unknown to the majority of physicists, who by reading only second-hand reports are led to the erroneous belief that Boltzmann dealt only with ideal gases; this paper clearly indicates that he considered mutually interacting molecules as well, with non-negligible potential energy, and thus, as we shall see in Chapter 7, it is he and not Josiah Willard Gibbs (1839–1903) who should be considered as the founder of equilibrium statistical mechanics and of the method of ensembles. In another paper, in 1887, Boltzmann discussed the mechanical analogies of the Second Law of Thermodynamics.

In 1886, deeply impressed by Hertz's experimental verification of the equivalence between electromagnetic waves and light, predicted by Maxwell's theory, Boltzmann spent considerable effort in redoing Hertz's experiments. These are documented in the last publication he wrote before leaving Graz (see Chapter 9).

FIG. 1.6. Ludwig Boltzmann at the age of 40, when he was a Professor in Graz and was studying the thermodynamics of radiation as well as what later became known as the theory of ensembles, after Gibbs.

At variance with what he did later, Boltzmann avoided discussions on philosophical aspects of science and knowledge during his Graz years.

In a relatively short time Boltzmann's approach to kinetic theory had become widely known, especially in Great Britain, as shown by the circumstance that H.W. Watson's little book on the kinetic theory, published in 1876, made use of Boltzmann's methods. He had already been celebrated as one the fathers of the kinetic theory of gases in a biography of Maxwell written in 1882. A scientific dispute with two famous British colleagues of his, P.G. Tait and W. Burnside, in 1885–7 contributed considerably to improving his relations with the British physicists. Since not many people dared to read his lengthy papers, it was through these discussions that Boltzmann laid the foundations for his international reputation, which grew perhaps earlier in England than in the German-speaking world. Scientific distinctions and awards kept accumulating.

Summarizing, we can say that during his second stay in Graz, Boltzmann became one of the great names of the physics of his times. This very circumstance may be considered the cause of one of the events which changed his life, as we shall see in the next section.

1.2 The crisis

In a few months the idyllic situation and the fruitful period of work described in the previous section suddenly changed. The problems started in January 1888 and got worse until they developed into a major physical and psychological crisis in the months between May and July that year.

What had happened? One can identify a series of unpleasant facts, none of which by itself is sufficient to explain a crisis of such proportions. Taken all together, however they can explain it, since they produced a pile-up of problems that Boltzmann was not able to overcome.

The first documented psychological crisis actually goes back to 1885, most probably related to the death of his mother. As such, it could be interpreted as a natural reaction for a man of 41 who had lost his father at the age of 15 and was deeply affectionate towards his only surviving parent. We must however report that in 1885 he wrote no scientific paper and there is no record of any letter written during that year.

A further problem originated in the previously mentioned fact that Boltzmann was elected to be the university Rector: an honour, but an onerous duty as well, for which Boltzmann was not prepared. On 22 November 1887, the pro-German students of Graz took away the bust of the Austrian Emperor from one of the halls where they used to have their parties and made anti-Habsburg speeches. Boltzmann, as Rector, was obliged to take disciplinary measures against the students, while he was closely watched by the Governor of Styria, the central administration in Vienna and the Emperor himself. This produced a state of tension in him, due to an excess of responsibility, especially because these activities lasted more than four months, continuing until the spring of 1888.

Another fact that, with its consequences, caused him problems was the death of Gustav Kirchhoff, which occurred in Berlin on 17 October 1887. Boltzmann commemorated him in Graz on 15 November and, in the last days of 1887 and the early ones of 1888, was invited to Berlin, where he was offered the opportunity to become his successor and a colleague of Helmholtz. He left immediately, visited the institutes and the laboratories, accepted the offer, and even chose his rooms in the Department. The relevant contract had already been signed by the Kaiser in March 1888, when Boltzmann asked for it to be cancelled. One may surmise that the rather formal manners of the Berlin academic world placed him in some difficulty, since he was accustomed to the relaxed atmosphere prevailing in Graz. For instance, Frau Helmholtz is reported to have said to Boltzmann when, after bargaining for his position at the Ministry, he was having a dinner with his colleagues: "Professor Boltzmann, I am afraid you will not feel at ease here in Berlin". But this is certainly no explanation for his change of mind.

Without doubt, as remarked by Höflechner [4], there must have been other reasons besides the atmosphere in Berlin. In fact, all the dealings related to his appointment in

Berlin should have remained secret, but they became known in the middle of January and had the consequence that Heinrich Streintz made his move: he asked for more space and more money, since he thought that these requests should not offend his Director or anybody else, since Boltzmann was to leave for Berlin.

On 1 February, Albert von Ettingshausen was appointed to a temporary post of Professor of Physics at the Polytechnic (*Technische Hochschule*) of Graz. This school originated in 1811 and reached a level close to that of a University from about 1872. In 1888 there was a vacant chair of physics and Ettingshausen was thought to be the best candidate for it. He was indeed appointed to that chair a few weeks later. When Ettingshausen left, Boltzmann lost his main support in his own Department, quite a severe blow in his situation.

When he came back from Berlin, he realized what he had done and started to think that he had been too daring. It was not normal, at least in those days, to accept an appointment in a different country without first asking for permission. As a consequence, he did not have the courage to inform the Austrian authorities that he had already accepted the proposal of an appointment in Berlin. But, although no official move was made, his intention was no secret to anybody. And Austria did not want to lose one of its most famous scientists. Intense negotiations began, to persuade him to stay in Graz, where he was still the Rector of the University. This forced him into a sort of double game, something that, as we already know from his pun on integrals and intrigues, was not exactly his forte and soon brought him to a nervous breakdown. During the spring of 1888, Boltzmann found himself facing a painful dilemma, since he did not dare to clarify his difficult situation. As underlined by Höflechner [4], the situation is now very clear from a couple of official letters that he wrote to the Prussian and Austrian administrations, as well as from official letters and coded telegrams with new salary offers and other written documents, preserved in the files of those administrations [5].

Boltzmann remained in a horribly confused and painful state for months. When he received the letter of official appointment to Berlin in March, he tried to exploit his myopia, saying that he would have difficulty in occupying the chair in Berlin because of this defect; but the answer from Berlin was that they would receive him with sincere understanding and attention and that they did not want to dispense with him. Thus Boltzmann's proposal to refuse the appointment did not have the desired effect. In the end, Boltzmann was obliged to break his pledge and write to renounce the chair in Berlin, without ever having begun his service.

A passage in the second letter sent by Boltzmann to explain his position [5] reads as follows:

By starting my activity in Berlin, I would enter into a new arena, mathematical physics. In the last 15 years I have lectured on only the basic notions of mathematical physics and the introductory concepts of differential and integral calculus [...] So far, I have however neglected almost completely many broad and significant chapters of mathematical physics. When I was in Berlin, it seemed to me, in my initial enthusiasm, much easier to remedy this inadequacy of mine. Now however, on the point of actually starting this new activity, I realize that it would strain my eyes too much. On the other hand, my conscience would not permit me to start in a new job, in this position of high responsibility, without complete experience in the whole area for which I have been appointed.

This letter appears very strange to us, if we think of the well-deserved fame that Boltzmann enjoyed for his familiarity with all subjects of the research carried out in his days. One may however think, in agreement with the interpretation of Höflechner, that the words of the letter that we have just quoted reflect a very serious and intimate aspect of Boltzmann's personality. They give us the portrait of a man full of anxieties and desire for perfection: when he was not able to reach the level of the picture of himself that he had developed, he was gripped by feelings of fear, suffering, and depression.

On 9 July 1888 the Kaiser cancelled the appointment.

1.3 Restlessness

While it is easy to describe in a few words the events that occurred between January and July 1888, it is not so simple to realize the change that they must have introduced into Boltzmann's life and personality. It was then in fact that his neurasthenia, his tendency to develop a manic-depressive syndrome, was observed. The time of tranquillity and beauty had ended; that of dissatisfaction and restlessness had begun. Boltzmann began to hesitate in his decisions, to look for changes of place and university. This may appear strange if interpreted from a simplistic psychological viewpoint: since he had escaped the grand temptation of Berlin, why should he not remain in Graz? But this is exactly an indication of his syndrome: he had refused Berlin for reasons that were not clear even to him and so he might have started to wonder whether he was not at the level of the great scientists, to whom, in a more or less conscious way, he might have compared himself. Perhaps, however, Graz was not important enough for him. These are presumably the thoughts that made him restless.

The death from appendicitis of his first son Ludwig in 1889, at the age of eleven, added new sorrow to the other problems. It is to this fact that the most traditionalist biographers attribute his depression. What is true is that Boltzmann reproached himself that he had not realized how serious Ludwig's case was, and had relied on the wrong diagnosis of a general practitioner. This sad event certainly contributed to increasing his sense of insecurity and isolation.

First of all he tried to leave Graz, which for him had become an unpleasant place for many reasons: the change in the attitude of his colleagues, provoked by his own behaviour, the idea that he deserved a more important position, the memory of his lost son. He even began to have problems with Streintz and his other colleagues and, already at the end of 1888, he was writing to Helmholtz to tell him that he was re-established in health and was again interested in the chair in Berlin which he had renounced a few months before.

He then informed all his colleagues in every university of his desire to leave Graz for another position. The person who understood the situation and exploited it was Eugen Lommel, professor of experimental physics in Munich, who, with the help of the famous chemist Adolf von Baeyer (1835–1917), the future Nobel prize winner, in 1890 obtained an appointment for Boltzmann to a chair of theoretical physics at the University of Munich. This was a new position there, and not very common elsewhere, and thus the faculty had to justify it. The carefully written argument pointed out that there was an "increasing separation of theoretical from experimental physics", due to their "difference in methods": "while experimental physics in its inductive work requires the knowledge and practice of experimental techniques which become increasingly complicated, theoretical physics uses mathematics as its main tool in its deductive process and demands intimate familiarity with all means of this quickly advancing science." The argument continued by pointing out that the rapid growth of physics would have the consequence that fewer and fewer physicists would be able to master both methods with the same perfection, or in other words, they would be obliged to specialize in one of the two branches. Boltzmann, because of his outstanding talent for theoretical research and "his most thorough mathematical education", was able "to develop further and to supplement the theories of Maxwell, Clausius, and Helmholtz." The philosophical faculty also mentioned his work in several areas which had earned him the reputation of being one of the best theoretical physicists.

Boltzmann was eager to go, exactly for the same reason that the faculty wanted him, and, by combining the salaries of two lapsed professorships, they were able to afford him.

When Boltzmann left Graz there was a solemn farewell party in the traditional academic style, on July 16. The new Rector, J.A. Tewes, and Boltzmann's colleague, H. Streintz, delivered speeches, in which they expressed the hope that Boltzmann would some day come back to work in Austria. Boltzmann replied with a speech, later inserted in his *Populäre Schriften*, entitled "On the significance of theories" [6]. Boltzmann began his reply by saying:

When some days ago I learnt of the plan for today's ceremony, it was at first my firm intention to ask you to refrain. For how, I asked myself, can an individual deserve being honoured in this way? Surely, all of us are just collaborators in a great enterprise, and everyone who does his duty in his post deserves equal praise. If therefore an individual is singled out from the community this can in my view never be aimed at him as a person but only at the idea that he represents; only by completely giving himself over to an idea can the individual gain enhanced importance.

Therefore I decided not to insist on my request only when I related all honours not to my own modest self but to the idea that fills my thought and action: the development of theory, for whose glory no sacrifice is too great for me; since theory is the whole content of my life, let it likewise be the content of my present words of thanks.

Since this talk is important in explaining Boltzmann's viewpoint on what is a theory we shall return to it in a subsequent chapter of the book (Chapter 10). Here we shall restrict ourselves to quoting the final part of the speech [6]:

If at the outset I have declared myself an advocate of the theory, I will not deny that I have myself experienced the evil consequences of its spell. Yet what should be more effective against this spell, what could drag us back more forcefully into reality than the living contact with so honourable a gathering as this present one? For this kindness that you have shown me I thank you all: first you, Rector, who organized this ceremony, next, the orator, colleagues and guests who followed his call, and finally the gallant sons of our alma mater, whose strong endeavours and noble enthusiasm were my support through 18 years. May Graz university grow and flourish and always be and remain what is highest in my view: a stronghold of theory!

In Munich Boltzmann was finally able to teach the subject that was dearest to his heart.

Having taught experimental physics for fourteen years, he used mechanical models to illustrate theoretical concepts in a very lively way. For instance, to visualize Maxwell's electromagnetic theory, he invented a machine called a *Bicykel* (the literal English translation as "bicycle" would be inappropriate, so we stick to the German name). This was a clever model used to illustrate, in mechanical terms, the mutual influence of two electric circuits. He had entrusted the building of the machine to Herr v. Gasteiger, master mechanic in Graz. On this apparatus, Ehrenfest in Boltzmann's obituary, written in 1906, comments that the clarity of movement and forces must have been an aesthetic pleasure for him. Arnold Sommerfeld (1868–1951), the great mathematical physicist who made important contributions to the early formulations of quantum mechanics and to the mathematical theory of electromagnetic waves, remembers this strange apparatus in a celebration lecture held in Vienna in 1944, on the occasion of the centenary of Boltzmann's birth:

This model, which even worked well, was made to Boltzmann's order; it was kept in my former department in Munich with all due reverence. However, it was used less for electrodynamics than for mechanics, i.e. to make it understandable how the differential gear in cars, which is completely analogous to Boltzmann's *Bicykel*, works.

We remark in passing that in the same talk Sommerfeld expresses the opinion that, for the atomistically oriented mind of Ludwig Boltzmann, quantum theory would have been the true playground.

There existed two specimens of the *Bicykel*, one at the University of Graz, the other at the University of Munich. In an inventory of the Department of Physics of Graz University dated 1914, the *Bicykel* is succinctly described in this way: "0 776: inexplicable apparatus with toothed gearing after Boltzmann". Both specimens were lost during the war. In his lectures on Maxwell's theory, Boltzmann wrote about his *Bicykel* "This instrument was made by Herr v. Gasteiger, mechanic; it is of first quality, and experiments carried out with it were entirely satisfactory."

A replica of the *Bicykel* was built by another master mechanic of Graz University, Herr Kurt Ansperger, on the occasion of an exhibition devoted to Boltzmann in 1985 [7].

In Munich Boltzmann used to meet once a week at the *Hofbräuhaus* to discuss academic questions over a glass of beer with some colleagues. Among them we find the mathematicians Walther Franz Anton von Dyck (1856–1934), a charming man and an enthusiastic teacher, expert in function theory, topology, and potential theory and one of the founders of the *Encyklopädie der mathematischen Wissenschaften*; and Alfred Pringsheim (1850–1941), famous for his studies of a power series on the points of the circle of convergence; the physicists Lommel and Leonhard Sohnke (1842–1937); the chemist Baeyer; the astronomer Hugo von Seeliger (1849–1924); and the cryogenics expert Carl von Linde (1842–1934), who invented the first apparatus to liquefy air. Urged on by this company, Boltzmann held lectures on mathematics, especially the theory of numbers. He spent four quite peaceful years, during which many students from all the countries of the world came to study under his guidance. At first, Boltzmann lived in Maximilian Strasse, which was convenient for both the university and the opera house, where he could listen to the works of one of his favourite composers, Richard

Wagner (1813–83). The *Evangelimann* by Wilhelm Kienzl was also greatly enjoyed by the Boltzmann family.

The only disadvantage was that at that time, the Bavarian government did not provide the university professors with a pension. Since his sight deteriorated more and more, Boltzmann started to worry about the future of his family. He remembered his father's early death and the example of the blind Georg Simon Ohm, who died without a pension in the most miserable circumstances. Already Frau Boltzmann would regularly read aloud scientific writings to spare her husband's eyes. In the introduction to the first volume of his treatise "Lectures on Gas Theory", which was to appear a few years later, he expresses this preoccupation: at the World Fair in Vienna in 1873 Professor Wroblewski had already asked him to write that book, but he had declined the offer then because he was afraid that he would soon lose his sight. It was certainly not the blunt answer "All the more reason for hurry!", however, that finally presuaded him to write the book in 1896, but rather the attacks to which the theory was exposed, as we shall see later.

Furthermore, already in 1892, he began to feel homesick; in a letter to Loschmidt in October of that year, we read that, yes he lived, "but certainly not better than in dear old Austria". Three months later his famous master, Josef Stefan, died and his colleagues in Vienna began manoeuvres to install Boltzmann in the place of his illustrious predecessor. Boltzmann hesitated and simultaneously tried to reinforce his position in Munich, while keeping the door open to the possibility of going back to Vienna. In the end the University of Munich persuaded him to stay by offering certain advantages, among which were a substantial increase in salary, a title, and an assistant. Soon after, however, Boltzmann informed his Viennese colleagues that he felt obliged to stay in Munich for just one year. If Vienna had materialized, he would have accepted.

We have an unusual witness for Boltzmann's years in Munich, Hantaro Nagaoka (1865–1950), one of the most widely known promoters of physics research in Japan, in an interesting letter written to *Toyo Gakuzgei Zasshi* (*Asian Journal of Science and Arts*) in April 1894. At that time Nagaoka was less than 30 years old, but two years later he became professor at the University of Tokyo. A translation of his letter from Japanese into English, due to Setsuko Tanaka, is available (see [5]). It contains some opinions on the city and the University of Munich as well as on Boltzmann's reputation and personality. The passage concerning Boltzmann is worth quoting. The first part explains that Nagaoka moved to Munich just because Boltzmann was lecturing there. This is a clear indication of how famous Boltzmann was. But it also strikes a note on Boltzmann's oddness:

Munich is called a city of art, just like our Kyoto. Though there is (normally) not much going on in science, since Professor Boltzmann was invited to give lectures here recently, I moved to Munich at the beginning of April to hear his lectures. Fortunately, he is going to give his reputed lectures on the "Kinetic Theory of Gases" and the "Application of Hamilton's Principle to Physics" in this summer semester. We hear that Professor Boltzmann was invited to Berlin University to succeed Kirchhoff, but he did not accept the invitation and chose to come to Munich. I do not know why he did so.

I think that no one can be as competent as he, perhaps, except for Helmholtz. His lectures are extremely transparent; he speaks lucidly, not like Helmholtz who speaks rather awkwardly. But he is a little odd fellow and sometimes ends up doing unintelligent things.

Nagaoka continues by explaining his interest in Boltzmann's lectures and also gives us a summary of their contents. In particular, he has some comment on the principle of equipartition of energy:

Professor Boltzmann is of Austrian origin. But he admires Maxwell and seems to have Maxwell's attitude in many respects. Such a situation is convenient for me since I was trained in the British tradition. I am especially interested in the Gas Theory which has been developed by him, Clausius and Maxwell. In particular, from the lectures of Boltzmann himself, I can clearly understand the Maxwell–Boltzmann doctrine of the distribution of energy which has recently been controversial.

At about the same time Nagaoka wrote to a certain Tanakadate (presumably not the famous volcanologist, who was only ten years old at that time) a letter in which he expresses similar opinions, though in different words. We quote again from the translation by Tanaka (see [5]):

Professor Bol is a man of bushy beard as you know. Students are much impressed by his features. But his lectures are surprisingly clear in contrast with those of Professor Hel. He appears to have an excellent brain, for he explained Hamiltonian functions or six-fold integrals without seeing any notes.

After informing his correspondent about the rather small number of students attending Boltzmann's lectures, Nagaoka says:

Professor Bol is gentle and honest, and has a personality to be loved by everybody rather in contrast with his features.

A laboratory building is under construction; it is about half the size of that in Berlin, but appears to shake less easily. Professor Bol guides only one student in experimental physics.

In 1894 Boltzmann became PhD *honoris causa* at the University of Oxford, where his ideas enjoyed great prestige.

In the spring of that year, his choice between Munich and Vienna changed every day, until he accepted the appointment in Vienna, where he arrived in the month of June. He was then the most important Austrian scientist and the preparations to receive him and persuade him to stay were impressive. The *Philosophische Fakultät* cancelled a chair of chemical physics, which had already been requested, because the money was needed to increase Boltzmann's salary. He was also promised the full pension in case of his becoming unfit to work, with the earlier periods of service in Austria being taken into account.

1.4 Scientific debates and travels

Boltzmann's initial euphoria over the impressive efforts made to bring him back to Vienna did not last long. After having been away from Vienna for 18 years, he did not find there as pleasant a group of friends and colleagues as in Munich. Especially when Ernst Mach, violently hostile to an atomic picture of nature, became professor of philosophy and history of science in 1895, Boltzmann found that he was not in the best of all possible places. We have already discussed Mach's philosophy and, in particular, the fact that he denied the existence of atoms. For Boltzmann's psychological weakness, it was too much to have quite a famous colleague who openly fought the very theory to which he had devoted his entire life. There was no open quarrel between Mach and Boltzmann, but it is clearly true that Mach's personality was quite different, distant and cool, sometimes displaying a condescending attitude towards Boltzmann in areas not related to physics. For example, in a letter to the philosopher Gomperz, Mach wrote: "Boltzmann is not malicious, but incredibly naïve and casual...,...he simply does not know where to draw the line. This applies to other things too, which are important for him."

In those years Boltzmann, who as we have seen, had always liked scientific discussions and had the attitude of a modern scientist in thinking that isolation damages the scientific progress, had to oppose implacably the enemies of the atomic theory of matter. And he was certainly not short of arguments. We shall talk about the theory of "Energetics" in Chapter 11. Here we shall just mention that it was a rather systematic in principle, though somewhat naïve in practice, application of Mach's ideas. If we cannot refer to things which we cannot observe, we may try to reduce everything to exchanges of energy, because this is what we can really measure and feel, in one way or another, and base on this idea the entire foundations of physics. This in short was the tenet of what was called energetics. Though this concept may sound reasonable (especially after Einstein's discovery of the equivalence of mass and energy), this programme was carried out in a way that actually impoverished the tools of theoretical physics and its ability to predict new phenomena.

We should not be surprised to see that Boltzmann was a strong adversary of energetics. For instance, in the lecture by Sommerfeld previously quoted, we can read the description of a debate that occurred in a meeting held in Lübeck in 1895 between Ostwald, who denied the existence of atoms, and Boltzmann:

The champion for Energetics was Helm; behind him stood Ostwald, and behind both of them the philosophy of Ernst Mach (who was not present in person). The opponent was Boltzmann, seconded by Felix Klein. The battle between Boltzmann and Ostwald was much like a duel of a bull and a supple bullfighter. However, this time the bull defeated the toreador in spite of all his agility. The arguments of Boltzmann struck home. We young mathematicians were all on Boltzmann's side; it was at once obvious to us that it was impossible that from a single energy equation could follow the equations of motion of even one mass point, to say nothing of those for a system of an arbitrary number of degrees of freedom. On Ostwald's behalf, however, I must mention his remark on Boltzmann in his book *Grosse Männer* (Leipzig, 1909, p. 405); there he calls Boltzmann "the man who excelled all of us in acumen and clarity in his science".

Whereas the physicist Georg Ferdinand Helm (1881–1923) is remembered only for having been a rather stubborn champion of energetics, Felix Klein (1849–1925) was a genial mathematician, particularly remembered for his contributions to topology, the theory of analytic functions, and the applications of group theory to geometry. He is the author of the famous Erlangen Programme and conceived the *Encyklopädie der mathematischen Wissenschaften*, a great enterprise that was realized under his direction. His name is attached to several mathematical objects, among which we shall recall only the Klein bottle, a typical example of one-sided surface (a sort of Möbius strip without boundaries).

Boltzmann suffered much from the fact that he felt isolated and thought that his ideas received little attention in Germany. Characteristic of the way he felt about this is the frequently quoted sentence contained in the first paper in which he answered the attack made by Zermelo on the H-theorem (see Chapter 5): "Zermelo's paper shows that my writings have been misunderstood; nevertheless it pleases me for it seems to be the first indication that these writings have been paid any attention in Germany."

When Boltzmann came to Vienna in 1894 as director of the physics institute, he normally would have had responsibility for directing laboratory courses, but he asked to be excused from this duty. Franz Exner, professor of physical chemistry, was obliged to conduct the physics laboratory course. This produced a curious if not chaotic circumstance, because Exner could not handle both duties.

Boltzmann soon realized that his return to Vienna was not a happy move. In a letter to Ostwald dated 13 December 1898, he made no secret that there were "far fewer students ready for scientific work" than in Germany, and there were few scientific meetings and societies and no scientific stimulation. One year later, his wife added that his activity in Vienna had the "character of a schoolmaster's drilling of candidates in secondary school education". This of course did not do justice to his talent or his aspirations. Similar concepts are expressed by Boltzmann in letters to his former colleagues in Munich. He also said that he was also dissatisfied with political conditions in Austria.

Among the many trips abroad that Boltzmann made to increase his scientific contacts, the three longest were those to the United States. In the first of these, in 1899, he was accompanied by his wife. They embarked in the steamer *Kaiser Wilhelm der Große* of North German Lloyd in Bremen and travelled to New York via Southampton and Cherbourg. He gave four talks on the principles and basic equations of mechanics at Clark University in Worcester (Massachusetts), which was celebrating its tenth anniversary. There he also received a PhD *honoris causa*.

During the trip the Boltzmanns wrote six letters to their children. The first was written on the ship and deals mainly with Henriette's seasickness, which prevented her from attending a nine-course dinner. They were quite impressed with New York: "The jostling of the electric trams and steam trains on, above, and below the street is quite splendid. They travel extremely fast. It is really quite dangerous." The couple found that Boston was terribly dusty. Other places they visited were Montreal, Buffalo, Washington, Baltimore, and Philadelphia.

In 1900 Boltzmann decided to accept an appointment as Professor of Theoretical Physics in Leipzig, where the physical chemist Wilhelm Ostwald had built up a great research centre and the faculty recommended him as the "most important physicist in Germany and beyond". Although when he had learned that this appointment was very probable, he had written to Ostwald that he had come out of his "depressed mood", the actual decision to accept the appointment strained him and caused him such a serious nervous breakdown that he had to spend a short time in a psychiatric hospital to recover.

Leipzig, too, did not suit him. Though Ostwald as a person was a friend of Boltzmann's, he was also the father of energetics, which, as we said, is based on Mach's philosophy and adverse to Boltzmann's ideas. The scientific debate between them became so bitter that Mach himself thought that the argument was becoming too violent, and proposed a reconciliation of mechanistic and phenomenological physics.

Fɪɢ. 1.7. Ludwig Boltzmann at the age of 58, when he was a Professor in Vienna.

Thus in Leipzig Boltzmann found himself in an even worse situation than in Vienna. The constant struggle depressed him, since he rather preferred teamwork. In addition, for some reason he did not like the lifestyle in vogue in Saxony and made one attempt at suicide. When asked by a colleague why he felt unhappy, he was unable to say why. One may speculate about the marshy climate or the North German Protestant customs, or some other reason. What is certain is that he asked the Saxon government to release him for "reasons of health". Even before leaving Vienna, however, he had started to negotiate a return.

In 1901 Mach retired because of a stroke. Since during his voluntary exile in Leipzig his chair had remained vacant, Boltzmann went back to Vienna in 1902. But the Establishment did not immediately forgive his previous flight. The Minister for Research and Education did not have an easy time in explaining the problems of Boltzmann's personality to the Emperor Franz-Joseph and silencing the rumours circulating in Vienna, according to which Boltzmann was mentally ill and would not be able to perform his duties as a professor. The Minister had to consult the psychiatrist and doctors who had taken care of Boltzmann before, and he had to give his word in writing that he would not try to leave Austria again. Fortunately he had by then become more a symbol of the scientific independence of Austria than a professor.

Life in Vienna soon began to show its problems too. In a letter to his assistant Stefan Mayer [5], Boltzmann expresses his preoccupation with the big cuts in the finances of the institute. In addition, he had to wait a couple of years before he could be re-elected as a full member of the Imperial Academy. He was offended by this circumstance, which was related to the fact that he had resigned from the academy on taking up his chair in Munich; the Emperor Franz-Joseph himself was against his immediate reappointment. In February 1903, his wife wrote to their daughter Ida, who had stayed in Leipzig to finish her gymnasium course: "Daddy gets worse every day. He has lost his faith in our future. I had imagined a better life here in Vienna." His sight had deteriorated to such a point that he had to pay a lady to read scientific papers for him and it was his wife who wrote down his own papers.

He suffered from bad attacks of asthma during the night and perhaps of angina pectoris as well. He was also disturbed by bad headaches caused by overwork. His teaching duties at that time were: five hours a week for a course of theoretical physics, a seminar on the same subject, and a course of at least one hour a week, every third semester. Since 1903, in addition, Boltzmann taught for two hours a week the philosophy course that had been given by Mach. It is easy to imagine that these heavy duties, together with his scientific activity, were more than his already fragile health could stand.

There had also been a misunderstanding over the control and use of the physics apparatus, because of the curious situation created by his previous stay in Vienna and the subdivision of duties between him and Exner. Boltzmann had to give up some of the experimental apparatus, but the ministry assured him that he would be given money to buy what he had lost and still needed. Boltzmann actually asked for and obtained money twice, in 1902 and 1903.

The lectures on philosophy were the most popular lectures by Boltzmann, who, when appointed to that duty, had imagined that philosophy was his true calling. The first one was a great success. Although the largest hall had been chosen, people were standing even on the steps. The hall was adorned with twigs of silver fir and he received enthusiastic ovations. All the newspapers reported the event. He received a great number of letters of approval. He was also invited to have an audience with the Emperor Franz-Joseph, who told him how happy he was that he had come back to Austria and how he had heard about the crowd present at his lectures. After two or three sparkling talks, however, his enthusiasm diminished and with this the audience as well. This resulted in a sense of failure. From the letters written in those years to Franz Brentano (1838–1917), who had had to leave his chair of philosophy in Vienna because he had been a Catholic priest and then married, one understands the importance that those lectures had for him. He also went to Florence to meet Brentano, with the attitude of someone who is going to see his psychiatrist.

He continued his evening social activity, referred to above. In spite of this, in his last years of life he was frequently still working at 5 a.m. In this circumstance too we can recognize a symptom of a manic-depressive syndrome. Those who suffer from it tend to wake up early and work intensively if they are in the manic phase, whereas the depressive

FIG. 1.8. Ludwig Boltzmann lecturing in California, as imagined in a cartoon by K. Przibram.

phase brings, together with sleepiness and a reluctance to act during the day, a lack of sleep in the early morning accompanied by unpleasant thoughts and plans for suicide.

Nevertheless, in October 1904 Boltzmann participated in a meeting in St Louis (accompanied by his son Arthur Ludwig, as on a cruise in the Mediterranean undertaken in 1901 with the intention of improving the health of the scientist). This time the trip, from Hamburg to New York on the packet steamer *Belgravia* of the Hamburg America line, lasted ten days and was extremely uncomfortable, especially because of the continual wailing of the foghorn that prevented sleep. They also visited Detroit and Chicago. The return took place on the *Deutschland*.

In 1905, travelling alone, Boltzmann went for the last time to the United States, where he gave 30 lectures at a summer school held at the University of California at Berkeley. He had been invited there together with Ostwald. Of this last trip he gave an entertaining description, entitled *Reise eines deutschen Professors ins Eldorado* [6] (a translation of this essay is included at the end of this book).

During his long stay in the United States in July 1905 he suffered from attacks of asthma; this had not happened on his previous journeys. He wrote about this to his assistant Stefan Mayer, whom we have already mentioned:

Fig. 1.9. A view of Duino with the New and Old Castle.

FIG. 1.10.   A cartoon by K. Przibram showing Ludwig Boltzmann during a lecture.

What I have seen is very interesting and I hope to see even more interesting things, but all of this exhausted and tired me. Above all because the climate of California is not so pleasant as one frequently hears. The changes from a hot subtropical heat to cold weather, from a dry weather to fog, are hard to bear for a European. In addition, the rain-water, kept in large cisterns from the winter, upsets my stomach. The wine one almost conceals as a schoolboy does his cigar. This is what they call freedom.

We can read about Boltzmann's travel to *Eldorado* in the local newspapers of the time. The *Daily Californian* (Vol. XXV, no. 17, p. 1) says:

Professor Boltzmann ranks very high among the physicists of the world and stands in the same class with Arrhenius, the Swedish physicist, and De Vries, the celebrated Dutch botanist, both of whom were members of the faculty in the last summer session.

and a little later:

The success of President Wheeler's efforts to secure eminent scientists and professors to instruct in the summer sessions has been greatly augmented by the acceptance of his invitation by Dr. Ludwig Boltzmann, professor of theoretical physics in the University of Vienna.

In his field of research Dr. Boltzmann has attained a pre-eminence that classes him with the world's greatest scientists. He has done for mathematical physics what Professor Arrhenius has done for physical chemistry.

After telling his readers that "Dr. Boltzmann is the greatest living exponent of the atomic theory", the newspaper gives further information about him:

Dr. Boltzmann is such a deep and abstruse thinker that the known mathematics is not advanced enough for some of his higher works, so that he has to evolve his own mathematical formulae.

A similar report can be read in the 29 June issue of *The Record*:

Dr. Boltzmann arrives for Summer Session—Will conduct an advanced course during the six weeks—Is one of the most prominent Physicists of Europe—Has been connected with several universities in Germany over thirty years.

Dr. Ludwig Boltzmann occupies the same rank in the world of scientists as Arrhenius and De Vries, the eminent physicist and botanist who were the stars of the Summer Session last year. He was invited to come to California last year, at the time when he came to this country to take part in the International Congress of Arts and Sciences at the St. Louis fair. He could not accept the invitation then, but promised to come this summer. This is his first visit to the Pacific Coast, his busy career having kept him on the other side of the Atlantic. During the last thirty years Professor Boltzmann has been connected with the universities of Graz, Munich, Leipzig and Vienna, his connection with the latter institution dating from 1902.

Boltzmann was very proud of his English, as we learn from his report on the journey. Yet the comments of people attending the lectures express some reservations about his command of the language. One comment has that it was "somewhat deficient, to put it mildly". "Oh! I learn 'heem' tonight" is another comment. Or more explicitly:

It is reported that if he had lectured in German most of his audience might have been able to follow him. But he did lecture, four times a week, in (so-called) English, on the subject "Mechanical Analogies of Thermodynamics with Special Reference to the Theorems of Statistical Mechanics".

According to Höflechner [4], "the folks in Berkeley till today are a little disgruntled about Boltzmann's behaviour in Berkeley, which was regarded as a mixture of manic ecstasy and the rather pretentious affectation of a famous German professor."

1.5 The tragic fate of a great scientist

Boltzmann's life had a tragic end in 1906. At that time he was spending a few days with his wife and youngest daughter in Duino, a village near Trieste, famous for its castle perched on a rock, at the point of a promontory on the Adriatic sea, having the sea on one side, deep forests of cork oak on the other. The castle is associated with the poet Rainer Maria Rilke, who was there several times and wrote his *Duino Elegies*. Empress Elizabeth (the famous "Sissi" of a film series), wife of Franz-Joseph, and the composer Franz Liszt also stayed there.

Boltzmann had never worried about his health, but had sacrificed it to his scientific activity. When however even that vacation in Duino did not bring any relief from his illness, in a moment of deep depression he committed suicide by hanging on 5 September 1906. The next day he should have gone to Vienna to start his lectures.

Is it possible to explain this sudden decision? How can it be reconciled with the humorous description of his trip to California, written a few months before? He was probably exhausted and disappointed by the visit to the United States. His lectures in Berkeley, as well as those on the philosophy of science in Vienna, did not prove as successful as he had anticipated. As we shall say in the next section, Boltzmann was considered a great teacher and his audiences got the impression that he was deeply interested in the subject of the lecture and happy to give it. But we know from his assistant Stefan Mayer that lecturing was painful for him and whenever there was a possibility of cancelling a lecture, he was more relieved than disappointed. When he gave a lecture he was like an actor and very ambitious to do his best. But, like an actor who feels that he has failed, he could also become very sad and depressed. In the case of his lectures on the philosophy of science there might have been uneasiness and fear as well.

A few last spasms of euphoria surfaced when he wrote the description of his trip to California. As is typical of the syndrome from which he was suffering, these spasms were followed by the deepest possible depression. We may also add that suicide was fairly common among Viennese intellectuals in those days, as we shall see in a later section of the present chapter.

There are a few documents and witnesses that allow us to reconstruct some aspects of the last months and the last hours of Boltzmann's life. They indicate some deep and some superficial reasons for his deed. Even superficial motivations should not be discarded, given the seriousness of the syndrome from which he was suffering. Some of them have already been made known to a wide public in a book devoted to irreversibility [8].

In May 1906 the Dean of the Philosophical Faculty of the Imperial-Royal University of Vienna wrote a letter to inform the competent Ministry that Boltzmann was suffering from a serious form of neurasthenia and had to abstain from any scientific activity [5].

Here is a particularly accurate description of his problems with his eyesight: "Those who knew Boltzmann will remember the pair of heavy highly-powerful spectacles resting on a deep groove in his nose. For many years his eyesight had been failing, and he found it increasingly difficult to complete the many researches which were on his mind." [9]

Let us now consider some impressions of the problems related to Boltzmann's physical and mental health. Referring to the latter part of the year 1905, an anonymous journalist says: "In the fall of last year the scholar decided to visit a mental hospital near Munich, but he left the asylum a short time later and returned to Vienna." [10]

Ludwig Flamm was a student of Boltzmann's and married his daughter Ida. He was one of the first to discuss an intuitive meaning of the 3-D cross-sections of Schwarzschild's space–time, the most widely known exact solution of the equations of Einstein's theory of general relativity. He says [11]:

I, myself, as a student was able to hear the last lecture which Boltzmann held on theoretical physics; it was in the winter semester 1905–1906. A nervous complaint prevented him from continuing his teaching activity. Together with another student I took and passed my oral examination in his Villa in Währing. On leaving, after the examination was over, we heard from the front hall his heartrending groans.

Alois Höfler says, referring to the year 1906: "When I visited him during the Easter holidays for the last time, he expressed his physical and mental suffering thus: 'I never would have believed that such an end was possible.'" [12]

Coming to the summer of the same year, we learn from no less than Ernst Mach [13] that

Boltzmann had announced lectures for the summer semester [1906], but had to cancel them, because of his nervous condition. In informed circles one knew that Boltzmann would most probably never be able to exercise his professorship again. One spoke of how necessary it was to keep him under constant medical surveillance, for he had already made earlier attempts at suicide.

When he was in Duino, where he had gone to satisfy an old wish of his wife Henriette ("in Venezien" we find in one of the early letters by Henriette to him), Boltzmann was, according to an article in the newspaper *Die Zeit* of 7 September 1906, "upset and nervous because he was anxious to return to Vienna. His condition otherwise seemed better. On the day of his death Boltzmann showed himself particularly excited. While his wife and daughter went swimming, he carried out the deed." [14]

Here is another piece of information provided by Frau Dr Lili Hahn: "He was very melancholy for a long time and did not want to send his suit to be cleaned because it would mean a further delay in returning to Vienna. After his wife left, taking the suit with her, he hanged himself." [15]

Of the manner in which he carried out the deed, we can learn from *Die Zeit* again: "He used a short cord from the crossbars of a window casement. His daughter was the first to discover the suicide." [14]

As for the place where Boltzmann hanged himself, there are at least two versions. It seems likely that it was his hotel room. Frau Auguste Dick, however, claimed that a colleague of Boltzmann's, the mathematician Mertens, reported that he committed suicide in the church of Duino [5].

Henriette outlived her husband for many years. She died on 3 December 1938 of a lengthy illness, aged 85.

Among the people who were shocked by the news of Boltzmann's death, was Erwin Schrödinger, then about 19 years old. He had expected, as W. Moore says in a recent biography [16], to begin his studies in theoretical physics within a few months under the great master. Schrödinger himself describes his feelings that autumn, when he entered the physics building. In 1929, when giving his inaugural talk at the Prussian Academy of Sciences, he said:

The old Vienna Institute, from which shortly before Ludwig Boltzmann had been torn away in a tragic fashion, the building where Fritz Hasenöhrl and Franz Exner and many other Boltzmann's pupils went in and out, engendered in me a direct empathy for the ideas of that powerful spirit. For me his range of ideas played the role of a scientific young love, and no other has ever again held me so spellbound.

Another person who was frustrated by Boltzmann's suicide was Ludwig Wittgenstein (1889–1951), who, when leaving Linz in 1906, had hoped to learn from him [17]. Wittgenstein is famous for his *Tractatus Logico-Philosophicus*, published just after World War I, in which, according to Janik and Toulmin [18], he tries to solve the problem of reconciling Hertz's and Boltzmann's physics and philosophy of science with the ethics of Kierkegaard and Tolstoy.

Concerning the death of Boltzmann in Duino, Moore [16] aptly quotes from the first of the *Duino Elegies* of Rainer Maria Rilke (1875–1926):

> *Aber die Liebenden nimmt die erschöpfte Natur*
> *in sich zurück, als wären nicht zweimal die Kräfte,*
> *dieses zu leisten*

(But lovers are taken back by exhausted Nature / into herself, as though such creative force / could never be re-exerted) [19].

This quotation probably renders the feelings of Boltzmann when he decided that life is harder and heavier than the weight of anything else. He was a deceived lover who had devoted himself to atomic theory, but his love was not reciprocated because his contemporaries were not able to understand his great vision. But if we look at his suicide with hindsight we can perhaps see him as a hero rather than as a lover and recall the two lines that just precede those quoted by Moore:

> *denk: es erhält sich der Held, selbst der Untergang war ihm*
> *nur ein Vorwand, zu sein: seine letzte Geburt*

(Consider: the Hero continues, even his fall / was a pretext for further existence, an ultimate birth.) [19]

1.6 Boltzmann as a teacher

Boltzmann was an excellent teacher. He had an amazing memory and always lectured completely without notes. He was a born teacher and his lectures were crystal-clear, witty and humorous, full of stimulating anecdotes. He livened up the lecture by using unusual expressions such as "gigantically small". In his philosophy lectures he also spoke about matters that were then exotic, such as multidimensional and curved spaces. A student rag referred to this in the following couplet [1]:

> *Tritt der gewöhnliche Mensch auf den Wurm, so wird er sich krümmen;*
> *Ludwig Boltzmann tritt auf; Siehe, es krümmt sich der Raum!*

(When a normal human being treads on a worm, it will curl up; / when Ludwig Boltzmann treads the stage: Look, space curls up!) This joke illustrates the enormous effect created by Ludwig Boltzmann's outstanding personality and his lectures.

In Vienna he had among his students Paul Ehrenfest (1888–1933), Fritz Hasenöhrl (1874–1915), Stefan Mayer, and one of the future discoverers of uranium fission by slow neutrons, Lise Meitner (1878–1968); in Graz, Svante August Arrhenius (1859–1927) and Walter Nernst (1864–1941). Hasenöhrl (whose most notable paper anticipated in 1904 the equivalence of mass and energy, though with an erroneous factor) writes about him: "He never exhibited his superiority. Anybody was free to put him questions and even to criticize him. The conversation took place quietly and the student was treated as a peer. Only later one realized how much he had learned from him." His colleague in Graz, H. Streintz, writes: "He gave advice in any difficult situation. He was not upset even if a student disturbed him at home when he was working. The great scientist remained available for hours for his student, always keeping his patience and good temper." On his lectures Lise Meitner said:

He gave a course that lasted four years. It included classical mechanics, hydrodynamics, elasticity theory, electrodynamics, and the kinetic theory of gases. He used to write the main equations on a very large blackboard. By the side he had two smaller blackboards, where he wrote the intermediate steps. Everything was written in a clear and well-organized form. I had frequently the impression that one might reconstruct the entire lecture from what was on the blackboards. After each lecture it seemed to us as if we had been introduced to a new and wonderful world, such was the enthusiasm that he put into what he taught.

The seminar talks by Boltzmann were also exceedingly stimulating, as witnessed, for example, by a statement in Ehrenfest's thesis.

After Boltzmann's death, lectures in theoretical physics were suspended for eighteen months, until Hasenöhrl was appointed to the vacant position. His inaugural lecture was a masterly synthesis of the statistical theories of Boltzmann and an exposition of the philosophy of the great master.

1.7 Boltzmann and inventions

Boltzmann was fascinated not only by scientific problems, but also by promising scientific inventions. Although he was an eminent theoretician, as we have mentioned and shall discuss in more detail in Chapter 9, he was also a down-to-earth scientist, with a special skill and interest in experimental work, perhaps inherited from his grandfather, who, as we said at the beginning of this chapter, was a musical-box maker.

He was deeply interested in technology and more than once praised the role of technology in the development of science, as one can see from a couple of passages in his *Populäre Schriften* [6] quoted in Chapter 12. In Boltzmann's view, science should not pride itself on the ideal character of its goals and look down somewhat contemptuously on technology and practice, since it owed its rise to a striving for satisfaction of purely practical needs. To this he added the fact that the successes of science would not have been so brilliant, had not science found in technologists very capable forerunners. Further, even logic is a consequence of technological achievements. If we had not attained them, we should not know how to infer. Only those inferences are correct that lead to practical success. It is clear that Boltzmann is making use of his Darwinian view of the development of our brain as a tool for producing world pictures. This organ, because of the great utility of these pictures for the preservation of the species, has developed in man to a particular degree of perfection. That this was not just a philosophical pose is shown by some facts to be mentioned presently.

When in 1902 Nernst invented his electric lamp (now practically forgotten, and made of a small cylinder of oxides of rare earths heated by a small platinum spiral), he sent some specimens of it to Boltzmann, who wrote these verses on one of them:

> *Da Du den sprödesten Stoff Dir gewählt*
> *Und ihn zwangst, den elektrischen Strom zu leiten,*
> *Schufst Du das glänzendste Licht*

(Since for yourself you chose the most refractory of material / and oblige it to carry an electrical current, / you're able to create the most glittering light.)

He also invited the members of the Viennese Physical Society to a home demonstration of the lamp. Fifty-five invitations were sent out; many sandwiches and 50 litres of beer were ready for the guests, but only seven came. This indicates that his enthusiasm was not always shared by his colleagues.

Boltzmann also encouraged Kress, who in 1880 had invented an aeroplane. In 1894 he even gave a lecture on this invention at the *Gesellschaft Deutscher Naturforscher und Ärzte* (German Society of Scientists and Physicians). In particular, Boltzmann predicted the superiority of the aeroplane over the dirigible. Unfortunately Kress shared the fate of many inventors, not just in Austria: he ran out of money and was forgotten.

Boltzmann was rather optimistic about the possibility of human flight, and made a comparison with one of his favourite heroes, Columbus. To make this comparison appear stronger he also quoted the verses by Schiller on Columbus [6]:

> *Zieh' hin, mutiger Segler, mag auch der Witz dich verhöhnen,*
>
> *Mag der Schiffer am Steuer senken die mutlose Hand,*
>
> *Immer, immer nach West, dort muß die Küste sich zeigen,*
>
> *Liegt sie doch schimmernd und liegt deutlich vor deinem Verstand.*
>
> *Mit dem Genius steht die Natur in ewigem Bunde.*
>
> *Was der eine verspricht, leistet die andre gewiß.*

(Advance, proud sailor, do not pay attention to the jokes about you. / Let the lazy sailor give up the tiller! / Advance toward the west, always toward the west! / It is there that the coast will appear. Look at it, shining in your head! / With the genius stays Nature in eternal bond; / what the former promises, the latter certainly keeps.)

Boltzmann himself, using his experimental skills, built an electric sewing machine for his wife, who used it to make clothes for all the family.

1.8 Ludwig Boltzmann and his times

Boltzmann lived in a rather unusual period for his country and Europe. For some centuries, Austria had been a leading country in Europe, rivalling England and France: Vienna was a great capital at the level of Paris and London. It was a metropolis for those days, since it numbered about two million inhabitants at the beginning of the century. But the positioning of Austria as well as of Europe as the centre of the world had begun to fade at the end of the eighteenth century. Two facts, not entirely unrelated—the American and French revolutions, both children of the highest point of European culture, the Enlightenment—had changed the world. The former provided the impetus for the formidable rise and expansion of the United States; the second, with the remarkable follow-up of Napoleon and his wars, changed Europe irreversibly. In 1815 the Congress of Vienna re-established the *Ancien Régime*, but this was an unstable solution that was bound to collapse sooner or later. The new map of Europe, so close to the one that existed before Napoleon, was the brainchild of a genial diplomat and statesman, Prince Metternich, who had succeeded in the seemingly impossible game of marrying an Austrian princess to Napoleon, being allied to him in the war against Russia, and then letting his country emerge as powerful as before, though controlling a different area (the Italian and Balkan peninsulas, Hungary, Bohemia, and part of Poland, rather than Germany). Prince Metternich was the obedient Prime Minister of Emperor Franz I and then the obedient executor of his will during the thirteen years of the reign of the frail Emperor Ferdinand. As Metternich himself, who followed the rules set by Franz I even if he was not always in agreement with them, said: "*J'ai gouverné l'Europe quelquefois, l'Autriche jamais*" [20] ("I ruled Europe sometimes, Austria never"). The goal of Franz I was obtuse preservation of the *status quo*, even when this meant retaining some policies of his predecessor, the enlightened Emperor Joseph II. "My realm", said Franz once, "resembles a worm-eaten house. If one part is removed, one cannot tell how much will fall" [20]. It is thus not surprising that the nineteenth century was marked by the emergence of wars of independence in various countries or changes of regime in some other countries like France. Austria was defeated in turn by France (1859) and Prussia (1866). The latter war ended Habsburg claims to hegemony in the German-speaking world; Austria became, at best, a second-rate power and the German Empire came into being. Italy in both wars was on the side of the winner, and so also gained her independence. In 1867 the Hungarian question had been solved by creating two distinct states, the Austrian Empire and the Kingdom of Hungary, united under the rule of the same dynasty. All this was over when Boltzmann began his scientific career, so that he enjoyed a period free from wars. There were still troubles with Slavs, Czechs, Italians, and Hungarians, but they were solved by peaceful methods. Thus for example two former enemies, the Emperor of Austria and the King of Italy, reconciled at last, could be found in the audience for a gala occasion, a performance of Donizetti's *Lucia di Lammermoor*, followed by a ballet, at Venice's La Fenice theatre (which was devastated by fire in January 1996).

The real explosive move—the annexation of Bosnia-Herzegovina to the Austro-Hungarian empire, which was to cause the bitter reaction of Serbian patriots and lead, after the assassination of Archduke Franz-Ferdinand, the heir of the Emperor Franz-Joseph, to the outbreak of the First World War in 1914—occurred in 1908, after Boltzmann's death.

When Boltzmann died, Robert Musil (1880–1942), who had previously taken a degree in engineering, had probably already started work for his thesis on Mach, which was to gain him his PhD (Berlin, 1908). Some of his vivid and affectionate descriptions of Austria and Vienna in his monumental novel *Der Mann ohne Eigenschaften* (The man without qualities) [21], where he provided an ironic analysis of the ills of the age, unmasked false attitudes to life, and made an attempt to apply scientific precision of thought to social and spiritual experiences, may allow us to reconstruct the impressions that Boltzmann might have had about the ambience of the time. After all, it is a character of Musil's, Ulrich, presumably a portrait of the author, who proposes that in ethics everything flies hither and thither like molecules according to the kinetic theory of gases, producing very stable average results and very strange agreements. Progress is then the statistical consequence of random flights of ideas, and the behaviour of a single person is of no consequence for the march of history.

Thus we shall freely use Musil's descriptions. Occasionally we shall cite his own words from the English translation [21], but otherwise we shall paraphrase his sentences.

Musil coined a special name, Kakania, for his own country. This name combines two meanings. On one hand it is derived from the initials K.K. (*ka-ka*) (*kaiser-königlich*, imperial-royal), which preceded the name of all the major institutions of the Empire. On the other hand, familiarity with German-speaking children's language indicates that it also has a meaning that could be translated as "Shitia" or "Crapland".

Here is how he talks of Franz-Joseph:

The Emperor and King of Kakania was a legendary old gentleman. Since that time a great many books have been written about him and one knows exactly what he did, prevented or left undone; but then, in the last decade of his and Kakania's life, younger people who were familiar with the current state of the arts and sciences were sometimes overtaken by doubt whether he existed at all. The number of portraits one saw of him was almost as great as the number of inhabitants of his realms; on his birthday there was much eating and drinking as on that of the Saviour; on the mountains the bonfires blazed, and the voices of millions of people were heard vowing that they loved him like a father. Finally, an anthem resounding in his honour was the only work of poetry and music of which every Kakanian knew at least one line. But this popularity and publicity was so overconvincing that it might easily have been the case that believing in his existence was rather like still seeing certain stars, although they ceased to exist thousands of years ago.

In the Austro-Hungarian Empire, a relic of the past but also a model of perfect administration, one tried to slow down the speed of progress. It was just the opposite of what people knew of American cities, where everybody appeared to run or to stop with a chronometer in his hand. We saw what impressed Boltzmann most: aerial and underground trains, pneumatic mail, elevators, trains, and cars were moving people and news horizontally and vertically, leaving just the possibility of saying a few words between one kind of transportation and another, before being sucked into and swallowed by the perpetual motion of an advanced society. Questions and answers had to engage like gear teeth, activities were grouped together in certain places, one ate while moving, whereas amusements were grouped in other areas and people lived in still different places, where big buildings acted as containers for the family, the gramophone, and the soul. To be happy one had to reach an objective, it did not matter what.

This was of course New York, and when thinking of America, one thought of New York; the remainder of the country was for a European a somewhat foggy picture, from which a few colourful images relating to cowboys and the Wild West emerged to give an additional touch of unreality to the New World.

The United States of America (still called "the revolting colonies" by Queen Victoria of England) was lagging behind in basic science because, in spite of the presence of a few exceptional physicists (Benjamin Franklin, Joseph Henry, Josiah Willard Gibbs), people were more interested in inventions and practical applications of science (Samuel Morse, Thomas Alva Edison, Alexander Graham Bell) than in developing its cultural and fundamental aspects in an original fashion.

In Europe and especially in Austria, the marvellous new world developing beyond the Atlantic was known about, but there was no hurry to imitate that kind of life; when away from his country, an Austrian still thought of it as a collection of white roads, wide and comfortable, where one marched or moved with a stage-coach, meandering in all directions like canals of a pre-established order, like ribbons of that pale ticking used for uniforms, encircling hills and valleys in a bureaucratic embrace. We can see this also in the description of Boltzmann's trip to California, when he compares the views of the mountains of America to those that populate the Austrian countryside.

And what a variety of countryside! Seas and glaciers, the Carso caves and Bohemian wheat fields, nights on the Adriatic sea with restless chirping crickets and Slovakian villages where smoke came out of the chimneys as if from the nostrils of a snub nose, while the village crouched between two small hills as if Earth had slightly opened her lips to warm her child. The first cars could be seen and there were thoughts of conquering the air, but not so actively. Occasionally a ship set sail for Latin America or the East Indies; but not so often. Austrians did not cherish imperialist ambitions, especially since the failure to establish Maximilian of Habsburg, brother of Franz-Joseph, as Emperor of Mexico and his execution by the republican troops of Benito Juarez in 1867. People were happy to find themselves in the middle of Europe, where the ancient axes of the world met. The words "colony" and "overseas" reached the ears as faraway and unfamiliar things. There was some luxury, but not so refined as in France. There was some sport, but not so active as in England. As for military ambitions, Musil puts it this way: "One spent tremendous sums on the army, but just enough to assure one of remaining the second-weakest among the Great Powers."

Similarly, Vienna was a little smaller than other metropolises of the world, but a little larger than the usual great cities. Like any European metropolis, it was made up of irregularities, alternations, falls, intermittencies, collisions of things and events, and, in between, points of deep silence; of rails and virgin lands, of a great rhythmic beat and of the eternal disagreement and upsetting of any rhythm; and, as a whole, it looked like a bladder boiling inside a pot and stuffed with houses, laws, regulations and historical traditions. It was however a great cultural centre. In general, one can say that the cultural role of the whole German-speaking world was increasing to the point of rivalling and almost outshining France and Great Britain.

Austria, not the small country of today but a great Empire, even if no longer of the size that it had been under Charles V, was administered—with shrewdness, circumspection, and ability directed to cautiously smoothing every corner—by the best bureaucracy in Europe, which could be criticized in only one respect: to this bureaucracy, genius and spirit of enterprise in a person not authorized to possess such gifts by his or her high birth or government post appeared impertinent and presumptuous. Nobody, after all, likes to see an unauthorized person laying down the law! Then a genius was always thought to be a blockhead; but it never happened, as in other countries, that a blockhead could be taken for a genius.

To use the words of Musil [21]:

It was *kaiser-königlich* (imperial-royal) and it was *kaiser- und königlich* (imperial and royal) to every thing and person; but esoteric lore was nevertheless required to be sure of distinguishing which institutions were to be referred to as *k.k.* and which as *k.u.k.* On paper, it called itself the Austro-Hungarian Monarchy; in speaking, however, one referred to it as 'Austria'—that is to say, it was known by a name which it had, as a state, solemnly renounced by oath while preserving it in all matters of sentiment, as a sign that feelings are just as important as constitutional law, and that regulations are not the really serious thing in life. By its constitution it was liberal, but its system of government was clerical. The system of government was clerical, but the general attitude to life was liberal. Before the law all citizens were equal: not everyone, of course, was a citizen. There was a Parliament, which made such vigorous use of its liberty that it was usually kept shut; but there was also an Emergency Powers Act, by means of which it was possible to manage without Parliament. And, each time that everyone was just beginning to rejoice in absolutism, the Crown decreed that there must now again be a return to parliamentary government.

The hostility of one man to another had been perfected into an extremely refined etiquette and grown to become a collective feeling, to such a point that one distrusted even oneself and one's own destiny with stubborn arrogance. People acted—sometimes even to the highest levels of passion and its consequences—in a way completely different from what they thought, or they thought in one way and acted in another. Unprepared observers misunderstood this as courtesy or as a weakness that they called "the Austrian characteristic". But they were wrong, says Musil, as it is always wrong to explain a country's outward show by means of the characteristics of its inhabitants. In fact, beyond the characteristics related to profession, nation, state, class, geography, sex, consciousness, unconsciousness, and privacy, there is a tenth characteristic: the fantasy of unfilled space, which allows everybody all things but one: to take seriously what his other nine characteristics express, or, in other words, what forbids him to obtain exactly what could fill that space. This space is certainly difficult to describe, but certainly in Italy it has a colour and form different from in England, because different colours and forms stand out, but it is the same in both places, an empty invisible space, within which reality, like a small town of building blocks, is abandoned for fantasy.

Thus Austria went on, as today the entire world seems to do, with its negative freedom, with a sensation that one's existence does not have sufficient reason. Things, even unusual things, happened and events and blows of fate became as light as feathers or thoughts. Austria, like the West nowadays, was full of geniuses. And this was probably the cause of its fall.

Boltzmann, who anticipated in many respects the cultural interests and disappointments of the intellectuals of the twentieth century, accepted the situation prevailing in his country, although his heroes were certainly not emperors and kings.

From the political viewpoint, Boltzmann may be considered to have been, according to E. Broda [22], a democratic radical and a resigned republican. For this reason he admired the America of his days (apart from the difficulty in finding wine), as one can see from the aforementioned account of his trip to California. There he maintained that George Washington's and his followers' struggles had not a local patriotic, but rather a historic universal, meaning. Then he went on, quoting his beloved Schiller, who once said: "Another thousand fellows like me, and Germany will become a republic that will make Rome and Sparta look like nunneries." But he also objected to this quotation: "Clearly that did not come to pass. A few thousand fellows like you? The world has not produced even one." "But ideas do not die", he went on, with the American example in his sight: "The republic, beside which Rome and Sparta were nunneries, does exist beyond the ocean, and how colossal it is and how it grows!", and concluding, quoting Schiller again: "Freedom breeds colossi" (see "A German professor's journey into Eldorado" later in this book).

It is true that Vienna was then quite a liberal place, with the limitations shown for example by the case of Brentano that we have already quoted. Back to the great heart of the Austro-Hungarian Empire, with its sun by then setting, had flown Slavs, Hungarians, Italians, and Jews, who had no problem with living together with Austrians. There was also wide contact between the landed aristocracy, the very rich, frequently Jewish middle class, whose children thronged the crowds of intellectuals, and the Austro-Marxist socialists. But the power was, and had to stay, in the hands of the aristocracy, headed by the Habsburg family and strictly tied to a very rigid Church, to the military caste, to the highest level of bureaucrats, and to a centralized financial power. When it was a case of assigning important positions in the arts and the sciences, the chosen person had to be already integrated or to adapt himself. He could be a conservative or a liberal, but certainly not a radical [22].

It is an interesting question whether, in the last years of his life, Boltzmann realized the importance of the new cultural developments. He was certainly aware of the new trends in arts, as witnessed by a passage which we shall quote in Chapter 10, where he mentions quite different methods of dramatic poetry from those of Goethe and Schiller, as well as impressionism, secessionism, and plein-airism in painting, and the music of the future. Earlier in his life, perhaps distracted by technical and scientific novelties, he had seen only classical painters, poets imitating Goethe and Schiller, architects building houses in Gothic and Renaissance styles, i.e. artistic conformism. But, as a miracle that repeats itself over and over in the history of culture, after years of spiritual sluggishness, a small ascent was under way. From a view of the world as smooth as oil, a refreshing fever was arising everywhere in Europe. Men stood up to fight against the past. Practice and theory were meeting, great minds were emerging. They were very different from one another. They loved the superman and the underdog; they adored sun and health, but also the fragility of consumptive girls; they professed the cult of the hero and socialism; they were believers and sceptical, they loved naturism and refinement, they were sturdy and morbid; their dreams featured old walks near castles, autumnal parks, glassy fishponds, precious gems, hashish, disease, and demons, but also prairies, unbounded horizons, forges and rolling mills, naked wrestlers, revolutions of enslaved workers, and destruction of society. These were certainly contradictions and very antithetic war-cries, but they had a common inspiration: whoever had wanted to analyse that period would have found a nonsense, something like a squared circle made of wooden iron, but in reality everything was of a piece and had a glimmer of meaning. That illusion, materialized in the magic date of the turn of the century, was so strong that some were enthusiastically hurtling over into the new century, whereas others were lingering in the old one, as in a house from which one was obliged to move anyway, but the two attitudes did not look so different. This movement should not be overestimated, because it took place in the thin and variable layer of intellectuals, with no permanent influence on the bulk. Yet something was passing through the tangled mass of faiths and beliefs, like when several trees bend under the same gust of wind: a spirit of faction and reform, the happy consciousness of a new principle, a blooming, a dawn, a small resurrection, things that may occur only in the best times.

In Habsburg Austria, sceptical and unorthodox ideas could arise only outside the official structure described above, as a result of the fact that Vienna had become a melting pot of different cultures, typically in modest apartments or in cafés at the end of the last century and the start of the present one. One may recall Sigmund Freud (1856–1939) and psychoanalysis, Arnold Schoenberg (1874–1951) and dodecaphonic music, *Jugendstil* (Art Nouveau), centred around the journal *Ver Sacrum* ("Sacred Spring") and the advanced literary criticism and pacifism of a Karl Kraus (1874–1936) with his satirical fortnightly, *Die Fackel* ("The Torch"), exposing corruption wherever he found it. Ludwig Wittgenstein (whom we have already mentioned), the neo-positivistic movement, later to become famous by the name Vienna Circle, or a genius of literature such as Musil could come to life only in such a cradle. Perhaps less widely known among the names we have mentioned is that of Kraus, but he was a real character of Viennese life at the end of the nineteenth and in the first third of the twentieth century. To illustrate his importance, we may mention that he was much admired by Wittgenstein [17] and that Schoenberg presented him with a copy of his great musical textbook *Harmonielehre*, with the following inscription: "I have learned more from you, perhaps, than a man should learn, if he wants to remain independent." [23]

Boltzmann loved nature and the arts, but his object in life was always his scientific world. His involvement may be characterized well by the words that he put at the beginning of his book "Principles of mechanics" [24]:

> *Bring' vor, was wahr ist;*
> *Schreib' so, daß klar ist*
> *Und verficht's, bis es mit dir gar ist*

(Bring forward what is true; / write it so that it's clear; / defend it to your last breath!)

His philosophical standpoint, realistic and materialistic, which we shall examine in Chapter 10, might have found a response among the socialists. Indeed, his opinions on the role of theory and on evolutionism (which Lenin [25] approvingly quotes, with the result that Boltzmann was considered a hero of scientific materialism in the former Soviet Union) remind us in a singular way, as remarked by Broda [22], of those expressed by Engels in his "Anti-Dühring", though there is more than one reason to believe that Boltzmann not only did not know them, but was even unaware of Engels's existence. The concept of class struggle is completely absent in his writings. For him, science and philosophy were a single entity, embraced by the very broad term (in Boltzmann's use) of "Mechanics". Sometimes this was not without some irony, as in the following passage [26]:

The enthusiastic love of freedom of men like Cato, Brutus and Verrina arises from feelings that had grown in their souls by purely mechanical causes and we again can also explain mechanically that we live contentedly in a well-ordered monarchical state and yet like to see our sons reading Plutarch and Schiller and draw inspiration from the words and deeds of enthusiastic republicans. This too we cannot alter; but we learn to understand and bear it. The god by whose grace kings rule is the fundamental law of mechanics.

(We remark that no great man named Verrina seems to have existed. This appears to be one of the not unusual and strange mistakes in Boltzmann's *Populäre Schriften*, presumably related to his sight problems. Either it is a misprint or Boltzmann had in mind a reference to the orations of Cicero against Verres, known as the *Verrinae*).

In the imagination of most people, Vienna means Strauss waltzes, cafés, delicious pastries. But the words: "Ah, Vienna, City of Dreams! There is no place like Vienna" are put into the mouth of a madman in the third volume of Musil's novel [21]. A few weeks after the army of Franz-Joseph was defeated by Prussia at Sadowa (July 1866), Strauss diverted the minds of the Viennese bourgeoisie from that disastrous event. We are used to regarding the waltz as the typical expression of the Viennese enjoyment of life. But there are many reports that indicate that many of his contemporaries considered his music as African, erotic, hedonistic, demonic, bacchantic, and dangerous (see e.g. [27]).

So, beneath the glittering surface was a society whose members were incapable of opening themselves to others. To quote Musil again, "the notion that people who live like that could ever get together for the rationally planned navigation of their spiritual life and destiny was simply unrealistic; it was preposterous." [21]

Correspondingly the suicide rate was very high. The most famous case, which has become the subject of so much romantic literature, occurred in 1889, when Crown Prince Rudolf took his life and that of his mistress, Baroness Maria Vetsera. The list of people who committed suicide in those times is very long and, in addition to Boltzmann and the case just quoted, includes three elder brothers of Ludwig Wittgenstein, Otto Weininger, who had published a genial book entitled *Sex and Character* a few months before his tragic act, Georg Trackl (a lyric poet), Otto Mahler (brother of Gustav, and himself a talented musician), Alfred Redl (in 1913), and Eduard van der Null, who could not bear the criticism of the Imperial Opera House which he had designed.

1.9 A poem by Ludwig Boltzmann

This section is concerned solely with the poem *Beethoven im Himmel* mentioned in the first section of this chapter. The German original appears in ref. [5]. It was presumably written in the last five or six years of Boltzmann's life. The English translation we are going to quote is by F. Rohrlich [28]. As Rohrlich himself says, he "had to sacrifice the rhyme in favour of sense and meter" and even then he "had at times to resort to a somewhat free translation. The rhyme is preserved only in the last paragraph to indicate the full flavour of the poem."

Boltzmann called it a *Scherzgedicht* (jocular poem). That is the way it starts. But one cannot avoid noticing certain facts. It speaks of a journey of Boltzmann's soul, detached from the body, to heaven. It is not necessary to be an expert in psychology to see in this a (perhaps unconscious) desire for death. And in fact, in the last part we also see an example of sudden change of humour (passage from Carnival to Lent!): the thing on the earth most missed by the souls in heaven is pain. And examples of pain and sorrow are explicitly mentioned.

Let us examine the poem in more detail. It starts:

> With torment that I'd rather not recall
> My soul at last escaped my mortal body.
> Ascent through space! What happy floating
> For one who suffered such distress and pain.

The last words constitute more than an explicit mention of unhappiness. But they might be considered a casual reference to the pains of human life. This interpretation would seem to be justified by the next sentences, which have rather a light tone, but pain and sorrow are the main theme of the poem, as indicated by its second part.

Boltzmann continues by saying that he passes near other worlds, to which he pays hardly any attention because he has a higher goal: heaven! When he arrives there he hears a wonderful harmony: choirs of angels are singing, but Boltzmann tells them that their songs appear monotonous to him:

> They laugh: "A truly German soul you are!
> Your art of music causes envy here.
> Begin the song 'God praise eternity'
> So he will see what we can do up here.
> But watch it so in unison we'll be!"

The last sentence seems to have been familiar in Boltzmann's home, when he played with his son. The poem continues with the angels singing "a mighty choral hymn". And Boltzmann recognizes Beethoven's style, though the piece is new to him. Then he questions the angels, who tell him that it was composed by Beethoven "upon the Lord's command" and is the best among their songs. Boltzmann now expresses a desire to meet Beethoven, so that his travel will have had a good purpose. Thus he is led to meet the soul of the great musician, who, after a few words of welcome, asks whether the piece pleased him. Boltzmann is confused and keeps silent. Beethoven encourages him to speak. Boltzmann confesses that he has not found the same beauty in the angels' hymn as he was used to on earth. Beethoven agrees with Boltzmann's judgement and says that he has stopped writing music, because he is not as successful as he used to be. Only for the Last Judgement had he agreed to write the part for trumpets. "The Lord would be embarrassed otherwise."

Why has Beethoven been robbed of his creative spark? The mightiest tone is absent, the tone provided by pain! It is pain that "rings with might and resonates like steel, / and when it grips you every fiber shakes." Then Beethoven quotes examples of pain that make us feel human beings:

> "What force commands a mother love her child?
> It's without doubt the nameless agonies
> She suffered all through many nights eternal
> When she and God alone have watched the child.
> Have you not wept together with your wife?
> If not, you missed that bond that joins forever;
> It is the pain that you together shared
> Whose memory will linger as your angel.

The saint who suffers pain and grief
Redemption's rays illuminate his way.
No man achieves a hero's worldly fame
Who has not forced himself with all his power;
And as it caused his aching heart to tremble
His valiant deed will live in song immortal.

The Lord himself when he among us dwelt
Was He a king, a rich man or the like?
He was a human's son beset with pain!"

Beethoven affirms that the warmest mode of life is pain and that he built his music on this foundation. But in heaven he lacks inspiration, because there is no sorrow. And here are the concluding lines:

In shock almost I gazed upon his face.
"How truly wondrous are these worldly ways!
Just hours ago I begged for death again,
'Oh spare my heart the suffering and the pain'
But here in heaven pain is what one yearns
Oh human heart your ways one never learns."

1.10 Boltzmann's personality

The account of Boltzmann's life, with its ubiquitous *leitmotiv* of humour and depression, and the related enthusiasm for his scientific theories, always on the borderline between complete success and profound failure, shows that this great thinker had an unusual weakness, as a consequence of a very human sensitivity. We have seen his almost childish fascination for scientific problems and inventions, his tender, sensitive personality in his relations with his wife and his children, his socializing, entertaining character in his relations with colleagues and students, his broad education and his wide interests: music, literature, nature, and of course science in all its aspects.

There is no choice nowadays, after reading his letters [5], other than to adopt a picture of his most intimate aspects that was first pointed out by Höflechner [4]. It is a picture of one of the greatest heroes of physics, that, to use the words of Höflechner himself, "will seem provocative, derogatory, and a disparaging treatment of his personality", because "our picture of Boltzmann is [...] hazy and ambivalent, just as his personality was very ambivalent and difficult to understand for his contemporaries."

This gives us a clue to the fact that much of his work is only known through someone else's presentation, while his original arguments and papers appear to have been forgotten. He was in a sense a disturbing presence. He wrote lengthy papers, and was always ready to fight for his priority and the correctness of his viewpoint.

The important feature of Boltzmann's personality was psychological instability, so well characterized by his own joke about his birth in the night between Carnival and Lent. It was this instability that eventually led him to commit suicide. His letters [5] and even some sentences in his scientific papers and books give us a portrait of a man full of anxieties and desire for perfection, a portrait borne out by the first part of his scientific life, when he was an ingenious, successful, and fortunate professor and everything worked out perfectly for him. As a consequence of his early success, he perceived himself a great man, misunderstood by his contemporaries, but he occasionally felt that he was not able to maintain the level of this picture. Because of this weak side, to our eyes he seems closer to a normal human being than to a great hero of physics history. When these weaknesses showed up, feelings of fear, suffering, and depression took over. He tried to react to this situation by travelling, by meeting new people, by making his theories more widely known; hence the restless development of his last eighteen years. In this part of his life he did not behave in a way that would result from rational and normal thinking. Some of his moves, as has been stressed by several authors, might have been the result of his wife's worries about her future because of the increasing problems with his physical and psychological health, but, even granted this, one could argue that he should have thought more carefully about the consequences of his own decisions.

The evolution of his attitude can be almost read in his photographs, from that showing a young professor who looks at us with friendly eyes, sure that we understand his importance, to those of the late period, where he appears, besides the inevitable changes due to ageing, as if he were afraid of the world around him: he does not look at us, nor at the surrounding world, but, with the bewildered eyes of a man who has been defeated by life, at something indefinite but frightening. The verses of his beloved poet Schiller certainly inspired him in his youth and gave him the courage and faith to attack the most difficult problems of atomism and solve them in a profound way. They were still with him in his last years, but as the only surviving connection with a world that he felt extraneous and hostile.

We may see his suicide as the inevitable conclusion of his life, as inescapable as Fate in a Greek tragedy: the last, extreme move. To use the words of Höflechner [4], "he, this time without breaking his pledge, switched universities for the last time and entered the one great university of the immortal intellectual giants of science."

2

Physics before Boltzmann

2.1 From Galileo and Newton to the early atomic theories

What was the state of physics when Boltzmann began to contribute to its development? It was in a phase characterized by two conflicting aspects: on the one hand nobody had doubts about the basic structure erected by Newton after the sagacious innovation in both concepts and methods proposed by Galileo; on the other hand the new discoveries in the new areas of thermodynamics and electromagnetism were slowly transforming the imposing edifice in such a way that its inherent contradictions were sooner or later bound to come to light.

This chapter aims to give an account of the status of science before Boltzmann started his work. It will not be a balanced description, because we shall keep in mind those aspects which are more closely related to Boltzmann's work, and underline the difficulties that were hidden beneath the great successes and spurred new developments. Even with these constraints, some interesting material had to be omitted, in order to keep the size of this chapter within acceptable bounds. We shall certainly mention the basic contributions of Volta (who with his battery changed not only physics but chemistry as well), Oersted, and Faraday, who discovered the phenomena upon which the modern production, distribution, and use of electric power are based; but we cannot dwell in detail on these topics. We shall say something more about thermodynamics because the theoretical discoveries spurred by the need to improve the efficiency of steam engines and electric motors are closely related to Boltzmann's main line of research.

Boltzmann is in a sense a typical witness and perhaps the most significant scientist to consider in order to illustrate the debates that were taking place on the new developments occurring during his lifetime. On the one hand he was the inflexible continuer of the great tradition of the mechanical interpretation of nature; on the other hand he saw, so to speak, the protean subject of his researches changing its appearance in his own hands.

He seems to have been clearly conscious of this circumstance. In his address to the meeting of natural scientists in Munich (September 1899) "On the development of the methods of theoretical physics in recent times" [1], given when Boltzmann was fifty-five, after examining the situation that he had found at the beginning of his career as a scientist, he says, without reserve:

This was the stage of development of theoretical physics when I began my studies. How many things changed since then! Indeed when I look back on all these developments and revolutions

I feel like a monument of ancient scientific memories. I would go further and say that I am the only one left who still grasped the old doctrines with unreserved enthusiasm—at any rate I am the only one who still fights for them as far as I can. I regard as my life's task to help to ensure, by as clear and logically ordered an elaboration as I can give of the results of classical theory, that the great portion of valuable and permanently usable material that in my view is contained in it need not be rediscovered one day, which would not be the first time that such an event had happened in science.

I therefore present myself to you as a reactionary, one who has stayed behind and remains enthusiastic for the old classical doctrines as against the men of today; but I do not believe that I am narrow-minded or blind to the advantages of the new doctrines, which shall receive due justice in the next section of my talk, so far as lies within my power; for I am well aware that like everyone else, I see things subjectively tinged through my own spectacles.

Let us then examine the body of scientific knowledge which was already old and classical at the end of the nineteenth century. Relying on the previous work of many scholars, Galileo Galilei (1564–1642) and Isaac Newton (1642–1727) had created a system which is the starting point of modern science. Newton added his own law of action and reaction, which supplies a precise meaning to the vague concept of force. He was able to obtain particularly profound and successful results in the study of the motion of the celestial bodies, by modelling them as point masses, i.e. mathematical points endowed with the additional property of inertia, expressed in a quantitative way by the concept of mass. On the strength of the experimental evidence supplied by Kepler's laws, between two given point masses he assumed an attractive force, directed along the straight line that joins them, and inversely proportional to the square of their distance apart. Assuming that the same force acts between any two given point masses constituting any material body, and applying the laws of motion obtained from the observations made on terrestrial bodies, Newton succeeded in deducing the motion of all celestial bodies, the weight of any material object in the neighbourhood of the Earth's surface, and the theory of tides.

Given these exceptional successes, Newton's successors strove to explain further natural phenomena by using the same method, i.e. merely introducing suitable modifications and additions. Thus by reviving an old assumption, introduced by Democritus in the fifth century BC, they imagined material bodies as aggregates of countless atoms. In this respect they were partly following Newton himself, who in his *Opticks* [2] says:

It seems probable to me that God in the beginning form'd matter in solid, massy, hard, impenetrable, movable particles, of such sizes and figures and with such other properties, and in such proportion to space, as most conduced to the end for which he form'd them; and that these primitive particles being solids are incomparably harder than any porous bodies compounded of them; even so very hard, as never to wear or break in pieces; no ordinary power being able to divide what God himself made one in the first creation.

His successors, however, began to think of the atoms as of point masses, between which at least another force, in addition to Newton's attraction, should act. The true nature and properties of this force were, however, mysterious.

The followers of Galileo and Newton were busy for more than a century applying their ideas and tools. It was attempted to explain all physical phenomena in terms of the basic laws so clearly formulated by Newton. This gave birth to a certain trend in physics, one that was later named mechanics; it was the natural development, following in Galileo's and Newton's footsteps, of the atomism of the ancient Greek philosophers. According to this view of the world, the physical universe is made of matter whose primary elements follow intelligible laws, written in mathematical terms. In short it was, with the addition of the most sophisticated methods of calculus, the full development of the programme outlined by Galileo in the famous passage of *Il Saggiatore*:

Philosophy is written in this grand book—I mean the universe—which stands continually open to our gaze, but it cannot be understood unless one first learns to comprehend the language and interpret the characters in which it is written. It is written in the language of mathematics, and its characters are triangles, circles and other geometrical figures, without which it is humanly impossible to understand a single word of it; without these one is wandering about in a dark labyrinth.

This explains why mathematics played such an important role in the beginnings of modern science. The elementary ideas of calculus, which were developed simultaneously with the elaboration of physical concepts, are necessary for a reasonable understanding of the content of this book.

The most systematic treatise on this early atomic theory, which produced some interesting results, is due to Roger Joseph Boscovich (1711–87). His extensive output concerns practically every aspect of the science and culture of his time, but his treatise [3], which is an attempt to understand the universe by means of a unified model, is perhaps the highest point of his researches. In this book the force between point masses is imagined to be repulsive at certain distances, attractive at others, as seemed appropriate to explain the known phenomena (see Fig. 2.1). The basic facts to be explained were: the impenetrability of particles, the fact that solid bodies have an equilibrium volume at low pressures, the fact that such bodies change into liquids with larger equilibrium volume, and the well-established circumstance that any two pieces of matter have an inverse-square gravitational attraction.

Let us stop a little while to consider the life and work of this little-known scientist, who was, however, widely quoted in the nineteenth century. Boscovich is a typical example of an eighteenth century physicist. If we devote some space to him, it is not only because he was the first who tried to develop a systematic atomic theory in the framework of Newton's mechanics, but also because he exerted a great influence on great physicists of the nineteenth century, such as Faraday, Oersted, Lord Kelvin. In spite of this and the fact that an English translation of his treatise appeared in 1922, he was practically forgotten by great scientists who lived at the beginning of the twentieth century, as testified by the following sentence by the Nobel Laureate Max Born (1882–1970) [4]: "Lord Kelvin quotes frequently a Father Boscovich as one of the first to use atomistic considerations to solve physical problems; he lived in the eighteenth century, and there may have been others, *of whom I know nothing*, thinking on the same lines." (italics mine).

The rather ambitious theory proposed by Boscovich tried to summarize what was known at his time and thus is an early example of an attempt at unifying different physical phenomena under a single principle. In addition, this principle is based not on matter but on a system of forces filling all space and making the separate existence of

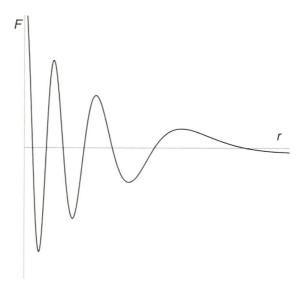

FIG. 2.1. The interatomic force according to Boscovich.

matter almost unnecessary. It is thus related to the world-view of Immanuel Kant, who had advocated the construction of a system of physics based on attractive and repulsive forces (*Naturphilosophie*): atoms, if they exist at all, are merely centres of force and have no separate existence.

The importance of the work of Boscovich lies in the fact that he was the first to propose something that may sound trivial today, but in his day was not. It is sufficient to look at some other studies of the eighteenth and early nineteenth centuries to realize that their authors seemed to think that the nature of atoms ought to be different in the gaseous, liquid, and solid states. For instance, the great Pierre Simon Laplace (1749–1827) assumed that short-range attractive forces between atoms accounted for phenomena such as surface tension, capillarity, and the cohesion of solids and fluids. He also made use of the theory of caloric particles (see below) to explain the long-range repulsion that he thought necessary to account for the behaviour of gases, but he did not formulate his theory quantitatively in such a way that it could be tested experimentally. In particular, his derivation of the ideal gas law, ignoring the basic ideas of Bernoulli (see next chapter), clashed with a previous theorem due to Newton, according to which, in any such explanation the long-range repulsive force must be inversely proportional to the distance. Laplace's force is exactly the opposite of what we assume today between atoms (apart from gravitation), i.e. that the interatomic force is attractive at large distances and strongly repulsive at short distances. Yet Laplace's theory provided the best explanation of capillarity for many years and was superseded only by van der Waals's in 1873.

Let us return to Boscovich. He was born in Dubrovnik (called Ragusa at that time) in Dalmatia (which is now part of Croatia but was then part of the Republic of Venice), the son of a merchant. Although his mother was of Italian ancestry and he never wrote anything in Serbo-Croat (the common language of a large part of Yugoslavia, which is written in the Latin alphabet by Croatians and in the Cyrillic alphabet by Serbians: the one difference, in addition to religion, between two peoples that would otherwise be one), the family of Boscovich is of purely Serbian origin. He was proud of his Slav nationality; in a dispute with D'Alembert, who called him an Italian, he said: "we will notice here in the first place that our author is a Dalmatian, and from Ragusa, not Italian; and that is the reason why Marucelli, in a recent work on Italian authors, has made no mention of him". Also in a poem on eclipses, which he dedicated to the king of France, Louis XV, Boscovich (Rudjer Josif Bošković in Serbo-Croat) pays a tribute to his native town and native land in the dedicatory epistle. He was a recognized scientific authority of his time; he was a member of the Jesuit order and taught mathematics at the Roman College until 1760. In that year we find him in Paris, where he was greatly honoured on the strength of his scientific reputation, though his religious order had been banished from France. He did not feel easy in Paris, however, and in the same year he moved to London, where he was warmly welcomed and made a member of the Royal Society. He was later commissioned by the Society to proceed to California to observe the transit of Venus; since he was unwilling to go, he was instead sent to Constantinople (now Istanbul) for the same purpose. He did not arrive in time for the observation, and when he did arrive, he fell ill. He thus had to remain seven months in Constantinople. He then reached Warsaw, after a journey through Thrace, Bulgaria, and Moldavia, which he described in his *Giornale di un viaggio da Costantinopoli in Polonia*. In 1762 he returned to Rome, but shortly afterwards he was appointed to a chair at the University of Pavia; however, in 1764 the building of the astronomical observatory of Brera had already begun in Milan in accordance with Boscovich's plans, and in 1770 he was appointed its director. Unfortunately, only two years later he was deprived of office by the Austrian government (under whose rule Lombardy came at that time) in favour of another Jesuit, Father Lagrange. His position was further complicated by the fact that the Society of Jesus was suppressed by Pope Clement V. Boscovich then moved to Paris again, where he became a French subject and the director of the Optical Institute of the Royal Navy of France.

In France he had disputes with D'Alembert, Rochon, and Laplace, but had as his devoted friend and admirer the famous astronomer Lalande, from whom we learn many details about his character and habits. In particular, as we have seen, he wrote poems, and Lalande has a precise comment on this ability: "Boscovich wrote verses in Latin only, but he composed with extreme ease. He hardly ever found himself in company without dashing off some impromptu verses to well-known men or charming women. To the latter he paid no other attention, for his austerity was always exemplary."

During a journey to Italy to supervise the printing of his five volumes of optical and astronomical works, Boscovich fell ill and died in Milan.

Boscovich had also an intensive public life, as a diplomat and adviser to the Pope on such different topics as the land reclamation of the Pontine Marshes, the measurement of the size of the Italian peninsula, and the stability of the dome of St Peter's Church in Rome.

After Boscovich, mechanics did not disappear; actually it flourished in the nineteenth century, but it took several diverging paths that were to lead to a more composite view of nature, a process that has continued in the twentieth century with a clearer understanding of the limitations of the classical views.

In spite of the importance of Boscovich in the history of science, his views were insufficient to explain known phenomena, and Boltzmann explicitly rejected them in a letter published in *Nature* (in English). The aim of this letter was to clarify his views on the existence of atoms and the role of hypotheses in physics, with particular concern for the kinetic theory of gases. Here are some sentences from Boltzmann's paper [5]:

For a long time the celebrated theory of Boscovich was the ideal of physicists. [...] If this theory were to hold good for all phenomena, we should be still a long way off what Faust's famulus hoped to attain, viz. to know everything. But the difficulty [...] would be only a quantitative one; nature would be a difficult problem, but not a mystery for the human mind. [...] this simple conception of Boscovich is refuted in every branch of science, the Theory of Gases not excepted. The assumption that the gas-molecules are aggregates of material points, in the sense of Boscovich, does not agree with facts.

Here Boltzmann is referring to the fact that an atom cannot be a simple object, as was amply known in his time from spectroscopy. It was the study of this structure that paved the way to the theory of elementary particles in the twentieth century. These are the bricks from which one builds atoms and may derive a force between atoms of the kind imagined by Boscovich.

We must also remark that Boscovich was the first to assert strict determinism, though conceding some room for free will. It is worth while to quote the following passage in his treatise, published in 1763 [3, p.141]:

Any point of matter, setting aside free motions that arise from the action of arbitrary will, must describe some continuous curved line, the determination of which can be reduced to the following general problem. Given a number of points of matter, & given, for each of them, the point of space that it occupies at any given instant of time; also given the direction & velocity of the initial motion if they were projected, or the tangential velocity if they are already in motion; & given the law of forces expressed by some continuous curve, such as that of Fig. 1, which contains this Theory of mine; it is required to find the path of each of the points, that is to say, the line along which each of them moves. [...] Now, although a problem of such a kind surpasses all the powers of the human intellect, yet any geometer can easily see thus far, that the problem is determinate, & that such curves will all be continuous [...] & a mind which had the powers requisite to deal with such a problem in a proper manner & was brilliant enough to perceive the solutions of it (& such a mind might even be finite, provided the number of points were finite, & the notion of the curve representing the law of forces were given by a finite representation), such a mind, I say, could, from a continuous arc described in an interval of time, no matter how small, by all points of matter, derive the law of forces itself; [...] Now, if the law of forces were known, & the position, velocity & direction of all the points at any given instant, it would be possible for a mind of this type to foresee all the necessary subsequent motions & states, & to predict all the phenomena that necessarily followed from them.

This should be compared with the celebrated passage by Laplace (written in 1814) [6, p.3]:

Nous devons donc envisager l'état présent de l'universe comme l'effet de son état antérieur, et comme la cause de celui qui va suivre. Une intelligence qui pour un instant donné connaîtrait toutes les forces dont la nature est animée et la situation respective des êtres qui la composent, si d'ailleurs elle était assez vaste pour soumettre ces données à l'analyse, embrasserait dans la même formule les mouvements des plus grands corps de l'universe et ceux du plus léger atome; rien ne serait incertain pour elle, et l'avenir comme le passé serait présent a ses yeux.

(Thus we must consider the present state of the universe as the effect of its previous state, and as the cause of its following one. An intelligence which could know, at a given instant, the forces by which nature is animated and the respective situation of the beings who compose it—and also was sufficiently vast to submit these data to analysis—would embrace in the same formula the movements of the greatest bodies of the universe and those of the lightest atom; for it, nothing would be uncertain and the future, like the past, would be present before its eyes.)

2.2 The first connections between heat and mechanical energy

Alongside the standard Newtonian approach, the current usage of calculus had shown, under the impetus of Galileo's laws of falling bodies and the treatment of collisions and of the motion of the compound pendulum by Christiaan Huygens (1629–1695), the importance of another quantity, which later came to be universally named energy. Whenever a point mass subjected to a force covers a path of given length along the direction of the same force, one observes an increase in a quantity which depends upon the mass and the velocity and is nowadays called kinetic energy. This quantity and its increase clearly show up when a force acts in the same fashion on all particles of a body, as in the case of weight. The same quantity tends to decrease if only some of the particles are subjected to certain forces, as occurs in the case of friction or a collision. In all processes of the latter kind a certain amount of heat arises. Now up to the end of the eighteenth century, notably in the work of chemists, heat was treated as a substance, named caloric by the great French chemist Antoine Laurent Lavoisier (1743–1794), who made the first attempt to introduce the methods and concepts of physics laid down by Galileo and Newton into chemistry. As a substance, heat could be neither created nor destroyed, and was thought to be weightless and self-repulsive, whereas it was assumed to be attracted by ordinary matter, within which it was able to penetrate to occupy the spaces between the minute particles of such matter in such a way as to hold them apart against their gravitational attraction. Friction was simply assumed to squeeze out this latent caloric, liberating it and resulting in the heating ordinarily experienced. This theory provided at least a qualitative explanation for all the known properties of heat.

In the last decade of the eighteenth century Benjamin Thompson (1753–1814), who had been born in America but later became Count Rumford in Germany, was boring out cannons for the Elector of Bavaria. In this process there is an enormous expenditure of mechanical energy as well as a great output of heat. Rumford became convinced that there was a direct relation between the loss of mechanical energy and the heat produced. He then proposed to assume that heat, previously considered to be a material substance, is none other than an irregular motion of the most minute particles of a body, which collide one against another. This motion cannot be observed directly, because the particles themselves are invisible; this microscopic motion however is perceived by the particles of our nerve terminations and generates the sensation of heat. In other words, what he proposed was to extend to all phenomena the principle of conservation of mechanical energy.

One can trace a rudimentary idea of this principle already in Galileo's *Discorsi e dimostrazioni matematiche intorno a due nuove scienze* (1638), where he avails himself of considerations which imply the concept of energy, in order to justify the law of inertia with a limiting argument based on the motion of falling bodies. His argument is founded on the fact that a body falling along whatever curve (in the absence of friction) must reach exactly the same speed that it would need to climb again to its initial height; otherwise it would be possible to construct perpetual motion, possibly using a reversal of the body's motion.

It is not easy to explain his considerations, which are a mixture of bold assumptions and subtle arguments, but we shall try. One of Galileo's lesser-known discoveries is that the basic laws, those of motion in the absence of friction, must be time-reversible. This result is of course restricted to the case of gravity, the only force (external to the body acted upon) that he ever considered. He started from the simple example of a pendulum. If one observes an oscillating pendulum, one may wonder why it moves and why eventually it stops. The motion is of course produced by gravity and should persist for ever, if friction were not present. Thus we can consider, as Galileo did, the idealized case of no friction. Then it is obvious that the two halves of the pendular motion are symmetric and, for a given height, the speed is the same at the two positions corresponding to that height. The same situation should occur if a point mass moves on two inclined planes, each of which is the mirror-image of the other with respect to a vertical plane (to avoid problems with the corner where the planes meet, we can imagine smoothing it off). Then Galileo assumed that the motions in the horizontal and vertical directions were independent (something trivial for us) and concluded that if we change the slope of one of the planes, the speed must remain the same at the same height. In fact, let us assume that the speed reached in the fall is greater than the body would need to climb again to its initial height on the other plane. Then it would be possible to increase its speed continuously, by interrupting the climb when it has reached the height from which it started and letting it go down and up with a similar arrangement repeated periodically. If on the contrary the speed reached in the fall is lower, we could again construct perpetual motion with a continuously increasing speed, by reversing the motion and using the assumed symmetry upon time reversal.

The body under consideration can then climb on an inclined plane of an arbitrary slope and, the less inclined the plane, the further the body will go. The passage to the limiting case of a horizontal plane and hence of indefinite motion with a constant velocity is left to the imagination of the reader. This omission is perhaps not fortuitous, but intentional rather, in order to avoid considering an infinite plane on a spherical Earth.

After Galileo there had been developed, as obvious consequences of his discoveries and principles, the concept of mass, assumed to be constant, and the first theories about collisions, even if restricted to the case of head-on collisions. In relation to these studies, René Descartes (1596–1650) introduced momentum, and maintained that this quantity was conserved on the grounds of metaphysical-theological speculations; the lack of the idea of velocity as a vector, however, prevented him ever arriving at correct results (in particular, he did not account for the sign of the relevant velocity component in the case of a head-on collision).

Huygens, in a paper written in 1669, whose proofs appeared only in his *De motu corporum ex percussione*, published in his *Opuscula posthuma* (1703), showed that he had understood not only that the sum (with the correct sign) of momenta preserves the same value before and after an elastic collision, but also that the same holds for the product of each mass by the square of the corresponding velocity. It is remarkable that the first of these laws is deduced from the second through Galileo's relativity principle: in fact, Huygens makes use of a thought experiment and imagines himself to be on a beach and to observe a collision occurring on a moving ship to arrive at this deduction. The first scientist to publish results on the conservation of energy was however Gottfried Wilhelm Leibniz (1646–1716), who had learned a lot from Huygens, in a paper published in 1695 [7], where the mistake of Descartes is criticized. Leibniz called kinetic energy *vis viva* (live force), as opposed to *vis mortua* (dead force), i.e. the usual force. It is also to be remarked that Leibniz used the term *potentia motrix*, which includes for the first time the potential energy (at least in the case of weight): *in natura conservari summam potentiae motricis*. This is his answer to the objections raised against his conceptions [8]:

I had maintained that the *vis vivae* are conserved in the world. It has been objected that in a collision two soft or inelastic bodies would lose their [live] force. I answer that things are not so. It is true that the bodies as a whole lose it as far as their total motion is concerned, but their parts acquire it, because the collision strength creates an inner agitation. Thus this loss is only apparent. The forces are not destroyed, but dissipated among the most minute parts. There is no loss, but everything goes as if somebody wanted to change a coin into smaller pieces.

In a letter of 15 January 1696 to the Marquis de l'Hospital, Leibniz almost appears to augur a programme of a future theory of the structure of matter:

You see that the principle of equality between cause and effect, i.e. the exclusion of perpetual motion, is at the bottom of my measurement of the force. This is preserved, because of this principle, in an immutable equality, according to which there is always conservation of the quantity required to produce a fixed action, to lift a weight to a fixed height, to stretch a spring, to communicate a fixed speed, and this without the possibility that, during such action, the least gain or the least loss may occur, though, without any doubt, a portion of this force, a portion that one should never neglect to take into account, is absorbed by the imperceptible particles of the body itself or of the surroundings. On the contrary, there is no indication that momentum is conserved in nature. As for the bodies that we can observe in nature, experience contradicts the hypothesis of the conservation of momentum, and our reason does not exhibit any argument to admit such conservation in the imperceptible parts of matter, where we must always assume the same actions to occur as in the perceptible and visible objects, with the single exception of their sizes. But, as far as these objects are concerned, the opinion that I maintain here is not obviously based on experiments referring to collisions, but on principles which account for these very experiments and allow us to judge on cases for which no experiments have been performed or laws formulated as yet and the only and unique source of these principles is the equality between cause and effect.

Leibniz's considerations were generalized and systematized by Johannes Bernoulli (1667–1748), who introduced the term "energy", and by Leonhard Euler (1707–83), whereas Daniel Bernoulli (1700–82) was the first to re-establish a connection with Galileo and Huygens, avoiding any metaphysical consideration.

Daniel Bernoulli, about whom we shall say more in the next chapter, was the originator of a new trend in a period in which scientists were busy improving Newton's methods and making them more precise and elegant, as one can see in the remarkable work of Joseph Louis Lagrange (1736–1813), who provided a systematic technique for writing the equations of motion for arbitrarily complex systems composed of point masses and rigid bodies which are acted upon by given forces.

2.3 The springtime of thermodynamics

The concept and term "work" emerge in connection with the great development of applied mechanics that occurred in France at the beginning of the nineteenth century thanks to Lazare Carnot (1753–1823), the "organizer of victory" in the French Revolution, Gustave-Gaspard de Coriolis (1792–1843), and Jean-Victor Poncelet (1788–1867).

One might expect that the experimental work of Rumford and all the theoretical considerations which we have just described should have established the viewpoint that heat is none other than a form of energy, but the caloric theory had a strong hold on the scientists of the eighteenth and early nineteenth centuries. When Rumford pointed out that the production of heat in his cannon-boring experiments was practically inexhaustible, the supporters of the caloric theory replied that there is always enough caloric stored in the metal and its surroundings to be transferred from the latent to the free state.

We must emphasize the fact that geologists, and in particular James Hutton (1726–97), had opposed the theory of gradual cooling of the earth proposed by Georges-Louis Leclerc, Count Buffon (1707–1788). Although Buffon's arguments and laboratory experiments were impeccable, given the knowledge of his days, Hutton remarked that a "subterraneous fire" must "exist in all its vigour at this day" [9]. Today we know that he was right, in spite of the objections of many illustrious and lesser-known critics until the twentieth century (the most famous being Lord Kelvin). The source of energy is provided by radioactivity.

We do not know whether Hutton was influenced by the steam engines of his friend James Watt (1736–1819), and how much in turn the debate on his ideas influenced the subsequent development of physics.

However, thanks to the spur of the industrial revolution and the manufacture of steam engines, the problem emerged from academic debate and became a practical necessity which was widely felt and had to undergo an experimental test. The consequence of the theory that the heat produced must always be exactly proportional to the kinetic energy lost was confirmed by the famous experiments of James Prescott Joule (1818–89) in the years 1838–43. He was born in Manchester, the heart of the first industrial development in England, and had begun with a study of the efficiency of the electric motor, utilizing the new energy source supplied by the electric battery invented by Alessandro Volta.

The principle of equivalence between energy and heat was enunciated by Julius Robert von Mayer (1814–78), a physician, who had already tried to publish a paper on this subject in 1841; the paper was rejected by Poggendorff's *Annalen der Physik und Chemie*, because, in spite of his great intuition, Mayer's physics was bad and he used personal jargon that was certainly unpalatable to most physicists. It was retrieved and published in 1881 after his death. In 1842 he produced a new paper [10], which shows a great improvement in his physics and his ability to master the concepts of what we would call kinetic and potential energy. He gave quite an accurate value for the proportionality constant between mechanical work and heat, on the basis, as he disclosed only three years later [11], of the difference between the specific heat capacities of a perfect gas. It is remarkable that Mayer's interest in this subject was stimulated by his observation of the colour of the venous blood when he was serving as a ship's doctor on a trip to the East Indies, and later on he studied the role of energy in biological phenomena. Here he was naturally led to introduce the horsepower as a unit of power and, when studying the role played by oxidation reactions in the muscles, to state that the blood, "a slowly burning fluid, is the oil in the flame of life". He also tried to explain heat production in the sun and correctly concluded that chemical reactions do not suffice; not having nuclear reactions at his disposal, he proposed a very curious and objectionable explanation, based on the fall of an enormous number of meteors on the sun. The paper was not published by the Academy of Sciences in Paris, to which it was sent in 1846, but was worked over and enlarged in an essay published in 1848 [12].

Returning now to atomic theories, the scientists of the early nineteenth century had precise opinions. We can summarize them by using Boltzmann's words [1]:

It was further presupposed that in solid bodies every particle oscillates about a certain rest position and that the configuration of these rest positions determines the solid shape of the body. In liquids molecular motions are so vigorous that the particles creep past each other; vaporization occurs by total detachment of particles from the surface of bodies, so that in gases and vapours the particles mostly fly off in straight lines like bullets from a gun. This accounts in a natural way for the occurrence of the three states of aggregation of bodies, as well as for many facts of physics and chemistry. From many properties of gases it follows that their molecules cannot be material points. It was therefore supposed that they are complexes of such points, perhaps surrounded by layers of aether.

In addition to ponderable atoms that make up bodies, the presence of a second kind of matter was in fact assumed, made up of subtler atoms, the luminiferous ether, and people succeeded in explaining (after the theory of the corpuscular nature of light favoured by Newton's followers had been abandoned) almost all phenomena concerning light through the assumption of transverse regular vibrations of the ether, according to the bold theory of Augustin Jean Fresnel (1788–1827). Some difficulties still remained, such as the complete lack of longitudinal waves in the luminiferous ether that not only occur in all ponderable bodies but actually predominate in solids. Fresnel's ether had to be assumed ethereal and rigid at the same time, without having the mechanical properties of any known body!

In 1796, when Lazare Carnot was a member of the Directory of France, his elder son, Nicholas Léonard Sadi Carnot (1796–1832), was born. A few years later Lazare was Minister of War and he often took his little son Sadi with him when he visited Napoleon.

Once Napoleon was amusing himself throwing stones near a group of ladies (including his wife) who were on a boat on a lake, splashing them. The little boy ran up and shouted at him: "You, beastly First Consul, stop teasing those ladies!" Napoleon laughed and so there was no serious consequence for the history of science. Until he was 24, Sadi Carnot devoted his life to studying and serving as an officer in the army. He then retired on half-pay and started to study concrete problems of physics and economics. When Lazare died in 1823, Sadi and his younger brother Hippolyte set up home in a small apartment. There Sadi wrote his pioneering work [13] which paved the way to a new discipline, thermodynamics. This remarkable little book was partly expressed in the language of heat as a substance, but this idea was used in a very cautious and critical way. The circumstance that Carnot subsequently accepted the mechanical theory of heat is widely documented by some notes printed in an edition published posthumously in 1878. In fact he caught cholera during an epidemic and died at the age of 36. In these notes we can read (see the English translation quoted in [13]):

We may be allowed to express here a hypothesis concerning the nature of heat.

At present, light is generally regarded as the result of a vibratory movement of the etherial fluid. Light produces heat, or at least accompanies the radiant heat and moves with the same velocity as heat. Radiant heat is therefore a vibratory movement. It would be ridiculous to suppose that it is an emission of matter while the light which accompanies it could only be a movement.

Could a motion (that of radiant heat) produce matter (caloric)?

Undoubtedly no; it can only produce a motion. Heat is then the result of a motion.

Then it is plain that it could be produced by the consumption of motive power and that it could produce this power.

When reading this passage, a reader of today cannot avoid thinking that similar thoughts might have been going on in the head of Einstein, about 80 years later, when he arrived at the opposite conclusion, that light is made of particles, soon to reach the supreme synthesis that one cannot distinguish between matter and energy (see Chapter 12).

Carnot's remarkable argument in his little book [13] was aimed at showing that the motive power of heat is a universal function of the temperature. His reasoning was based on the simple periodic process (cycle) that now bears his name. The basic point of the Carnot cycle is that one needs at least two thermal sources, at different temperatures, to perform work and, given the temperatures, there is a maximum amount of work that can be performed.

The process devised by Carnot is described in almost any book of thermodynamics. We shall give a brief summary of it. Carnot considers a gas, air for example, enclosed in a cylinder equipped with a piston that can change the volume of the gas, facilities for thermally insulating the cylinder, and two heat reservoirs, A and B. The temperatures of these reservoirs will be denoted by T_A and T_B, and T_A will be assumed to be higher than T_B. One can then let the gas undergo adiabatic (= no exchange of heat) and isothermal (= constant temperature) transformations of volume and pressure. In Fig. 2.2 the changes in the pressure p are plotted as a function of the volume V. The volumes corresponding to particularly important stages of the cycle are marked with the letter V

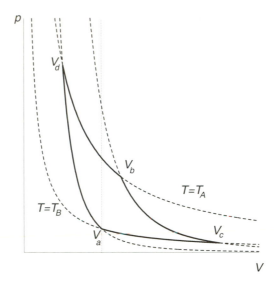

FIG. 2.2. Carnot's cycle in the pressure–volume plane.

with a subscript, such as V_a, i.e. by the corresponding value of the volume. Since we follow well-defined transformations (drawn as full lines), there is no need to show the corresponding values of the pressure. This kind of picture is not contained in Carnot's memoir [13] and was first used by Clapeyron [14].

Let the gas be first in contact with A. By moving the piston, one can let the gas expand isothermally, so that its volume changes from say V_a to say V_b. In Fig. 2.2 the representation of the transformation starts from the intersection of the straight line $V = V_a$ with the upper full line. Now the gas is thermally insulated and is allowed to expand adiabatically until it reaches the temperature T_B (a gas cools when expanding, if no heat is supplied to it). The gas will now occupy a volume V_c (please note that, whereas V_b was arbitrary, V_c is now fixed by the data we have already mentioned). Now we reduce the volume by an isothermal process at the temperature T_B, until the volume has again become V_a (now on the lower curve). Now an adiabatic compression takes place until the temperature again becomes T_A (the volume will have some value, say V_d). Finally we let the gas undergo an isothermal transformation that brings it to the volume V_b again (passing through the initial state, of course). We have thus closed the cycle and can continue.

Carnot remarks that the cycle is reversible, i.e. it can run in the opposite order, with all the obvious changes (e.g. expansions will be replaced by compressions). Then we can repeat the cycle a certain number of times forward and the same number of times backward, until we return to the state from which we started. The two sequences of operations neutralize each other. Thus the *caloric*, he says, cannot produce a greater quantity of motive power than that which we obtained from it by our first sequences of operations. Then he concludes that the motive power of heat is fixed solely by the temperatures T_A and T_B. There is obviously a hidden assumption, which Carnot considers in a footnote, i.e. that after the transformation the gas will contain the same heat that it contained at first. He adds [13]:

This fact has never been called in question. It was first *admitted without reflection*, and verified afterwards in many cases by experiments with the calorimeter. To deny it would be to overthrow the whole theory of heat to which it serves as a basis. For the rest, we may say in passing, *the main principles on which the theory of heat rests require the most careful examination. Many experimental facts remain almost inexplicable in the present state of this theory.* [Italics mine.]

This footnote is probably related to his new idea that heat must be a form of energy and not of matter, as indicated by his posthumous notes quoted above.

After our description, which follows quite faithfully Carnot's argument, one cannot avoid asking the question: why did he not start from the beginning of one of the four processes that make up the cycle? Why did he not omit the change from V_a to V_b, or why did he not start from the volume that we called V_d? As underlined by M. Klein [15], one cannot answer these questions outside the historical context.

Clapeyron [14] presented Carnot's work in a simpler way and avoided what appears to be a clumsy aspect of the original argument. He started from the volume V_d. Was Clapeyron cleverer than Carnot? Not at all. In fact his proof depends in an essential manner on the assumption that heat is a substance, the aforementioned caloric, whereas Carnot's does not.

If one rejects the caloric theory, one is left with the problem of defining the point of the cycle used as a starting point by Clapeyron. This can be done, with difficulty, by using the laws of adiabatics [15], or, as J.C. Maxwell did in his little book *Theory of heat* [16], by omitting the first change from V_a to V_b. This is the clearest and most elegant procedure of establishing the Carnot cycle. Yet it was devised by one of the most gifted physicists who ever existed, half a century after Carnot's discovery and after the First Law (see below) had been clearly stated. We cannot avoid repeating with Klein [15]: "The analysis of what seemed to be an awkward and perplexing aspect of Carnot's *Réflexions* only reinforces one's respect for his critical insight."

The importance of the Carnot cycle is that an apparatus following it constitutes the simplest tool for extracting work from heat or doing work to reduce the temperature (as in a refrigerator). It has also great interest in theoretical considerations and in establishing the efficiency of thermal engines and refrigerators. In fact it gives an example of ideal efficiency, once the two temperatures T_A and T_B have been established.

The subject of thermodynamics was born from a synthesis of Carnot's work and the results of the experiments performed by Joule. The birth of the new discipline is due to the work of Rudolph Clausius (1822–88) in Germany and of William Thomson (later Lord Kelvin of Largs, 1824–1907) in Great Britain. This subject, without mentioning the atomic structure of matter, drew several consequences from two grand laws, the conservation of energy (the First Law) and the existence of irreversible processes (the Second Law).

The First Law starts from the fact that in any physical system there are two kinds of energy (for simplicity, we ignore the possible presence of electric and magnetic fields), mechanical and thermal. Their sum may change because one performs work on the system or supplies heat to the system. Here, as is traditional, the sign of the work and/or heat may be negative: in this case, in ordinary language (as opposed to the mathematical language usually adopted in physics), we would say that the system performs work and/or supplies heat. In the same way, having a negative amount of money in one's bank account means owing the corresponding positive amount to the bank. The First Law simply states that the change in total energy equals the work performed on, plus the heat supplied to, the system (measured in suitable units).

The Second Law indicates that not all the processes compatible with the First Law can actually occur. Whereas one can easily perform work to heat up the system, it is not always enough to supply heat to increase the mechanical energy. At least two heat sources at different temperatures are needed, as shown by Carnot's argument (sometimes one of the sources may be naturally supplied by the environment). Essentially the Second Law states that heat can never pass from a colder to a warmer body without some other related change occurring at the same time (see also the next chapter).

The new ideas were propagated in a rapid fashion thanks to the lively and elegant exposition by Thomson [17]. The conservation of energy in the atomic model found its final treatment in the hands of Hermann von Helmholtz (1821–94), who, like Mayer, started from physiological considerations, and about whom we spoke in detail in the previous chapter. In his fundamental work of 1847 [18] he explicitly introduced the concept of potential energy. His great predecessors had been Galileo, Huygens, Leibniz, and D. Bernoulli . The new thermodynamic ideas were easily extended to continuous media. The mechanics of continua had been developed especially by Leonhard Euler and Augustin Louis Cauchy (1789–1857). The addition of thermodynamic concepts led to a discipline that is today frequently called thermomechanics of continua.

Thermodynamics, which can be regarded as a limitation of our ability to act on the mechanics of the minutest particles of a body, was not enough to complete our description of nature. Greater changes were occurring in physics and this did not escape the attention of Helmholtz. In a lecture devoted to a wide audience [19], he remarked that the successes of mechanics had amazed and intoxicated scientists to such a point that mechanics was thought to be able to solve any problem. This juvenile self-confidence led not only to naïve attempts at interpreting new phenomena, but also at constructing marvellous automata which tried to imitate human abilities (so well indeed that some of the constructors languished for some time in the jails of the Spanish Inquisition, because they were suspected of black magic). These ingenious people set themselves a serious, not a futile, goal, chosen with boldness and pursued with remarkable insight. Their creatures, which nowadays would be regarded as complicated toys or historical curiosities, enriched mechanics but revealed its limitations. Thus, says Helmholtz [19], we do not try to construct machines that perform the thousands of acts typical of a man, but rather require that one machine performs a single act and replaces thousands of men. The study of thermodynamics certainly enriched the available tools, but another theory, that of electromagnetism, was still needed.

2.4 Electricity and magnetism

Knowledge of electric and magnetic phenomena, after the experiments performed by Charles-Augustin de Coulomb (1736–1806) in the most strict Newtonian orthodoxy, was enormously widened thanks to Luigi Galvani (1737–1798), Alessandro Volta (1745–1827), Hans Christian Oerstedt (1777–1851), André Marie Ampère (1775–1836), and many others, among whom one should recall Georg Simon Ohm (1787–1854) for his celebrated law relating the electric current to the difference in the electrostatic potential at the extremes of a conducting wire. The last seal to the experimental picture came from the fundamental contributions of Michael Faraday (1791–1867). It is worth quoting Boltzmann's words [1] about him: "The latter, using rather limited means, had found such a wealth of new facts that it long seemed as though the future would have to confine itself merely to explaining and practically applying all these discoveries."

We can add that in those days Poggendorff's *Annalen der Physik* used to publish translations into German of the most important papers written in other languages. Faraday was the author who received most attention; his papers filled the equivalent of nearly three volumes, more than twice the space given to any other foreign physicist.

Faraday was born near London and was the son of a blacksmith. His family was too poor to afford the expenses of his education and, at the age of 13, he took a job as an errand boy in a bookshop, where one year later he became an apprentice bookbinder. Faraday read many of the books that he was binding and became ardently interested in science. He started to make experiments and, perhaps without knowing that William Nicholson (1753–1815) and Anthony Carlisle (1768–1840) had performed the electrolytic decomposition of water about ten years before, rediscovered electrolysis and formulated its basic laws. Then, at the age of 22, he began to look for a job. It is thus that, towards the end of the year 1812, Sir Humphry Davy, the celebrated chemist (1778–1829), whose lectures Faraday was attending during his apprenticeship, received a letter in which the writer expressed his desire to obtain employment in Davy's scientific laboratory. The letter had an enclosure: a neatly written copy, still preserved at the Royal Institution in London, of notes that the young man had made of Davy's own public lectures. At Davy's recommendation, after an interview, Faraday was in the following spring appointed to a post in the laboratory of the Royal Institution, where he was to remain for the whole of his active life, first as an assistant, then (on the death of Davy) as Director of the laboratories, and from 1833 onwards as the occupant of a chair of chemistry which was founded for his benefit. It is perhaps interesting to remark that the Royal Institution had been established at the end of the eighteenth century under the auspices of Count Rumford, whom we encountered earlier.

It had long been thought of particular electric and magnetic fluids as the cause of electromagnetic phenomena. Ampère succeeded in explainingmachine by means of molecular electric currents, thanks to which the assumption of magnetic fluids became unnecessary, and Wilhelm Weber (1804–91) perfected the theory of electric fluids by completing it in such a way that all the electromagnetic phenomena known up to then could be explained in a simple way. To this end he imagined the electric fluids as consisting of minute particles just as ponderable bodies and the luminiferous ether, whereas forces quite analogous to those between the other substances were assumed to act between these electric particles with only one important modification, that the forces acting between any two particles were further to depend on their relative velocities and accelerations.

In 1831 a physics chair became vacant in Göttingen. The most influential scientist in Göttingen was Karl Friedrich Gauss (1777–1855) (*Princeps mathematicorum*) and his opinion was of course asked. He named and ranked five physicists and then proposed a trio for the faculty recommendation to the government in Hanover. One of them was Weber, then aged 27, but Gauss clearly indicated a preference for C.L. Gerling, a good experimenter and lecturer, and this preference did not escape the attention of the curator of the university, the intermediary between the faculty and the government. The curator transferred to Weber a laudatory sentence used by Gauss for Gerling, and Weber got the appointment. Sometimes bureaucrats are not as bad as scientists depict them; at least, they were not in the year 1831 in Hanover.

In a short while, Weber became a friend of his great colleague. They started to work together on the earth's magnetism; the experimental results obtained by Weber surpassed Gauss's expectations. Later Weber became involved in political trouble. In fact, in 1837, the union of Hanover with the British Empire, which had continued since the accession of the Hanover dynasty to the British throne, was dissolved by virtue of the Salic law, according to which a woman could not accede to the throne in countries where this law held. Thus when Princess Victoria became Queen of England (where this law of French origin was of course not applied), her uncle Ernest-Augustus succeeded to the crown of Hanover. He revoked the free constitution enjoyed for some time by Hanoverians, and ordered the civil servants, including the Göttingen professors, to take the oath of loyalty to his own person. Seven professors, including Weber and the Germanists Jacob and Wilhelm Grimm (famous for their anthology of folk tales), refused to do that; in fact, they argued, they had already sworn an oath of office on the state's constitution and would not dishonour it. Much to Gauss's grief, Weber was deprived of his professorship. His dismissal was not accompanied by exile, as was the case for some of the other six of his colleagues. Eventually, in 1843 he accepted a chair in the University of Leipzig, which he occupied until 1849. It was in this period that his principal theoretical researches in electricity were made. In 1848, the political situation in Hanover having changed, Weber agreed to go back to Göttingen, where he actually returned in the spring of 1849.

Before the work described in the previous section and the first part of the present one, physicists had assumed, besides the tangible substances, the existence of a caloric substance, a luminiferous substance, two electric fluids, two magnetic fluids, and so on. At the stage where we have arrived in our description, it was sufficient to consider the ponderable matter, the luminiferous ether and the two electric fluids. It was thought that each of these substances was made up of atoms, and the task of physics seemed confined to determining the law of action of the force acting at a distance between any two atoms and then to solving the equations that followed from all these interactions under given initial conditions.

It is obvious (at least a posteriori) that this attempt to follow Newton's scheme at all costs, when interpreting the experiments of Oersted and Faraday, was highly artificial and was not destined to stand for long. The new avenue had been already opened up by Faraday, who admitted to not being able to judge the theories of potential of Laplace and

Siméon Denis Poisson (1781–1840) because of his own lack of mathematical education. He, however, in the words of James Clerk Maxwell (1831–79) [20, Preface, X], "in his mind's eye, saw lines of force traversing all space, where the mathematicians saw centres of force attracting at a distance: Faraday saw a medium where they saw nothing but distance: Faraday sought the seat of the phenomena in real actions going on in the medium, they were satisfied that they had found it in a power of action at a distance impressed on the electric fluids."

Let us consider for example the simplest of the new electrodynamic phenomena, the action of an electric circuit on a magnetic pole lying in the same plane; this action reveals itself by a force normal to this plane, as follows immediately from the interpretation of the well-known experiment by H.C. Oersted. This force is not easily interpreted according to the mechanistic viewpoint, which should consider it as the resultant of elementary forces acting between the circuit elements and the magnetic pole. Such forces, known from Newtonian theory (universal attraction, Coulomb's electrostatic and magnetostatic forces, intermolecular forces according to Boscovich and Laplace), are directed along the straight line joining the two interacting elements and satisfy the principle of action and reaction. But the Newtonian model had penetrated to such a degree into the marrow of scientists that even men of the stature of Laplace, Ampère, and Weber tried to find elementary laws of the aforementioned kind, without worrying about the difficulties that have been hinted at. These elementary laws, differing one from another, led to results in agreement with experiment when they were used to obtain laws which referred to a complete circuit rather than to its elements.

The first attack on the scientific system that we have just illustrated was directed against its weakest side, Weber's theory of electrodynamics. This theory is, in the words of Boltzmann [1]:

so to speak the flower of the intellectual work of that gifted enquirer, who has earned immortal merit on behalf of electric theory by his many ideas and experimental results recorded in the system of electrodynamic units and elsewhere. However, for all its ingenuity and mathematical subtlety, it bears so much the stamp of artificiality, that there can surely never have been more than a few enthusiastic followers who believed unconditionally in its correctness.

Against this theory Maxwell developed his own work. While giving frank recognition to Weber's achievements, he set out to develop Faraday's conception in a mathematical form.

Maxwell was the son of a landed proprietor in Kirkcudbrightshire and, in contrast to Faraday, felt at home with mathematics. At the age of 14 he won the Edinburgh Academy's mathematics medal for a paper showing how to construct a perfect oval curve with pins and thread. A few years later he presented to the Royal Society two papers, one "On the theory of rolling curves" and another "On the equilibrium of elastic solids". Both papers were read before the Society by somebody else, because, according to George Gamow [21] "it was not thought proper for a boy in round jacket to mount the rostrum there". After finishing his studies in Edinburgh, Maxwell enrolled as a student at Trinity College, Cambridge, of which college he became a fellow in 1855, when he was 24. In 1855–6, not long after his election to a fellowship, he read a paper to the Cambridge Philosophical Society on the first of his attempts to form a mechanical concept of the electric field. In 1856 he was appointed to the chair of natural philosophy at Marischal College in Aberdeen, where he remained until in 1860 he was appointed to a professorship at King's College London. But in 1865, his desire to study in solitude, and perhaps his uneasiness at the rather formal college, brought him back to his country estate at Glenlair, where he composed his famous *Treatise* [20]. He was called back to Cambridge in 1874 as the first Director of the then newly created Cavendish Laboratory. He died in Cambridge in 1879 (the same year as Einstein was born), at the age of 49, after a short but painful illness, almost certainly cancer.

After his first attempt at a mathematical description of Faraday's ideas, he slightly modified his concept of electromagnetism and introduced an additional term in one of his equations, the so-called displacement current. Maxwell's papers are innovative in two respects. On the one hand he warned against regarding any given view of nature as the only correct one just because a series of its consequences had been confirmed by experience; on the other hand he introduced a completely new theory of electromagnetic phenomena.

As regards the first point, Maxwell shows by many examples how a group of phenomena can often be explained in two totally different ways. Both modes of explanation represent the entire set of facts equally well. Only on adding new and hitherto unknown phenomena does the advantage of one method over the other show up, but the former may have to give way to yet a third after further facts have been discovered. Whereas it was perhaps less Newton than his successors who claimed to have recognized the true nature of things through an explanation that brought them back to the Newtonian theory, Maxwell clearly stated that his own theory was to be regarded as a mere picture of nature, a mechanical analogy, which at that moment allowed one to summarize all the phenomena in the most comprehensive way. This position of Maxwell was extremely influential for the further development of physical theories, because his theoretical ideas were immediately followed by practical successes.

As regards the second point, Maxwell proposed to conceive of any electric or magnetic body as acting only on the neighbouring particles of a medium that fills the whole of space (the ether; nowadays we would say the field); these ether particles in turn act on their neighbours in the medium until the action has propagated itself to the next body. This contact action (rather than force at a distance) was already known from the theory of continuous media and permitted, among other things, the most characteristic properties of the elementary forces to be re-established, i.e. the principle of action and reaction to be satisfied.

The phenomena known until then were equally well explained by both theories, but Maxwell's prevailed over the old one. According to his theory, as soon as it was possible to produce sufficiently rapid electric motions, they would produce in the medium waves that exactly obey the laws of light-waves. Maxwell therefore surmised that within the particles of luminiferous bodies there are constant rapid motions of electricity and that the oscillations thereby provoked in the medium are precisely light. The medium transmitting electromagnetic effects thus becomes identical to the luminiferous ether already previously required. One may thus give it the same name as before, although it must have many other properties in order to serve as the support of electromagnetic actions. The existence of other electromagnetic waves, differing from light, was experimentally proved in 1888 by the German physicist Heinrich Hertz (1857–94) and this led, after the pioneering experiments of Guglielmo Marconi (1874–1937), to the development of radiocommunication techniques, which represented the first step toward one of the major branches of the development of the so-called Third Industrial Revolution that we are witnessing today.

We should remark that Maxwell was not the first to notice the possible connection between a dynamical theory of electricity and magnetism on the one hand and the wave theory of light on the other. In fact the constant that appears in his equations and turns out to be equal to the speed of light in vacuum had popped up in the works of Weber and Carl Neumann, but they tended to stress the superficiality of the analogy, based on a numerical (and with the data then available, not so accurate) coincidence. The only scientist who had preceded Maxwell in stating that this circumstance was not a coincidence was the great mathematician Bernhard Riemann (1826–66) (better known for his work on the functions of complex variables and especially for his researches on curved spaces, which paved the way to Einstein's theory of general relativity). He explicitly stated that the electrical actions propagated with the speed of light and wrote a wave equation for the electric potential. Although it is not easy to interpret Riemann's views (he withdrew his paper from the *Annalen* and it was published only after his death), he certainly had a grand view of a field theory, in a sense more modern than Maxwell's. In fact Maxwell's remark about Riemann [20], that he avoided talking about the medium through which the wave propagated, is not a criticism in our eyes, since the view that ether does not exist has prevailed in the twentieth century.

As we shall see in the next chapter, Maxwell was the most original innovating scientist in the study of the kinetic theory of gases as well, and paved the way for Boltzmann's great discoveries, which we shall discuss in detail in the next few chapters.

When Boltzmann started his university studies, the main emphasis in the university courses was still on the mathematical methods of the French school, which had given an elegant form to Newtonian physics. But, as Boltzmann says in his obituary of Kirchhoff [22], "then theoretical physics struggled to descend from heaven to earth". The mixture of mathematics from the French school and the new, partly empirical approach of Maxwell ignited the spark of Boltzmann's inspiration and he always interpreted continuum models as the idealized representation of an atomistic reality [23]: "The reason why Laplace, Poisson, Cauchy and others started from atomistic considerations is evidently that in those days scientists were as yet more clearly conscious that differential equations are merely symbols for atomistic conceptions so that they need to make the latter simple."

Thus Boltzmann always saw the properties of macroscopic continua as a result of a study of atomic models. The opposite view is perhaps that given by Euler, who liked to define continuous media (which he simply calls materials because he does not use atomic models) starting from their macroscopic (idealized) properties. This is a typical sentence of his [24]: "The ideal materials must be defined; their properties must then be explored; and the ambit of their applicability to specific physical materials must be established subsequently, through the comparison of the detailed predictions of the theory with the results of the measurements." The idea of discarding atomic ideas in favour of continua gained ground in the nineteenth century under the influence of

Mach and lead to the concept of "energetics" (see Chapter 11). Representatives of both viewpoints (macroscopic materials regarded as continua to start with and the same bodies constructed as limits of atomic structures) are still with us.

3

Kinetic theory before Boltzmann

3.1 Early kinetic theories

As early as 1738 Daniel Bernoulli [1] advanced the idea that gases are formed of elastic molecules rushing hither and thither at high speed, colliding and rebounding according to the laws of elementary mechanics. Of course this was not a completely new idea, because several Greek philosophers had asserted that all bodies are made up of particles in motion even when the bodies themselves appear to be at rest. We have seen that Newton and Boscovich shared the same view.

The first atomic theory is actually credited to Democritus of Abdera, who lived in the fifth century BC. We can echo here what Maxwell said in a popular lecture, if we modify the number of centuries:

We do not know much about the science organisation of Thrace twenty-*three* centuries ago, or of the machinery then employed for diffusing an interest in physical research. There were men, however, in those days, who devoted their lives to the pursuit of knowledge with an ardour worthy of the most distinguished members of the British Association; and the lectures in which Democritus explained the atomic theory to his fellow-citizens of Abdera realised, not in golden opinions only, but in golden talents, a sum hardly equalled even in America.

The atomic theory was also maintained by other philosophers such as Leucippus (fifth century BC) and through Epicurus (341–270 BC) it was transmitted to the Romans. The most complete exposition of the view of the ancients is the famous poem of Lucretius (99–55 BC), *De rerum natura* ("On the nature of things"). At first the atomic doctrine might be regarded as a relic of the old way of conceiving a magnitude as being given by an integer (in terms of some basic unit), i.e. the concept that led to Zeno's paradoxes, of which that of Achilles and the tortoise is a celebrated example. These paradoxes arose when space began to be regarded as a continuum but time was still regarded as made up of a finite number of instants, until Aristotle pointed out that time, like space, is divisible without limit. It is then easy to attempt to apply similar arguments to matter and praise the concept that matter is a continuum as more scientific than atomism.

The arguments of Lucretius are worth recalling. He thought that the atoms do not fill up space, because there is vacuum in between. If it were not so, there could be no motion:

> *Quapropter locus est intactus, inane, vacansque.*
> *Quod si non esset, nulla ratione moveri*
> *Res possent; namque, officium quod corporis exstat,*
> *Officere atque obstare, id in omni tempore adesset*
> *Omnibus: haud igitur quicquam procedere posset,*
> *Principium quoniam cedendi nulla daret res.*

(Hence there is a space untouched, vain and void. If this did not exist, things could not move at all; in fact, the essential property of a body, to oppose and to obstruct, would belong always to any of them: nothing could then proceed, because nothing would start to yield. *De rerum natura*, I, 334.)

The continuum school began with Anaxagoras, who maintained that the single parts of bodies (*homoeomerias*, the ancestors of infinitesimal elements) are similar to the whole, that there is no vacuum, and that all motion is like that of a fish in water:

> *Cedere squamigeris latices nitentibus aiunt*
> *Et liquidas aperire vias, quia post loca pisces*
> *Linquant, quo possint cedentes confluere undae.*

(They claim that fluids yield to the moving fishes and open their liquid roads, so that fishes may leave spaces after, where the waves that have yielded may flow. *De rerum natura*, I, 372.)

The atomic theory proposed by Epicurus is striking in some modern aspects, but in addition to the fact that it was only qualitative and not quantitative, it contained some basic flaws. The most remarkable one appears to be due to ignorance of the first principle of mechanics, the law of inertia. Thus the perpetual motion of the atoms in a vacuum was thought to be due to their weight. This assumption was related to the idea of a flat earth and hence to the belief that the downward direction is absolute. In addition, according to Lucretius, the speed of the atoms is greater than the speed of light, because, he argues, the atoms of light move in the air, whereas the material atoms move in a vacuum. Finally, Lucretius introduces the *clinamen* (deflection), of which there is no evidence in the writings of Greek philosophers; it is a random, occasional deviation from the vertical. This strange assumption has a twofold purpose: to allow the atoms to meet and hence form material objects, rather than keeping on falling, and to give room for free will in animals and humans.

In medieval times some Arab thinkers accepted the atomic theory, which on the contrary was fiercely attacked by the scholastic theologians in the West, who maintained that it conflicted with the dogma of transubstantiation. During the Renaissance period, ideas related to atomism occur in the writings of Giordano Bruno (1548–1600), Galileo Galilei (1564–1642), and Francis Bacon (1561–1626). Subsequently the French philosopher Petrus Gassendi (1592–1655) considered the idea of the atomic constitution of matter as a basic point of his philosophy.

Later, Descartes held that it is the basic property of space to be occupied by matter and denied the existence of atoms. Leibniz, on the other hand, regarded his monad as the ultimate element of everything.

Thus, from the start, modern science found itself confronted by two opposite views of reality, which can be regarded as a continuum or as made up of discrete particles. Sometimes both descriptions can be applied with the same results. The continuum theory makes use of a small number of continuous quantities, such as the density, the bulk velocity, and the pressure of a gas, whose evolution in time is ruled by essentially irreversible laws, which imply dissipation and growth of entropy (see below). These laws are in general asymmetric with respect to the time variable, the "time-arrow" pointing from the past into the future, i.e. with an orientation coinciding with that of increasing entropy. The atomic theory, on the other hand, conceives matter as discontinuous and made up of an extremely large number of particles, moving according to the time-symmetric laws of mechanics.

Since, as we have just said, these two descriptions frequently yield the same results, we must explain the fact that regular physical processes, described by continuous variables, emerge at the macroscopic level of everyday life from the extremely complex motions of an enormous number of particles and the fact that the passage to a continuum description is accompanied by the break in the time-symmetry and leads from reversible microscopic motion to irreversible macroscopic phenomena.

This problem is the main theme of Boltzmann's scientific work and we shall develop it in this and subsequent chapters.

In the previous chapter we referred to the theory of Boscovich, which may be taken as an example of the purest monadism. His theory was somewhat qualitative.

As mentioned above however, it was Daniel Bernoulli (1700–82) who introduced the idea of the atomic structure of matter as we understand it today. He explained the origin of pressure and gave birth to the kinetic theory of gases. The new idea was that the mechanical effect of the impact of these moving molecules when they strike against a solid constitutes what is commonly called the pressure of the gas. In fact, if we were guided solely by the atomic hypothesis, we might assume the pressure to be produced by the repulsion of the molecules.

Daniel Bernoulli belonged to a large family of mathematicians and mathematical physicists (their name is sometimes spelt Bernouilli). They descended from a Protestant family, which fled from Antwerp in 1583 to escape a massacre of Huguenots by Catholics. They first moved to Frankfurt and then to Basel. They became rich merchants. One of them, Nicholas the Old, was the founder of the family of mathematicians. They are so numerous that their first names are not enough to tell one from the other. In fact three of them were named Nicholas, three Bernoulli Johann (or Jean), two Daniel, and two Jakob (or Jacques). There is also some confusion among the historians of mathematics about the respective positions in the family tree, especially concerning the three Nicholases. Here we shall say a few words only about Jakob I (the first mathematician in the family, 1654–1705), Johann I (1667–1748), and Daniel I (1700–82). Since they are all the first to bear a typical name in the family, we shall simply use their first names.

Nicholas the Old wished that his first son had made a clerical career, but an irresistible calling had Jakob turning toward science. He was practically an autodidact and travelled through Switzerland and France first, through England and the Netherlands later. In 1687 he obtained a chair at Basel University, where he attracted many students. He was an admirer and correspondent of Leibniz and was elected to the Academies of Paris and Berlin. He was the founder of the calculus of variations and made important contributions to analytic geometry and probability. He also seems to have been the first to use the term "integral calculus", although this priority was later claimed by his brother and student, Johann.

The tenth child of Nicholas the Old must in fact have been in equal degree a great mathematician and a nasty person. Suspicious and envious, he quarrelled with both his brother and teacher Jakob and his son Daniel. His father had wanted Johann to become a merchant, but he melancholicaly saw his wishes unfulfilled again. Johann became first a medical doctor, but later he devoted himself to mathematics under the guidance of Jakob. He was first a professor in Gröningen (where his son Daniel was born) and later succeeded his brother in Basel. Johann was a remarkable mathematician and contributed greatly to diffusion of the differential and integral calculus in Europe. His fellow-citizens had a veneration for him, though they certainly had some reservations about the second line of the following verse, written by Voltaire as an epigraph for Johann's collected works:

> Son Esprit vit la vérité
>
> Et son Coeur connut la justice
>
> Il a fait l'honneur de la Suisse
>
> Et celui de l'humanité

(His Spirit saw the truth / and his Heart justice. / He has honoured Switzerland / as well as humankind.)

In spite of these reservations, the following words are carved on his tombstone:

> HOC SUB LAPIDE REQUIESCIT
>
> VIR QUO MAIOREM INGENIO BASILEA NON TULIT
>
> SAECULI SUI ARCHIMEDES
>
> NON ILLIS EUROPAE LUMINIBUS
>
> CARTESIIS, NEWTONIS, LEIBNITZIIS
>
> MATHEMATUM SCIENTIA SECUNDUS
>
> JOANNES BERNOULLI

(Under this stone lies such a man as Basel never had a greater, the Archimedes of his century, not second in mathematical sciences to those leading lights of Europe, the Descarteses, the Newtons, the Leibnitzs: Johann Bernoulli.)

John experienced with his son Daniel the same problem as his father had had with him. He wanted Daniel to become a merchant, but the young man was first a medical doctor and then a mathematician. This rebellion was the prime cause of the implacable dissension soon to show up between father and son. The discord became an open war in 1734, when the Paris Academy divided between them the prize for a competition (Daniel won prizes of this kind ten times). Johann did not hesitate to embezzle some results published in his son's *Hydrodynamica* [1]. This circumstance certainly stirred up the flames of the fire between them.

Among the many Bernoullis, Daniel must be rated as the most profound in the application of mathematics to physical problems. In fact he shares with Euler the honour of having founded mathematical physics. He was appointed professor of mathematics in St Petersburg at the age of 25, but he found that the place left much to be desired from the viewpoint of civilization. Then he returned to Basel, where he was in turn an anatomist, a botanist, and a physicist. He contributed to the theory of probability and its applications to such different topics as marriage and smallpox vaccination, to differential equations, and to hydrodynamics (with the widely known theorem named after him). There is an interesting anecdote of which he was proud. On a journey, during his youth, he introduced himself to a foreigner, with whom he had had an interesting conversation, by modestly saying: "I am Daniel Bernoulli", to which the foreigner replied "And I am Isaac Newton!".

Although Daniel Bernoulli's scheme [1] was able to account for the elementary properties of gases (compressibility, tendency to expand, rise of temperature during compression and fall during expansion, trend towards uniformity or space homogeneity), no definite opinion could be passed on it until it was investigated quantitatively. The actual development of the kinetic theory of gases accordingly took place much later, in the nineteenth century.

For the sake of simplicity, when we shall not state otherwise, the molecules of a gas will in this book be thought of as perfectly homogeneous, rigid, elastic spheres that move according to the laws of classical mechanics. Thus if no external forces, such as the earth's gravity, act on the molecules, each of them will move in a straight line unless it happens to strike another sphere or a solid wall. Bodies of this kind are usually called billiard balls, for obvious reasons. We shall also assume that the spheres are perfectly smooth (frictionless) so that any possible rotational motion about an axis through their centre can be completely neglected (since it is unchangeable); the centre's velocity will be called the velocity of the molecule, without any further specification. Only in Chapter 8 when we deal with polyatomic molecules shall we depart significantly from this gas model.

Although the rules generating the dynamics of these systems are easy to prescribe, the associated phenomena are not so simple. They are actually rather difficult to understand, especially if one is interested in the trend of the system over long periods of time (the so-called ergodic properties) or in the case when the number of spheres is very large (kinetic and hydrodynamic limits). Both aspects of the dynamics of hard spheres are relevant when dealing with a gas, but we shall usually be more interested in the problem of outlining the behaviour of this system when the number of particles is very large. This because there are about 2.7×10^{19} molecules in a cubic centimetre of a gas at atmospheric pressure and a temperature of $0°C$.

Given the vast number of particles to be considered, it would of course be a hopeless task to attempt to describe the state of the gas by specifying the so-called microscopic state, the position and velocity of every individual sphere; we must have recourse to statistics. A description of this kind is made possible because in practice all that our typical observations can detect are changes in the macroscopic state of the gas, described by quantities such as density, bulk velocity (e.g. the wind's velocity), temperature, pressure, and viscous stresses, and heat flow, which may be related to the suitable averages of quantities describing the microscopic state. A simple example, whose basic idea is contained in Bernoulli's work [1], is provided by an elementary calculation of the pressure in a container at rest at a uniform temperature. This calculation is outlined in Appendix 3.1, since it is the prototype of all calculations that can be performed in kinetic theory.

At this point however, a fundamental question must be considered. If we knew the exact position and velocity of every molecule of the gas at a certain instant of time, the further evolution of the system would be completely determined, according to the laws of mechanics. Even if we assume, as permitted, that at a certain moment the positions and velocities of the molecules satisfy certain statistical laws, we are not at all entitled to expect that at any later time the state of the gas will conform to the same statistical assumptions. This is the case for the assumptions embodied in eqn (A3.1) of the Appendix. This equation expresses the equality between the averages of the squares of the three orthogonal components of the velocity if the gas is in an equilibrium state. Although this seems statistically obvious, we must prove that this is what mechanics predicts. In the example just mentioned, it turns out that mechanics readily provides the required justification. However, things are not so easy if we take a step forward and ask how we can guarantee that the previous statistical assumption will be of practical importance, i.e. will actually be satisfied for a gas in equilibrium in a container. And questions obviously become much more complicated if the gas is not in equilibrium, as is for example the case for air around a vehicle moving at speed.

Questions of this kind have been asked since the dawn of the kinetic theory of gases; today this matter is quite well understood and a rigorous kinetic theory is emerging [2–4]. The importance of these developments stems from the need not only to provide a sound foundation for such a basic physical theory, but also to exhibit a prototype of a mathematical construct central to the theory of non-equilibrium phenomena of macroscopic size. We shall resume this subject in Chapter 5.

Returning now to the prehistory of kinetic theory, we can remark that the theory proposed by Bernoulli was soon afterwards brought forward independently by George-Louis Lesage of Geneva (1724–1803), who however devoted most of his work to the explanation of gravitation as due to the impact of atoms. Then John Herapath (1790–1869), in his *Mathematical physics* published in 1847, made a much more extensive application of the theory. He was born in Bristol, the son of a maltster, and in spite of his scanty early education, managed to study mathematics and French by himself. Apparently without knowing Bernoulli's work, he went further and was the first to state the idea of an absolute temperature depending on molecular velocities, although he used a rather unusual concept of temperature (essentially the square of our absolute temperature). We shall not dwell here on the inaccuracies of his theory, or on his studies of the effect of air molecules on the motion of trains. He gave the first explicit value for the average speed of a molecule in a gas. James Prescott Joule (1818–89), to whom this priority is usually attributed, seems to have based his calculations on Herapath's. In any case, their results were astonishing: the average speed of a molecule in hydrogen turned to be about two kilometres per second, greater than any velocity that had been met in artillery practice!

Another British scientist who rediscovered elementary kinetic theory and was not acknowledged by his contemporaries is John James Waterston (1811–83). Very few people paid attention to his work even in Britain. Reference to his papers is most

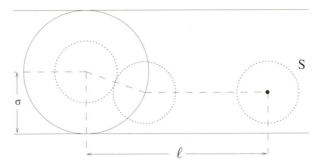

FIG. 3.1. The free path of a molecule between two collisions. The moving molecule S is represented as a point, the other molecules as spheres with double radius (the distance between the centres at the point of contact). The dotted circles represent the actual spheres.

remarkably absent from Maxwell's writings. Rankine did quote him twice in 1864 and 1869. His work was rediscovered, rather too late, by Lord Rayleigh in 1891.

Thus the glory of discovering the first ideas of modern kinetic theory shifted to Germany. A not particularly exciting paper by August Karl Krönig (1822–79) is, however, frequently quoted because it might have had the important role of drawing the attention of Rudolf Clausius (1822–88) to the subject. Clausius had just moved to the Eidgenössische Polytechnikum in Zurich when Krönig's paper appeared, but it seems plausible that he had done some unpublished work on the subject earlier.

Clausius took kinetic theory to a mature stage, with the explicit recognition that thermal energy is but the kinetic energy of the random motion of the molecules and the explanation of the First Law of Thermodynamics in kinetic terms.

As we know, Clausius had been the first to formulate the Second Law of Thermodynamics and to discover the hidden concept of entropy. His first paper on the kinetic theory of gases, entitled *Über die Art der Bewegung, welche wir Wärme nennen* (On the kind of motion we call heat), which appeared in 1857 in Poggendorff's *Annalen*, defined the scope of most nineteenth century work in kinetic theory. His earlier experience in atmospheric physics and in thermodynamics prompted him to find an answer to the naïve objections to the idea that molecules travel at very high speeds. Thus he laid down the programme and some concepts of the new theory. One of the first scientists to react to Clausius's first paper was the Italian chemist Stanislao Cannizzaro (1826–1910), who had revived the chemical atomic theory in its modern form based on the hypothesis of Amedeo Avogadro (1776–1856). His widely known *Sunto di un corso di Filosofia Chimica* (published in 1858 and reprinted as *Sketch of a Course of Chemical Philosophy*, The Alembic Club, Edinburgh, 1961) was distributed at a meeting in Karlsruhe in 1860 and quoted the new researches ("from Gay-Lussac to Clausius") to support Avogadro's views.

Clausius also introduced the concept of the mean free path (*normale mittlere Weglänge*), in 1858 [5]. The mean free path is the average distance covered by a molecule, i.e. a hard sphere whose diameter we shall denote by σ, between two subsequent collisions. We can consider a moving molecule that strikes any of the other molecules, which are

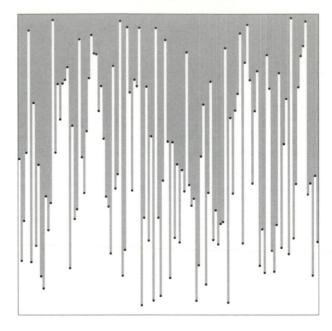

FIG. 3.2.   The possible free paths of a bullet in a wood (view of the cross-section in a horizontal plane with the usual conventions for north and south). There is exactly one tree in each strip having the width of the common diameter of the trees (black circles). The location of the tree in a given strip is randomly chosen. The grey areas denote the possible paths of a bullet and cover about half of the wood.

assumed to be motionless (this introduces no essential difference and simplifies the explanation). A collision occurs when the centre of the moving sphere arrives at a distance equal to σ from the centre of a sphere at rest. Then, in order to compute the length of the path between two collisions, we can consider the moving sphere S as a point and the sphere at rest as having a doubled radius σ (Fig. 3.1). Then if S travels a distance ℓ on average between two impacts, this means that there is only one molecule, i.e. S, in a cylinder of base $\pi\sigma^2$ and height ℓ, or $n\pi\sigma^2\ell \cong 1$, where $n = N/V$ is the number of molecules per unit volume. Hence $\ell \cong 1/(\pi\sigma^2)$.

To visualize this better, we can consider the problem of a man who fires a gun in a given direction at a given height in a forest where all the trees have the same diameter. Consider the mean range ℓ of the bullets inside the forest. In order to be able to make a statistical calculation, we can think of many men, occupying slightly different positions and each firing a bullet in the same direction, say from north to south (Fig. 3.2). In order to avoid the problem that some trees cannot be reached because the bullets will hit some other tree before being able to arrive near them, we shall consider an idealized forest such that if we subdivide it into many parallel strips from north to south with width equal to the common diameter of the trees, there is exactly one tree in each strip. The location of the tree in a given strip is randomly chosen. If we draw the sections of the trees on a piece of paper, we can shade grey the part of the strip corresponding to the path of the bullet. If there are N trees, we have N strips and the grey part will tend to cover half of the entire forest, if we neglect the small areas of the sections of the trees (see Fig. 3.2). This follows because of the random location of the trees: some paths will be longer, some shorter than the length of the forest from north to south, but the statistical average will be half this length. Then the area of the forest will be approximately $A = 2N\ell\sigma$, where ℓ is the average path of the bullet, and hence ℓ will equal $(1/2)A/(N\sigma)$.

The problem for molecules in three dimensions will be different for two reasons. First, in addition to the men staying outside the forest, we must consider various men staying inside, each with his back against a tree, and firing in the same direction; this will suppress the factor $1/2$ in the previous calculation. (The men outside are needed only if we have just a few trees in the direction of fire, as in Fig. 3.2; otherwise their role is negligible.) Second, the molecules are spheres and not circles like the sections of the trees drawn on the paper. Thus their section, rather than a segment (the diameter of a tree), will be a circle with area $\pi\sigma^2$ and the volume V occupied by the gas will replace the area occupied by the forest. Hence the formula obtained in the plane case will be replaced by $V = N\ell\pi\sigma^2$ and, as a consequence, ℓ will be equal to $V/(N\pi\sigma^2)$, as indicated above.

With this elementary argument we have related the length of the mean free path to the product $N\sigma^2$. Thus by measuring the mean free path we throw some light on the number (in a given volume) and size of the molecules of a gas. Both direct and indirect methods give a value for ℓ at atmospheric pressure of about a millionth of a centimetre. Thus we can obtain the product $N\sigma^2$ in a given volume. In order to obtain N and σ separately, we need to deploy more elaborate arguments, into which we shall not enter here. Suffice it to say that all the various methods which have been proposed (starting with Loschmidt in 1865) agree in giving for the number of molecules in a cubic centimetre a figure close to 10^{19} (i.e. 1 followed by nineteen zeros). As a consequence, σ turns out to be of the order of 10^{-8} centimetres, i.e. one hundredth of a millionth of a centimetre. This means that if we were to put ten million molecules side by side in a row, this row would be a millimetre long.

Now all these tiny bullets flying at extremely high speeds (of the order of the speed of sound) fortunately do not fly in the same direction; otherwise they would constitute a wind blowing at terrific speed, and how would we be able to stand? Actually they are not obnoxious but vital, because they allow us to breathe. The thrust of those hitting us in front is compensated by those hitting us on the back. They not only hit us or the walls of a house, but they collide against each other, quite frequently by our standards, though they cover approximately one hundred times their size between one encounter and another. When they collide, their paths are changed and they go off in new directions. Thus each molecule is continually getting its path altered. Hence, in spite of its high velocity it may be a long time before it reaches any great distance from the point whence it started. This can be checked by observing the diffusion of one gas into another, which can be made visible by colour or smelt with perfume.

Using an example first put forward by Maxwell, we can visualize what is going on among the molecules in calm air by observing a swarm of bees, when every individual bee is flying furiously, first in one direction then in another, while the swarm as a whole either remains at rest or moves slowly through the air. By a trick adopted by some owners to identify their bees when they tend to fly off to great distances, let us suppose that we throw a handful of flour at the swarm in such a way that it whitens only those bees that happened to be in the lower half of the swarm. If the bees still go on flying hither and thither in an irregular way, the floury bees will be found in continually increasing proportions in the upper part of the swarm, until they have become equally diffused through every part of it. Diffusion is of course caused not by the fact that we marked them with flour, but by their flying about; the only effect of the marking is to enable us to identify certain bees. But if we are not interested in distinguishing the bees, we could still be interested in counting the proportion of floury bees in a certain layers, a completely statistical property.

Now we have no simple way of marking a select number of molecules of air so as to trace them after they have diffused among others. However, we may easily communicate to them some property by which we can obtain evidence of their diffusion. We can impart a preferred motion in a given direction, or supply a certain amount of energy to a select part of the gas (by putting it in contact with a moving or hot wall); then this property will be transported by diffusion elsewhere; we thus create what is called a transport process, which is of both conceptual and practical interest. Here there is no practical way of seeing which molecule was close to the moving or hot wall. But if we are happy with just a statistical count, the analogy with the example of the floury bees will be clear.

In 1860, two years after Clausius had introduced the mean free path on the basis of this concept, James Clerk Maxwell (1831–79) developed a preliminary theory of transport processes such as heat transfer, viscous drag, and diffusion. He also remarked that the approach of assuming all the molecules at rest and one moving was too simplistic: one needed to compute how the molecules are distributed with respect to their speeds. He thus introduced the concept of a distribution function (a tool to compute the probability of finding a molecule in a given range of speeds). In the same paper he gave a (very) heuristic derivation of the velocity distribution function that bears his name [6]. He considered the velocity vector, ξ and projected this arrow that gives us both the speed and the direction of motion on three orthogonal Cartesian axes to obtain three numbers, ξ_1, ξ_2, ξ_3, the components of the vector ξ. Now, says Maxwell with remarkable intuition but not enough rigour, the existence of the velocity ξ_1 does not in any way affect that of the velocities ξ_2 and ξ_3, "since these are all at right angles to each other and independent", so that one can assume that the square of the speed (i.e. the square of the length of the arrow) has a probability distribution which is the product of the probability distributions of each of the squares of the three components. This leads, by a simple mathematical argument, to asserting that both the magnitude and each component will have a distribution given by the bell-shaped curve well known from statistics, first introduced by Gauss to describe the casual errors in experimental observations and commonly called Gaussian. In the context of kinetic theory the name Maxwellian is used, for obvious reasons.

3.2 The beginnings of modern kinetic theory and the problem of justifying the Second Law

Maxwell almost immediately realized that the concept of mean free path was inadequate as a foundation for kinetic theory and in 1867 developed a much more accurate method

[7], based on the so-called transfer equations, and realized the particularly simple properties of a molecular model, according to which the molecules are mass points (therefore not hard spheres) interacting at a distance with a repulsive force inversely proportional to the fifth power of the distance (these fictitious molecules are nowadays commonly called Maxwell molecules). In the same paper he gave a better justification of his formula for the velocity distribution function for a gas in equilibrium.

With his transfer equations, Maxwell had come very close to an *evolution equation for the distribution function*, but this last step [8] must beyond any doubt be credited to Ludwig Boltzmann (1844–1906), as we shall see more thoroughly in the next chapter. The equation under consideration is usually called the Boltzmann equation but sometimes the Maxwell–Boltzmann equation (to recognize the important role played by Maxwell in its discovery). We shall learn more of this work of Maxwell in the next chapter, when we shall discuss the Boltzmann equation.

In any case there remained the important unsolved problem of deducing the Second Law of Thermodynamics. As is well known from elementary physics, this principle is often subdivided into two parts, according to whether we consider just reversible processes or irreversible processes as well.

The modern idea of irreversibility in physical processes is based upon the Second Law of Thermodynamics in its most general form. The first to put forward this law was, as we saw in the previous chapter, Sadi Carnot in 1824 in a paper on the efficiency of the steam engine [9]. We also indicated that an extremely practical motive led him to that study, i.e. the problem of obtaining the maximum possible work from a given quantity of fuel. His analysis led him to a completely general theorem providing, as Émile Clapeyron [10] put it ten years later, "the common link between the phenomena caused by heat in solid bodies, liquids and gases". This was the idea of Carnot's cycle that we discussed in the previous chapter. Twenty-five years after Carnot's paper, William Thomson was still amazed to see this result and commented that "nothing in the whole range of Natural Philosophy is more remarkable than the establishment of general laws by such a process of reasoning" [11]. As we saw, Carnot was much ahead of his times and, although he mentioned caloric in his paper, made a minimal, cautious, and critical use of this theory and even rejected the theory of caloric in papers that were published only many years after his death. The version of Carnot's argument and his cycle became known to a wide number of scientists through the exposition by Clapeyron [10], who set out to transform what he called the "chain of difficult and elusive arguments" of Carnot into a more readily understandable mathematical discussion. Clapeyron made extensive use of the concept of caloric. Some details of this important aspect of the history of thermodynamics were given in Chapter 2. The reader is also referred to the paper of M. Klein [12], which has already been quoted in that chapter.

In the scientific literature before 1850 one finds scattered statements about something that is lost or dissipated when heat is used to produce mechanical work, but only in 1852 did William Thomson (later to become Lord Kelvin) assert the existence of "a universal tendency in nature to the dissipation of mechanical energy" [13]. The consequences of Thomson's Principle of Dissipation were elaborated by Hermann von Helmholtz, who two years later described the "heat death" of the universe, the consequence of the transformation of all energy into heat [14]. It is to be stressed that Clausius had already remarked in 1850 [15] that, although Carnot's argument can be reconciled with the equivalence of work and heat through a slight modification, something more than the impossibility of perpetual motion had to be invoked as a postulate. In fact neither the First Law (equivalence of heat and work) nor Carnot's argument shows any feature of irreversibility, whereas heat "always shows a tendency to equalize temperature differences and therefore to pass from hotter to colder bodies" [15]. A more accurate and succinct statement of the Second Law can be found in another paper by Clausius, published in 1855 [16]: "Heat can never pass from a colder to a warmer body without some other change, connected therewith, occurring at the same time".

The modern statement of the Principle of Dissipation is based upon the notion of entropy, introduced by Clausius in 1866 [17]. Although Clausius's formulation did not add any new physical content, the mere fact of choosing a new name for something which had previously been represented only by mathematical formulae and rather ponderous circumlocutions had an undoubted influence upon the subsequent development of the topic. Entropy derives from the Greek ἐντροπή, "conversion", "mutation", "evolution", but also "confusion" and "shame" (this latter meaning one finds in the writings of Hippocrates in the form ἐντροπία). The word *entropy* was intentionally chosen for its similarity to the word energy. However, instead of being conserved in an isolated system as energy is, entropy has the property of not decreasing (it does in fact usually increase, and this is the characteristic feature of irreversible processes).

In fact, Clausius showed that for every thermodynamic system there exists a function of the state of the system, its entropy, denoted by S. The latter is defined by a differential relationship for the increase dS in the entropy of the system in an infinitesimally short time interval, during which a time-*reversible* process occurs:

$$dS = \frac{d^*Q}{T}$$

where d^*Q is the heat supplied to the system and T is the absolute temperature. (The notation d^*Q emphasizes that Q is not a state function and hence, in mathematical language, d^*Q is not an exact differential). For processes that are *irreversible*, one can state only that the increase in entropy in a process leading from one equilibrium state to another is larger than the increase that would occur in a reversible process between the same two states.

As a matter of fact, this statement had been shown to be true on experimental grounds and was based on considerations about equilibrium states only. The task of first proving the statement in a more general form (for gaseous systems), was reserved for Boltzmann, who made use of the tools of kinetic theory, not at all unknown to Clausius, as we have seen.

We note that occasionally the remark is made that what Carnot called heat was actually entropy, which is conserved in his cycle. Although it would be preposterous to say that Carnot discovered entropy, we may understand why his argument worked and why Clausius and Thomson were able to rescue his cycle and incorporate it in the new discipline. Essentially he saw that there was something that was conserved in reversible processes; this was not heat or caloric, however, but what was later called entropy.

In the description of matter as a collection of molecules instead of a continuum, questions related to reversibility are presented for the first time in the invention, almost as a joke, of what is now known as "Maxwell's demon". As we have seen, Maxwell had already discovered his law of distribution of velocities in 1860 [6], and the fact that higher temperatures correspond to higher speeds in the random motion of the gas molecules was common knowledge. In 1867, during discussions with P.G. Tait and W. Thomson, he conceived a being whose faculties are so acute that it is capable of following the motion of each molecule. This "demon" is described for the first time in a letter from Maxwell to Tait in 1867 [18]. But its first appearance in public was in the *Theory of heat* by Maxwell, published in 1871 [19].

Imagine such a being, minute and extremely active, able to see the individual molecules and quick enough to manoeuvre them as would a tennis champion with tennis balls. Such a demon could succeed in overcoming the law of increasing entropy, by opening and closing a microscopic aperture built into the wall separating two compartments of a container full of gas, so that only the fastest molecules would pass through the aperture in one direction and only the slowest ones in the other. Given the size of the molecules and hence of the aperture, the work of opening and closing the gate is negligible. In this way the demon would produce an increase in temperature in one of the compartments, essentially without doing work, which would be prohibited by the Second Law. The essential point of this imaginary construction is that the law is not necessarily true at a molecular level. In fact, the physical existence of a demon having the characteristics assigned to it by Maxwell would be impossible. We could of course think of replacing it by a machine able to simulate its faculties. It has been calculated however that the device, given the complexity of the structure required of it, would increase its own entropy in order to decrease that of its environment.

The invention of the demon shows how clearly Maxwell saw to the heart of the matter and we can but be disappointed by the fact that his writings on the statistical character of the Second Law are penetrating but fragmentary. In this connection, M. Klein [20] appropriately quotes a passage of Tait concerning a particular feature of Maxwell's approach to science: "It is thoroughly characteristic of the man that his mind could never bear to pass by any phenomenon without satisfying itself of at least its nature and causes" [21].

The first attempts at explaining the Second Law on the grounds of kinetic theory are due to Rankine [22, 23]. He assumed atomic trajectories which do not change during a thermodynamic transformation. For this reason, he was criticized by Boltzmann [24], since this assumption is incompatible with the deformation that the system undergoes during the said transformation.

Boltzmann himself makes his first appearance in the field with a paper [25] in which he tries to prove the Second Law starting from purely mechanical theorems, under the rather restrictive assumption that the molecular motions are periodic, with period τ, and the awkward remark, which might perhaps be justified, that "if the orbits do not close after a finite time, one may think that they do in an infinite one". Essentially, Boltzmann remarks that temperature may be thought of as the time average of kinetic energy, while heat can be equated to the average increase in kinetic energy; if we compute the unspecified period from one of the relations and substitute the result into the other, it turns out that the heat divided by the temperature is an exact differential. This part of the paper appears to be a rather primitive justification of the first part of the Second Law; as for the second part, Boltzmann's argument belongs more to pure thermodynamics than to statistical mechanics and leads to the conclusion that entropy must increase in an irreversible process.

In 1866 Boltzmann was practically a nobody and the paper received little attention. It is thus not surprising to find that Clausius did not read it and published a completely analogous paper five years later [26]. Boltzmann was then ready to publish a comment [27] that quoted word for word about ten pages of his previous paper and concluded by saying: "I think I have established my priority. Finally I wish to express my pleasure because an authority like Dr Clausius contributes to the dissemination of the ideas contained in my papers on the mechanical theory of heat."

Clausius recognized Boltzmann's priority [28], apologizing because he had scarcely had time in the last few years to keep abreast of the subject. He also added that in his opinion, Boltzmann's arguments were not as general as his own. Boltzmann did not reply to this remark; by then, as we shall see in the next chapter, his scientific interests had taken another path.

As remarked by Klein [20], "Maxwell observed this and later disputes over the mechanical interpretation of the second law with detachment—and no little amusement." In Maxwell's words:

It is rare sport to see those learned Germans contending for the priority of the discovery that the 2nd law of $\theta\,\Delta$cs [a sort of abbreviation for thermodynamics] is the Hamiltonsche Princip, when all the time they assume that the temperature of a body is but another name for the vis viva of one of its molecules, a thing which was suggested by the labours of Gay Lussac, Dulong, etc., but first deduced from dynamical considerations by dp/dt [a *nom de plume* of Maxwell himself; for a detailed explanation of this strange choice, see Klein's paper [20]]. The Hamiltonsche Princip, the while, soars along in a region unvexed by statistical considerations, while the German Icari flap their waxen wings in nephelococcygia amid those cloudy forms which the ignorance and finitude of human science have invested with the incommunicable attributes of the invisible Queen of Heaven.

(from a letter of Maxwell to Tait on 1 December 1873 [18]).

Why was Maxwell so mocking about "those learned Germans"? Because the prize for which they were contending was an illusion. He knew already, as his discussion of the demon named after him shows, that if heat is motion, then the Second Law "is equivalent to a denial of our power to perform the operation just described [to transform heat into ordinary motion], either by a train of mechanism, or by any method yet discovered. Hence, if the heat consists in the motion of its parts, the separate parts which move must be so small that we cannot in any way lay hold of them to stop them" [19]. In other words, the Second Law expresses a limitation on the possibility of acting on those tiny objects, atoms, with our usual macroscopic tools.

As for Clausius, he always stuck to his more mechanical viewpoint; his use of probabilities seems to be essentially restricted to the use of mean values. It seems fitting to quote here Gibbs's words, in his obituary of Clausius: "In reading Clausius we seem to be reading mechanics; in reading Maxwell, and in much of Boltzmann's most valuable work, we seem rather to be reading in the theory of probabilities." [29]

Boltzmann had not discovered the simple but profound truth about the statistical nature of the second principle as yet. His painful path towards his own statistical approach and especially towards a clear understanding of the profound meaning of his own successful discoveries, will be described in the next few chapters. On the other hand, Maxwell seems also to have missed the importance of Boltzmann's main result, the so-called H-theorem, perhaps because he had classified it as a nebulous form in a region unvexed by statistical considerations. We shall return to this in Chapter 11. Here we just make a remark. *If* a scientist of the stature of Maxwell missed the importance of a result offering a way of measuring our inability to transform heat into ordinary motion, then we should be sympathetic to our contemporaries when they are unable to understand the meaning of Boltzmann's discovery. The reader of this book must be patient if he wishes to comprehend. The reason why we proceed slowly is due to the difficulties and pitfalls of the subject. In fact, we have now arrived at the point in this book when it is appropriate to say, with Virgil: *Paulo maiora canamus* (Let us sing of slightly greater things).

4

The Boltzmann equation

4.1 Irreversibility and kinetic theory

The problem of irreversibility came to the forefront in kinetic theory, as was hinted at in the previous chapter, with Boltzmann. In 1872 [1] he not only derived the equation that bears his name, but also introduced a definition of entropy in terms of the distribution function of the molecular velocities. He proved that as a consequence of his equation, the entropy that he had defined must always increase, or at least, in a situation of statistical equilibrium, remain constant. Most of this chapter will be devoted to the basic equation discovered by Boltzmann, leaving the aspects concerning irreversibility in kinetic theory to be discussed in Chapter 5. This choice is made with clarity of exposition in mind, even if it is slightly inaccurate from a historical viewpoint, because it is this second problem which seems to constitute the main motivation of Boltzmann's paper [1].

Before writing that paper, Boltzmann had learned to master Maxwell's techniques [2], which we have already alluded to. By 1868 he had already extended Maxwell's distribution to the case where the molecules are in equilibrium in a force field with potential [3], including the case of polyatomic molecules [4]. The energy equipartition theorem, which we shall discuss in more detail in Chapter 9, was also extended by him to the case of polyatomic molecules [5]. In the mean time he had established contacts with Kirchhoff and Helmholtz.

As remarked by M.J. Klein [6], Boltzmann interprets Maxwell's distribution function in two different ways, which he seems to consider as a priori equivalent: the first way is based on the fraction of a sufficiently long *time interval*, during which the velocity of a specific molecule has values within a certain volume element in velocity space, whereas the second way (quoted in a footnote to paper [3]) is based on the fraction of *molecules* which, at a given instant, have a velocity in the said volume element. It seems clear that Boltzmann did not at that time feel any need to analyse the equivalence, implicitly assumed, between these two meanings, which are so different. He soon realized however (footnote to paper [5]) that it was necessary to make a related, "not improbable" assumption for real bodies made of molecules that are moving because they possess "the motion that we call heat". This assumption, according to which the coordinates and the velocities of the molecules take on, in an equilibrium state, all values compatible with the assigned total energy of the gas, became later familiar as the *ergodic hypothesis*, the name given to it by Paul and Tatiana Ehrenfest [7]. We postpone a discussion on this subject to Chapter 7, devoted to equilibrium statistical mechanics.

Before embarking on the discussion of the basic paper of 1872, we remark that in 1871 Boltzmann felt ready for a new attempt to understand the Second Law [8], starting from the equilibrium law that he had obtained in his previous papers and illustrating the difference between heat and work. He equated, as he had done previously and was common after the work of Clausius and Maxwell, something that he denoted by T and called temperature, with the average kinetic energy per atom. He thus did not use the so-called Boltzmann constant, which was introduced by Planck much later. This identification, apart from a factor, is readily justified only if there is proportionality between thermal energy and temperature; this is the case for perfect gases and solids at room temperature. The concept of temperature is indeed rather subtle, because it does not have a direct dynamic meaning. In a more modern perspective, as we shall see below, the concept of entropy, introduced by Boltzmann in kinetic theory (together with thermal energy) appears more basic (though admittedly less intuitive) and temperature appears as a restricted concept, strictly meaningful only for equilibrium states.

These problems however do not enter in the case of equilibrium states, considered by Boltzmann, and we can accept his identification without objection. It is then clear that the total energy E, the sum of the kinetic and potential energies, will have the following average value:

$$\langle E \rangle = NT + \langle \chi \rangle, \tag{4.1}$$

where χ is the potential energy and $\langle q \rangle$ denotes the average of a quantity q, while N is the total number of molecules. It is then clear that one can change the value of $\langle E \rangle$ in two ways, i.e. by changing either the temperature or the average potential so slowly as to go through equilibrium states, to obtain

$$\delta \langle E \rangle = N \delta T + \delta \langle \chi \rangle, \tag{4.2}$$

where δ denotes an infinitesimal change. If we denote the heat supplied to the system in the process by $\delta^* Q$ and compute it as the difference between the increase in average total energy and the average work done on the system, we have:

$$\delta^* Q = \delta \langle E \rangle - \langle \delta \chi \rangle = N \delta T + \delta \langle \chi \rangle - \langle \delta \chi \rangle. \tag{4.3}$$

We remark that $\delta \langle \chi \rangle$ and $\langle \delta \chi \rangle$ are different; in fact, the operation of taking an average will depend on certain macroscopic parameters, typically temperature, which are allowed to change in the process under consideration. Thus it is not the same thing if we first average the potential and then look at the change in this average (which depends on temperature) or we first compute the change in χ (which does not depend on temperature) and then average it.

The expression in eqn (4.3) is not the differential of some state function Q, a circumstance underlined here by the presence of an asterisk affecting the symbol δ. Boltzmann showed however that, if we divide the expression under consideration by T, we obtain the exact differential of a function, which he identified with entropy. Thus he had obtained the essential content of the Second Law for reversible processes. He also proceeded to computing entropy explicitly for a perfect gas and for a simple model of a solid body, thus finding in the first case a result well known in thermodynamics, in the second case an expression from which he easily succeeded in obtaining the Dulong–Petit formula for specific heat capacities (see Chapters 8 and 12).

Even though somebody acquainted with the usual thermodynamic calculations may find it strange that the work performed on the system is due to the change in the potential rather than to the motion of a piston, the derivation by Boltzmann is impeccable if one grants that the equilibrium distribution is what is called nowadays Maxwell–Boltzmann, and is now more or less standard (see the appendix to Chapter 7).

4.2 The great paper of 1872

The paper by Boltzmann discussed in the previous section was an important step forward. However, it was a treatment that excluded irreversible phenomena, and it could not have been otherwise, since the said distribution holds only for equilibrium states. But Boltzmann was by then ready for the final step, i.e. the extension of the statistical treatment to irreversible phenomena, on the basis of a new integro-differential equation, hinted at in the title of the present chapter. As soon as he was sure of this result, he wanted to publish a short paper in Poggendorff's *Annalen* in order to ensure his priority for this discovery and subsequently to elaborate the results in complete form for the Academy of Vienna. Since Stefan was against publishing the same material twice, we are left with just the memoir of almost 100 pages presented to the Academy [1]. This may explain the strange title, "Further researches on the thermal equilibrium of gas molecules", chosen to present a wealth of new results.

The paper started with a critique of the derivation of velocity distribution in a gas in an equilibrium state, given by Maxwell [2], with an emphasis on the fact that that deduction had shown only that the Maxwellian distribution, once achieved, is not altered by collisions. However, said Boltzmann, "it has still not yet been proved that, whatever the initial state of the gas may be, it must always approach the limit found by Maxwell." [1]. When writing this statement Boltzmann had obviously in mind the spatially homogeneous case, to which the first part of the memoir is actually devoted.

On the basis of an "exact treatment of the collision processes", he obtained an equation for the distribution function, usually denoted by f or $P^{(1)}$, or simply P, i.e. the probability density of finding a molecule at a certain position x with a certain velocity ξ at some time instant t (we shall soon give a more precise definition of f).

In the first part of the memoir he restricted himself to the case when f depends only on time and kinetic energy. This equation may appear a little strange to the eyes of those who have in mind the version of the equation that can be found in more recent treatments, and not only because of the use of the letter x to denote kinetic energy. In fact the circumstance that he adopts this variable as an independent variable instead of velocity introduces several square roots into the equation; these are due to the fact that the volume element whose measure does not change with time during the evolution of the system, thanks to a theorem of Liouville (see Appendix 4.1), contains the volume element in velocity space $d\xi_1\, d\xi_2\, d\xi_3$. When transforming the variables in polar coordinates, one obtains, in addition to the solid angle element, the element $|\boldsymbol{\xi}|^2\, d|\boldsymbol{\xi}|$ or, in terms of the kinetic energy E_{kin} and apart from constant factors, $(E_{\mathrm{kin}})^{1/2}\, dE_{\mathrm{kin}}$ (see Appendix 4.5). Note, as a curiosity, that if one makes the same calculations in two dimensions, the factor with the power $1/2$ or, equivalently, the square roots of Boltzmann, disappear.

By means of his equation, Boltzmann showed not only that the Maxwell distribution is a *steady solution of the equation*, but that *no other such solution can be found*. This goal is achieved by introducing a quantity that turns out to be, apart from a constant factor, the opposite of the entropy; the possibility of expressing the entropy in terms of the distribution function, though in a certain sense not unexpected, does not cease to stand as a remarkable fact that must have produced a deep impression on Boltzmann's contemporaries. In fact, as remarked by the author himself, it implied an entirely different approach to the proof of the Second Law, which showed not only the existence of an entropy function for the equilibrium states, but also permitted one to study its increase in irreversible processes. As already stated, we shall deal with this basic aspect of the matter in the next chapter.

The paper continues with an alternative derivation based on a model with *discrete energies*, in such a way that the integro-differential equation for the distribution function becomes a system of ordinary non-linear differential equations. The use of discrete energies always appeared "much clearer and intuitive" [1] to Boltzmann, as we shall see in Chapter 6 as well. This statement of his may sound like a naïvety, but it might also indicate a surprising intuition about the difficulties of a rigorous proof of the trend to equilibrium, to which we shall return in the next chapter. As a matter of fact, these difficulties in the proof of the trend just mentioned disappear if one has to deal with a discrete, finite system of equations, since the unknown f is, at any time instant, a finite set of numbers instead of a function (in mathematical jargon, we would say that we are dealing with a finite-dimensional space, rather than with a function space); this simplification permits one to make use of a property already known in Boltzmann's days (the so-called Bolzano–Weierstrass theorem) in order to deduce the trend under consideration without particularly refined mathematical arguments.

Many historians of science have underlined the circumstance that these discrete models used by Boltzmann led Planck to the discovery of his energy *quanta*, as Planck himself acknowledged [9] and we shall discuss in Chapter 12.

Just a few pages of the voluminous memoir by Boltzmann concern the calculation of the transport properties in a gas. It is in these pages however that Boltzmann laid down his equation in the most familiar form for us, where the distribution function depends upon time t, upon velocity $\boldsymbol{\xi}$ and upon position $\boldsymbol{x}$ (the vector notation is of course anachronistic). His calculations show that the viscosity, heat conduction, and diffusion coefficients can be computed by means of his equation with results identical to those of Maxwell, for the so-called Maxwellian molecules that we already mentioned in the previous chapter. Boltzmann however warned his readers against the illusion of an easy extension of his calculations to the case of more complicated interaction laws.

In order to explain Boltzmann's contributions in this exceptionally important paper, we shall start from this last part. We shall use an approach anachronistic not only for the notation but also for the argument, because otherwise it does not appear to be possible to treat the subject in an understandable and short form.

As previously indicated, we shall imagine, unless we say otherwise, the molecules as hard, elastic, and perfectly smooth spheres. Not only will this choice simplify our presentation, but it is also in reasonable agreement with experience. Using more refined models would quantitatively improve this agreement, but would introduce several complications, without changing anything from a conceptual standpoint.

In order to discuss the behaviour of a system of N (identical) hard spheres it is very convenient to introduce the so-called phase space, i.e. a $6N$-dimensional space where the Cartesian coordinates are the $3N$ components of the N position vectors of the sphere centres x_i and the $3N$ components of the N velocities ξ_i. In this space, the state of the system, if known with absolute accuracy, is represented by a point having given values of the aforementioned coordinates. Whenever convenient, we shall talk of velocity space (a $3N$-dimensional space where the Cartesian coordinates are the $3N$ components of the N velocities) and of configuration space (a $3N$-dimensional space where the Cartesian coordinates are the $3N$ components of the N position vectors of the sphere centres x_i).

Let us denote by z the $6N$-dimensional position vector of this point in the phase space. If the state is not known with absolute accuracy, we must introduce a probability density $P(z, t)$ which gives the distribution of probability in phase space.

The probability of finding z in a region D of phase space at time t is

$$\text{Prob}(z \in D) = \int_D P(z, t)\, dz, \tag{4.4}$$

where dz denotes the volume element in phase space. Equation (4.4) gives us the exact definition of P.

Given $P_0(z)$, the value of P at $t = 0$, we can compute $P(z, t)$ ($t > 0$), provided we have an equation giving its time evolution. Here we meet the fact that the time evolution of the hard spheres is discontinuous; in fact, when two spheres collide, their velocities change instantaneously from certain values (those possessed when they approach each other) to certain other values (those possessed when they depart from each other). To deal with this difficulty, first we cancel the parts of the phase space corresponding to overlapping spheres and then we add suitable boundary conditions at the edge of the remaining domain. In the following we shall denote by Λ the subset of configuration space obtained by cancelling the region where the spheres partially overlap, i.e. Λ will be the set of positions permitted to the system. The time evolution of the initial state, given by a phase point z_0 will carry it at time t into another state associated with a point z_t. This point is then uniquely defined (except at the instant of collision), provided that the set $\Im_0$ of the phase space points which lead to triple and higher-order collisions and those leading to infinitely many collisions in a finite time are neglected. Today we are well equipped, from a mathematical standpoint, to discuss these assumptions, which may look suspicious; in fact, the probability of having an initial state of the aforementioned set $\Im_0$ is rigorously zero (in the sense that the volume of $\Im_0$ in phase space is zero) [11, 12].

The probability in eqn (4.4) is equal to the probability that the representative point was, at $t = 0$, in the region D_0 consisting of the points z_0 which will end up, in the time evolution, as points z of D at time t. Then:

$$\int_D P(z, t) \, dz = \int_{D_0} P_0(z_0) \, dz_0. \tag{4.5}$$

We can now exploit the fact that the set of points z belonging to D coincides with the set of points obtained, by time evolution, from the points z_0 belonging to D_0 and obtain (see Appendix 4.1) that, under rather broad conditions, $P(z_t, t) = P_0(z_0)$, i.e. the probability density remains unchanged along the trajectories of the system in phase space.

From this circumstance one can obviously obtain:

$$\frac{dP}{dt} = 0, \tag{4.6}$$

where the time derivative is of course taken along a trajectory. By writing this derivative in an explicit fashion, we obtain the so-called Liouville equation, which rules the time evolution of the function P (see Appendix 4.1).

The Liouville equation that we have just mentioned is a useful conceptual tool but cannot in any way be used in practical calculations because of the large number of real variables upon which the unknown depends (of the order of 10^{20}). This was realized (in a more or less explicit fashion) by Maxwell and Boltzmann when they started to work with the one-particle probability density, or distribution function $P^{(1)}(x, \xi, t)$. The latter, at variance with the function $P(z, t)$ which we have used so far, depends on just seven real variables, i.e. the components of the two vectors x and ξ and time t. In particular, Boltzmann wrote an evolution equation for $P^{(1)}$ by means of a heuristic argument, which we shall try to present to the reader in such a way as to show where extra assumptions are introduced.

Let us first consider the meaning of $P^{(1)}(x, \xi, t)$; it gives the probability density of finding one fixed particle (say, that one labelled 1) at a certain point (x, ξ) of the six-dimensional reduced phase space associated with the position and velocity of that molecule. It is thus clear that there is a simple relation between $P^{(1)}$ and P; in fact,

$$P^{(1)}(x_1, \xi_1, t) = \int_{\Omega^{N-1} \times \Re^{3N-3}} P(x_1, \xi_1, x_2, \xi_2, \ldots, x_N, \xi_N, t) \, dx_2 \, d\xi_2, \ldots dx_N \, d\xi_N \tag{4.7}$$

since $P^{(1)}$ is the probability of finding the first particle in a certain state, no matter what the states of the particles labelled $2, \ldots, N$ are. (In eqn (2.1) Ω denotes the region occupied by the gas in physical space (e.g. the interior of a vessel), which is described by each of the $N-1$ position vectors of $N-1$ molecules, $\Re$ denotes the real axis described by each of the $3N-3$ components of the $N-1$ velocities of the same molecules, and of course P is set equal to zero outside $\Lambda \times \Re^{3N}$.) Thus, in principle, the evolution of $P^{(1)}$ is contained in the Liouville equation; this remark will be useful later, but for now will be disregarded. As a matter of fact, the exact equation for $P^{(1)}$, which can be obtained from the Liouville equation, contains the two-particle function $P^{(2)}$ (see below); an equation containing just $P^{(1)}$ is an extremely important step in the treatment of the problem. Such is precisely the equation written by Boltzmann and bearing his name. Rather than starting from the Liouville equation, we shall try rather to write an equation for $P^{(1)}$ on the basis of its physical significance.

Note that in the absence of collisions, $P^{(1)}$ would satisfy the same equation as P, i.e. eqn (4.6), except for the fact that the time derivative is to be taken along a trajectory in a six-dimensional space rather than in a $6N$-dimensional one (we should thus take $N = 1$ in eqn (A4.11)). Accordingly the new effect that we must evaluate is the effect of collisions upon the time evolution of $P^{(1)}$. Note that the probability of occurrence of a collision is related to the probability of finding another molecule whose centre is exactly one diameter from the centre of the first one, whose distribution function is $P^{(1)}$. Thus, generally speaking, in order to write the evolution equation for $P^{(1)}$ we shall need another function, $P^{(2)}$, which gives the probability density of finding, at time t, the first molecule at x_1 with velocity ξ_1 and the second at x_2 with velocity ξ_2; obviously $P^{(2)} = P^{(2)}(x_1, \xi_1, x_2, \xi_2, t)$. Generally speaking we shall have

$$\frac{dP^{(1)}}{dt} = G - L. \tag{4.8}$$

Here $L\, dx_1\, d\xi_1\, dt$ gives the expected number of particles with position between x_1 and $x_1 + dx_1$ and velocity between ξ_1 and $\xi_1 + d\xi_1$ which disappear from these ranges of values because of a collision in the time interval between t and $t + dt$, and $G\, dx_1\, d\xi_1\, dt$ gives the analogous number of particles entering the same range in the same time interval. The count of these numbers is analogous to that made in Appendix 3.1 to compute the transfer of momentum from the molecules to a wall, provided we use the trick of imagining particle 1 as a sphere at rest and endowed with twice the actual diameter σ and the other particles to be point masses with velocity $V_i = \xi_i - \xi_1$. In fact, each collision will send molecule 1 out of the above sets and the number of its collisions will be the number of expected collisions of any other point mass with that sphere. Since there are exactly $(N - 1)$ identical point masses and multiple collisions are disregarded, $G = (N - 1)g$ and $L = (N - 1)l$, where the lower-case letters indicate the contribution of a fixed particle, say particle 2. For the details of the computation of g and l, see Appendix 4.2.

At this point we are ready to understand Boltzmann's argument. In a gas, even if rarefied, N is still a huge number and σ (expressed in units suitable for objects that we handle in everyday life, such as centimetres) is very small; to fix ideas, let us consider a box whose volume is 1 cm^3 at room temperature and atmospheric pressure. Then $N \cong 10^{20}$ and $\sigma = 10^{-8}$ cm. Then $(N - 1)\sigma^2 \cong N\sigma^2 = 10^4$ cm^2 = 1 m^2 is a sizeable quantity, while we can neglect the difference between x_1 and $x_1 + \sigma n$, where n is the unit vector directed as $x_2 - x_1$ (at the instant when the collision occurs, the magnitude of the latter vector is obviously equal to the molecular diameter σ). This leads us to think that the equation to be written can be rigorously valid only in the so-called *Boltzmann–Grad limit*, when $N \to \infty$, $\sigma \to 0$ with $N\sigma^2$ finite.

In addition, since the volume occupied by the molecules is about $N\sigma^3 \cong 10^{-4}$ cm^3, collision between two preselected molecules is a rather rare event. Thus two spheres that happen to collide can be thought of as two randomly chosen molecules. Hence it makes sense to assume that the probability density of finding the first molecule at x_1 with velocity ξ_1 and the second at x_2 with velocity ξ_2 is the product of the probability density of finding the first molecule at x_1 with velocity ξ_1 times the probability density of finding the second molecule at x_2 with velocity ξ_2.

These considerations may be understood in a more general scheme thought up to develop the following viewpoint: the fact that the continuum and molecular description have a common ground of validity leads to the idea that we can develop a method to permit rigorous deduction of the macroscopic equations starting from a microscopic model. The method is essentially based on the idea that the macroscopic behaviour arises as a limiting one from the microscopic laws and is essentially due to the extremely large number of molecules coming into play.

This is part of a more general vision of our understanding of natural phenomena at various levels of description. The coarser descriptions correspond to phenomena that we meet in everyday life, but we must descend to lower layers in order to understand the reasons behind certain types of macroscopic behaviour and obtain more accurate laws.

For the moment we accept the idea of statistical independence of two molecules that happen to meet in a dilute gas, but we shall return later to its justification. Thus we write (assuming *molecular chaos*):

$$P^{(2)}(x_1, \xi_1, x_2, \xi_2, t) = P^{(1)}(x_1, \xi_1, t) P^{(1)}(x_2, \xi_2, t) \tag{4.9}$$

for two molecules that are about to collide, or:

$$P^{(2)}(x_1, \xi_1, x_1 + \sigma n, \xi_2, t) = P^{(1)}(x_1, \xi_1, t) P^{(1)}(x_1 + \sigma n, \xi_2, t)$$
$$\text{for } (\xi_2 - \xi_1) \cdot n < 0 \tag{4.10}$$

Thus we can apply this *recipe* to the loss term L in the form (A4.15) of Appendix 4.2 but not to the gain term in the form (A4.16). It is possible however to apply eqn (4.10) (with ξ_1', ξ_2' in place of ξ_1, ξ_2) to the form (A4.20) of the gain term, because the transformation (A4.4) of Appendix 4.1 maps the hemisphere of the unit vectors n corresponding to molecules that are approaching onto the hemisphere of the same vectors corresponding to molecules that are separating.

If we accept all the simplifying assumptions made (more or less implicitly) by Boltzmann, we still obtain, for the gain and loss terms G and L, expressions which contain $P^{(1)}$ only: in fact the function $P^{(2)}$ is, like P, continuous at a collision, in spite of the fact that the molecular velocities undergo a discontinuous transformation. One can thus equate the value of $P^{(2)}$ after a collision to its value before the same collision (for different values of the velocities, of course, as explained in the appendices) and apply eqn (4.10) just to the distribution prevailing before the collision (see Appendix 4.2).

4.3 A critique of Boltzmann's approach

The Boltzmann equation is an evolution equation for $P^{(1)}$, without any reference to $P^{(2)}$ or P. This is its main advantage. However, it has been obtained at the price of several assumptions; the chaos assumption present in eqn (4.10) is particularly strong and requires to be discussed a little. The main difficulty in this area is to distinguish between probabilistic and dynamic arguments; it is in fact the use of probability that makes the results acceptable and avoids the paradoxes that arise from a purely dynamic interpretation.

The molecular chaos assumed in eqn (4.10) is clearly a property of randomness. Intuitively, one feels that collisions exert a randomizing influence, and this can be made convincing by further considerations, but it would be completely wrong to argue that the statistical independence described by eqn (4.10) is a consequence of the dynamics (as Boltzmann seemed to maintain in his first papers on his equation). It is quite clear that we cannot expect every choice of the initial value for P to give a $P^{(1)}$ which agrees with the solution of the Boltzmann equation in the Boltzmann–Grad limit (since the Liouville equation is linear, half of the sum of two initial data for P, each of which gives by assumption a $P^{(1)}$ in agreement with the Boltzmann equation, would not be able to give the same agreement, since the Boltzmann equation is non-linear). In other words, molecular chaos must be present initially and we can only ask whether it is preserved by the time evolution of the system of hard spheres.

It is evident that the chaos property (4.10), if initially present, is immediately destroyed, at least if we insist that it should be valid everywhere. In fact, if it were strictly valid at any point of phase space, the gain and loss terms, in the Boltzmann–Grad limit, would be exactly equal. Hence there would be no effect of the collisions on the time evolution of $P^{(1)}$. The essential point is that we need the chaos property only for molecules which are about to collide, i.e. in the precise form stated in eqn (4.10). It is clear then that even if $P^{(1)}$ as predicted by the Liouville equation converges nicely to a solution of the Boltzmann equation, $P^{(2)}$ may converge to a product, as stated in eqn (4.10), only in a way which is in a certain sense very singular. In fact, it is not enough to show that the convergence occurs for all points of phase space, *except those of a set of zero volume*, because we need to use the chaos property in a zero volume set (corresponding to the states of molecules which are about to collide). On the other hand we cannot try to show that convergence holds everywhere, because this would be false; in fact, we have just remarked that eqn (4.10) is, generally speaking, simply not true for molecules that have just collided.

How can we approach the question of justifying the Boltzmann equation without invoking the molecular chaos assumption as an a priori hypothesis? Obviously, since $P^{(2)}$ appears in the evolution equation for $P^{(1)}$, we must investigate the time evolution for $P^{(2)}$; now, as is clear, and will be illustrated in Appendix 4.3, the evolution equation for $P^{(2)}$ contains another function, $P^{(3)}$, which depends on time and the positions and velocities of three molecules and gives the probability density of finding, at time t, the first molecule at x_1 with velocity ξ_1, the second at x_2 with velocity ξ_2, and the third at x_3 with velocity ξ_3. In general, if we introduce a function $P^{(s)} = P^{(s)}(x_1, x_2, \ldots, x_s, \xi_1, \xi_2, \ldots, \xi_s, t)$, the so-called *s-molecule distribution function*, which gives the probability density of finding, at time t, the first molecule at x_1 with velocity ξ_1, the second at x_2 with velocity $\xi_2, \ldots$ and the sth at x_s with velocity ξ_s, we find that the evolution equation of $P^{(s)}$ contains the next function $P^{(s+1)}$, until we reach $s = N$; in fact $P^{(N)}$ is no other than P and satisfies the Liouville equation. It is thus clear that we cannot proceed unless we handle all the $P^{(s)}$ at the same time and attempt to prove a generalized form of molecular chaos, i.e.

$$P^{(s)}(x_1, \xi_1, x_2, \xi_2, \ldots, x_s, \xi_s, t) = \prod_{j=1}^{s} P^{(1)}(x_j, \xi_j, t), \tag{4.11}$$

where $\prod_{j=1}^{s} F_j$ (whatever F_j is) means that we must multiply s factors given by F_1, F_2, $F_3, \ldots, F_s$.

Thus the task becomes to show that, if true at $t = 0$, this property remains preserved (for any fixed s) and for molecules about to collide, in the Boltzmann–Grad limit. This is discussed in detail in Appendices 4.3 and 4.4.

There remains the problem of justifying the *initial chaos assumption*, according to which eqn (4.11) is satisfied at $t = 0$. One can give two justifications, one of them being physical in nature and the other mathematical; essentially they say the same thing, i.e. that it is difficult to prepare an initial state for which eqn (4.11) does not hold. The physical reason for this is that in general we cannot handle the single molecules, but rather act on the gas as a whole, if we act at a macroscopic level, usually starting from an equilibrium state (for which eqn (4.11) holds). The mathematical argument indicates that if we choose the initial data for the molecules at random, there is an overwhelming probability [1, 2] that eqn (4.11) is satisfied for $t = 0$. (We shall take this problem up again in the Appendix to Chapter 7, after developing the required tools.)

This clarification from a physical standpoint is due to the Ehrenfests [7], while the problems posed by a mathematically rigorous justification are at the moment only partly solved. The route to be followed will be briefly discussed in the next chapter.

A word should be said about boundary conditions. When proving that chaos is preserved in the limit, it is absolutely necessary to have a boundary condition compatible (at least in the limit) with eqn (4.11). If the boundary conditions are those of periodicity or specular reflection, no problems arise. More generally, it is sufficient that the molecules are scattered without adsorption from the boundary in a way that does not depend on the state of the other molecules of the gas [2].

It is appropriate here to comment upon Maxwell's transfer equations mentioned in the previous chapter. They were actually very close to the Boltzmann equation, although there is no hint that Maxwell thought that they could be used to study the time evolution of the distribution function. In fact they can be written as follows:

$$\int_{\Re^3} \frac{d}{dt} \left[\phi(\xi) P^{(1)}(x, \xi, t) \right] d\xi = \int_{\Re^3} \phi(\xi) G(x, \xi, t) \, d\xi - \int_{\Re^3} \phi(\xi) L(x, \xi, t) \, d\xi, \tag{4.12}$$

where ϕ is an essentially arbitrary function of ξ and G and L are the same as in eqn (4.8). Equation (4.12) is an equation for the average change in the function ϕ due to collisions. It is obvious that if ϕ is largely arbitrary, eqn (4.12) implies the Boltzmann equation, eqn (4.8), but this is the step that Maxwell missed.

5

Time irreversibility and the H-theorem

5.1 Introduction

Boltzmann not only showed that the equation bearing his name admits Maxwell's distribution as an equilibrium solution, but he also gave a heuristic proof that it is the only possible one. To this end he introduced [1] a quantity, which he denoted by E and was later (as here) denoted by H, defined in terms of the molecular velocity distribution. He then demonstrated that as a consequence of his equation, this function must always decrease in an isolated system or, at most, remain constant, the latter case occurring only if a state of statistical equilibrium prevails. His result is usually quoted as the "H-theorem" and indicates that H must be proportional to minus the entropy.

H was essentially defined by Boltzmann as the integral of $f \log f$ with respect to the kinetic energy, where "log" denotes the natural logarithm. By a detailed consideration of the properties of his equation (Appendix 5.1), he showed that the time derivative of H (Appendix 5.2) is never positive and vanishes if and only if the velocity distribution is Maxwellian. Since H is bounded from below (by its equilibrium value), it cannot decrease below this value and will hence tend to it, and at the same time f will tend towards a Maxwellian. In truth, a rigorous proof of this result is far from simple (see Appendix 5.3); many years elapsed before Carleman [2, 3] could give an unassailable proof of Boltzmann's "H-theorem". We recall from the previous chapter that Boltzmann had had recourse to a discrete model to make the result intuitive.

It is a remarkable fact that the proof can also be extended to polyatomic gases. In truth, the original proof by Boltzmann for this case is not completely general, as was pointed out for the first time by H.A. Lorentz in 1887 [4–7]; an "amended" version by Boltzmann based on the so-called "closed cycles of collisions" [5–7] did not really satisfy anybody. For a while, this aspect of the matter was forgotten until, using quantum methods, the required property was shown to follow from the property of unitarity of the so-called scattering matrix (S-matrix) used to describe collision phenomena in quantum mechanics [8, 9]. A satisfactory proof of the required inequality in the case of polyatomic molecules using purely classical methods was given only in recent times [10], but it seems to have escaped the attention of historians of kinetic theory [11]. We shall come back to this point in Chapter 8.

Boltzmann made his remarks on polyatomic gases and the importance of his result in the middle of a section, about one-third of the way through his paper [1], but he came back to this point at the end of the paper. The "complex molecule" in the gas might have been a macroscopic body interacting with its surroundings. Boltzmann thus thought that he had given a proof of the Second Law on a purely mechanical basis.

The *H*-theorem led to a strange situation, perhaps unique in the history of science: on the one hand, the Boltzmann equation had been successfully applied to a large number of physical phenomena; on the other hand, Boltzmann's ideas met with violent objections put forward by both physicists and mathematicians. These objections are usually formulated in the form of two paradoxes: *Loschmidt's paradox* and *Zermelo's paradox*.

Before examining the objections, it is important however to remark that the Boltzmann equation is, historically, the first equation to govern the evolution in time of a probability.

Now probability theory is very difficult to understand properly. Most people (even some mathematicians) think that it is something non-rigorous, non-scientific, and approximate, although it has now become one of the most profound and useful areas of mathematics. The fact is that most people think of the use of probability in the terms described in the following delightful little poem by the Danish architect and resistance leader during the Nazi invasion, Piet Hein. It is taken from his booklet *Grooks* [12] (a title that may sound more mysterious than probability; presumably it means epigrammatic poems):

A psychological tip

Whenever you're called on to make up your mind,
    and you're hampered by not having any,
the best way to solve the dilemma, you'll find,
    is simply by spinning a penny.
No—not so that chance shall decide the affair
    while you're passively standing there moping;
but the moment the penny is up in the air,
    you suddenly know what you're hoping.

In what follows, we shall first discuss the paradoxes, by stating the objections to Boltzmann's argument and Boltzmann's answer. We shall then deal with modern understanding and misunderstanding of the matter.

5.2 Loschmidt's paradox

Boltzmann's *H*-theorem is of basic importance because it shows that his equation has a basic feature of irreversibility: the quantity *H* always decreases in time (when the gas does not exchange mass and energy with a solid boundary). This result seems to be in conflict with the fact that the molecules constituting the gas follow the laws of classical mechanics, which are time-reversible. Accordingly, given a motion at a certain instant with certain molecular velocities, we can always consider the motion with opposite velocities (and the same molecular positions as before) at the same time instant;

the backward evolution of the latter state will be equal to the forward evolution of the original one. Therefore if H decreases in the first case, it will increase in the second case, which contradicts Boltzmann's H-theorem.

This paradox is mentioned by Thomson in a short paper which is seldom quoted [13]. It appeared in 1874 and contains a substantial part of the physical aspects of the modern interpretation of irreversibility, not only for gases but also for more general systems made up of molecules. Thomson notes that

the instantaneous reversal of the motion of every moving particle of a system causes the system to move backwards, each particle of it along its old path, and at the same speed as before, when again in the same position. That is to say, in mathematical language, any solution remains a solution when t is changed into $-t$.

Even without invoking thermodynamics, one would here be considering phenomena that were paradoxical from a common-sense point of view.

If, then, the motion of every particle of matter in the universe were precisely reversed at any instant, the course of nature would be simply reversed for ever after. The bursting bubble of foam at the foot of a waterfall would reunite and descend into the water; the thermal motions would reconcentrate their energy, and throw the mass up the fall in drops re-forming into a close column of ascending water. Heat which had been generated by the friction of solids and dissipated by conduction, and radiation, and radiation with absorption, would come again to the place of contact, and throw the moving body back against the force to which it had previously yielded. Boulders would recover from the mud the materials required to rebuild them into their previous jagged forms, and would become reunited to the mountain peak from which they had formerly broken away. And if also the materialistic hypothesis of life were true, living creatures would grow backwards, with conscious knowledge of the future, but no memory of the past, and would become again unborn.

He also remarks: "If no selective influence, such as that of the ideal 'demon', guides individual molecules, the average result of their free motions and collisions must be to equalize the distribution of energy among them in the gross. . ."

In other words, the impossibility of observing macroscopic phenomena that run backwards with respect to those that are actually observed is, in the last analysis, due to the large number of molecules present even in macroscopically small volumes.

Josef Loschmidt, to whom the paradox is usually attributed, mentioned it briefly in the first of four articles devoted to the thermal equilibrium of a system of bodies subjected to gravitational forces [14]. His intention was to demonstrate that the heat death of the universe (which seems to follow from the Second Law of Thermodynamics) is not inevitable. He sought to

destroy the terrifying nimbus of the second law, which has made it appear as a principle of destruction for all living creatures in the universe; and, at the same time, to open up the comforting prospect that the human race does not depend upon coal or the sun to transform heat into work, but may have an inexhaustible supply of transformable heat.

During the course of an attempt to demonstrate (in contrast to Maxwell's assertions) that the temperature of a gas at rest in a gravitational field depends upon height, he notes that in any system "the entire course of events will be retraced if, at some instant, the velocities of all its parts are reversed."

We have met Loschmidt in Chapters 1 and 3. From the latter chapter we know that he was a convinced atomist, from the former that he was a good friend of Boltzmann. We recall the fact that Loschmidt had proposed that he and Boltzmann should grind some spheres made of sulphur crystals while they were queuing to buy tickets at the Burgtheater. Here we may add that in a letter written to Boltzmann by his wife we learn that Loschmidt had been entranced by her coffee-pot and had bought another for himself. Probably the two friends had discussed the objection together. This may explain how, in spite of the obscure arguments of Loschmidt, Boltzmann immediately grasped the essential point.

He published a paper [15] in which he pays a handsome tribute to his critic because the doubt about the demonstration of the H-theorem "is ingeniously thought out and seems to be of great importance for a correct understanding of the Second Law". Then he proceeds to state the paradox in a much clearer form ("because the original statement by Loschmidt may seem difficult to understand for physicists due to its rather philosophical formulation"). Finally Boltzmann gives a thorough discussion of the paradox, concluding in a manner similar to that of Thomson.

We remark that in this paper Boltzmann explicitly recognizes the probabilistic nature of the Second Law:

We must make the following remark: a proof, that after a certain time t_1 the spheres must necessarily be mixed uniformly, whatever may be the initial distribution of states, cannot be given. This is in fact a consequence of probability theory, for any non-uniform distribution of states, no matter how improbable it may be, is still not absolutely impossible. Indeed it is clear that any individual uniform distribution, which might arise after a certain time from some particular initial state, is just as improbable as an individual non-uniform distribution; just as, in the game of Lotto, any individual set of five numbers is as improbable as 1, 2, 3, 4, 5. It is only because there are many more uniform distributions than non-uniform ones that the distribution of states will become uniform in the course of time. One therefore cannot prove that, whatever may be the positions and velocities of the spheres at the beginning, the distribution must become uniform after a long time; rather one can only prove that infinitely many more initial states will lead to a uniform one after a definite length of time than to a non-uniform one. Loschmidt's theorem tells us only about initial states which actually lead to a very non-uniform distribution after a certain time t_1; but it does not prove that there are not infinitely many more initial conditions that will lead to a uniform distribution after the same time. On the contrary, it follows from the theorem itself that, since there are infinitely many more uniform than non-uniform distributions, the number of states which lead to uniform distributions after a certain time t_1 is much greater than the number that leads to non-uniform ones, and that the latter are the ones that must be chosen, according to Loschmidt, in order to obtain a non-uniform distribution at t_1. [15]

We remark that the use of terms such as "ordered states" or "uniformity" may be confusing, since there are various levels of description at which one can consider "order". When talking about minima in the amount of order, or about an unusually ordered state, we should never forget that there are always constraints, given by the level at which we want to describe a system. We shall make further comments on this point in Section 5.4; here we can remark that if we look at individual molecules, all the states are ordered. This also remains true if the state is not known with perfect accuracy, but we insist on a probability referring to the positions and momenta of all the molecules (and satisfying the Liouville equation). It is only when we pass to a reduced description, based on the one-particle distribution function, that we lump many states into a single state and we can talk about highly probable (disordered) states; these are states into which, in the reduced description, an extremely large number of microscopic states are lumped together.

5.3 Poincaré's recurrence and Zermelo's paradox

There is another objection that can be raised against the H-theorem when presented as a rigorous consequence of the laws of dynamics. This paradox goes under the name of the famous mathematician Zermelo, who was then a young assistant of Max Planck at the Institute of Theoretical Physics in Berlin. He stated it in 1896, but he was not the first to introduce this argument. Even if we neglect a statement due to the philosopher Friedrich Nietzsche (1844–1900) [16], based on the fact that, as he thought, the world has always existed and cannot have a final state (otherwise this would have been already reached) and thus it cannot but repeat the same events infinitely many times, we find it for the first time in a short paper by Poincaré [17].

The famous French mathematician is concerned with obstacles met by the "mechanistic conception of the universe which has seduced so many good men". In fact, he says, "a theorem, easy to prove, tells us that a bounded world, governed by the laws of mechanics, will always pass through a state very close to its initial state". After noting the contradiction with the Second Law, he goes on to say (apparently ignoring Boltzmann):

I do not know if it has been remarked that the English kinetic theories can extricate themselves from this contradiction. The world, according to them, tends at first toward a state where it remains for a long time without apparent change; and this is consistent with experience; but it does not remain that way forever, if the theorem cited above is not violated; it merely stays there for an enormously long time, a time which is longer the more numerous are the molecules. This state will not be the final death of the universe, but a sort of slumber, from which it will awake after millions of millions of centuries. According to this theory, to see heat pass from a cold body to a warm one, it will not be necessary to have the acute vision, the intelligence, and dexterity of Maxwell's demon; it will suffice to have a little patience.

Thus Poincaré was fighting against the idea that everything can be reduced to the motion of atoms, or, in the verses of the previously quoted Piet Hein [12]:

Atomyriades

Nature, it seems, is the popular name
for milliards and milliards and milliards
of particles playing their infinite game
of billiards and billiards and billiards.

The theorem to which Poincaré alludes in the passage quoted above had been presented by him three years earlier in a famous paper on the three-body problem [18] and he had shown it in the particular case of a system governed by three first-order differential equations. In 1896 Zermelo [19], starting from this memoir of Poincaré's [18] and ignoring the latter's short paper [17], gives a short proof of the recurrence theorem for a system with any (finite) number of degrees of freedom. Then he applies it to the kinetic theory of gases with remarks similar to those of Poincaré. His conclusion is that

it is in any case *impossible* on the basis of the *present* theory to carry out a mechanical derivation of the second law without specializing the initial state. It is likewise impossible to prove that the well-known velocity distribution will be reached as a stationary final state, as its discoverers Maxwell and Boltzmann wished to do. I have not given a detailed examination of the various attempts at such a proof by Boltzmann and Lorentz, since because of the difficulties of the subject I would rather explain as clearly as possible what can be proved rigorously and what seems to be of greatest importance, and thereby contribute to a renewed discussion and final solution of these problems.

Zermelo did not have to wait too long for Boltzmann's answer, which was rather sharp [20]. As in his answer to Loschmidt, he mentioned that there was no question of proving the Second Law on a mechanical basis. On the contrary, the general validity of the principle was questioned because kinetic theory showed that its content was not completely certain but only highly probable. This statement is followed by a rather bitter sentence:

Zermelo's paper shows that my writings have been misunderstood; nevertheless it pleases me for it seems to be the first indication that these writings have been paid any attention in Germany. Poincaré's theorem, which Zermelo explains at the beginning of his paper, is clearly correct, but his application of it to the theory of heat is not.

The main point underlined by Boltzmann is that, at variance with Zermelo's opinion (but in agreement with the above-mentioned paper of Poincaré), the states that after a sufficiently long time lead to an equilibrium state are in an overwhelming majority. Concerning the recurrence time, Boltzmann remarks that it must be enormous. He remarks that, for a gas of 10^{18} particles in a box of 1 cm^3 the recurrence time (in seconds) should be given by a number with 10^{18} decimal figures, a quite inconceivable number (the presently estimated age in seconds of the Universe has just 17 figures). He continues [20]:

Thus when Zermelo concludes, from the theoretical fact that the initial states in a gas must recur— without having calculated how long a time this will take—, that the hypotheses of gas theory must be rejected or else fundamentally changed, he is just like a dice player who has calculated that the probability of a sequence of 1000 one's is not zero, and then concludes that his dice must be loaded since he has not yet observed such a sequence!

The papers by Zermelo and Boltzmann which have been just discussed are followed by two other papers by the same authors [21, 22], where they essentially repeat their arguments and viewpoints. However, after suggesting that one must be careful in applying a theory ("our thought pictures") beyond the domain where it has been experimentally tested, in his second reply Boltzmann indicates the two possible alternative statements on the state and evolution of the universe which must be taken into account in order to obtain a coherent view. It is worth while to quote the whole passage [22]:

One has the choice between two kinds of pictures. One can assume that the entire universe finds itself at present in a very improbable state. However, one may suppose that the aeons during which this improbable state lasts, and the distance from here to Sirius, are minute compared to the age and size of the universe. There must then be in the universe, which is in thermal equilibrium as a whole and therefore dead, here and there relatively small regions of the size of our galaxy (which we call worlds), which during the relatively short time of aeons deviate significantly from thermal equilibrium. Among these worlds the state probability increases as often as it decreases. For the universe as a whole the two directions of time are indistinguishable, just as in space there is no up and down. However, just as at a certain place on the earth we can call "down" the direction toward the centre of the earth, so a living being that finds itself in such a world at a certain period of time can define the time direction as going from less probable to more probable states (the former will be the "past", the latter the "future") and by virtue of this definition he will find that this small region, isolated from the rest of the universe, is "initially" always in an improbable state. This viewpoint seems to me the only way in which one can understand the validity of the Second Law and the heat death of each individual world, without invoking an unidirectional change of the entire universe from a definite initial state to final state.

Boltzmann makes several other important remarks. He starts by rejecting a possible "Machian" objection to his theory of dead universes [22]:

The objection that it is uneconomical and hence useless to imagine such a large part of the universe as being dead in order to explain why a small part is living—this objection I consider invalid. I remember only too well a person who absolutely refused to believe that the sun could be 20 million miles from the earth, on the grounds that it was inconceivable that there could be so much space filled only with aether and so little with life.

After remarking that giving up the statistical description would impoverish our theoretical understanding of the universe, and discussing the behaviour of H *versus* time (the so-called H-curve; see Chapter 6), Boltzmann concludes his paper [22] with the following remark:

The Poincaré theorem is of course inapplicable to a terrestrial body which we can observe, since such body is not completely isolated; likewise, it is inapplicable to the completely isolated gas treated by the kinetic theory, if one first lets the number of molecules become infinite, and then the quotient of the time between successive collisions and the time of observation [becomes zero].

5.4 The physical and mathematical resolution of the paradoxes

Who won the battle? Zermelo or Boltzmann? The physicists decided to follow Boltzmann's view, especially since atoms were beyond any doubt shown to exist (two years after Boltzmann committed suicide in 1906). Before that, Albert Einstein had explained Brownian motion by means of Boltzmann's ideas and, even earlier, Max Planck had been forced to use statistical concepts to explain the spectrum of black-body radiation (see Chapter 12). In 1930 the statistical justification of the Second Law was a sound fact for any physicist.

Pure mathematicians, however could not accept Boltzmann's arguments; a *theorem* may be true or false, but not *probably true*. In other words: either the H-theorem is a true theorem which can be applied to real gases, or it is not a theorem, and then what are we talking about? Thus Zermelo had won, according to pure mathematicians.

On the other hand, it was a Pyrrhic victory. Boltzmann had not succeeded in proving a mathematical theorem, because he lacked the mathematical language suitable for expressing his ideas. It has been remarked, by the way, that this is true for Poincaré and Zermelo as well, whose theorems are not much more rigorous than Nietzsche's statement [16], since measure theory, needed to express the recurrence theorem in a rigorous way, was not introduced by Lebesgue until 1902. However, this is a rather pedantic remark, because the essential properties required in the proof of Poincaré's theorem are the invariance of the volume of an evolving set in phase space and the fact that the volume of the union of a finite number of subsets equals the sum of the individual volumes.

About 1950 however, people began to consider the possibility of formulating a non-paradoxical theorem, even though it would have been very difficult to prove. In 1949 Harold Grad wrote a paper [23] which became widely known because it contained a systematic method of solving the Boltzmann equation approximately. In the same paper however, Grad made a more basic contribution to the kinetic theory of gases. In fact, he formulated a conjecture on the validity of the Boltzmann equation, reminiscent of the passage quoted at the end of the previous section and of the passages in Boltzmann's lectures on gas theory that we quoted in the previous chapter. In Grad's words:

From the preceding discussion it is possible to see along what lines a rigorous derivation of the Boltzmann equation should proceed. First, from equilibrium considerations we must let the number density of molecules, N, increase without bound. At the same time we would like the macroscopic properties of the gas to be unchanged. To do this we allow m to approach zero in such a way that $mN = \rho$ is fixed. The Boltzmann equation for elastic spheres, (2.37) has a factor σ^2/m in the collision term. If σ is made to approach to zero at such a rate that σ^2/m is fixed, then the Boltzmann equation remains unaltered. [...] In the limiting process described here, it seems likely that solutions of Liouville's equation attain many of the significant properties of the Boltzmann equation.

Here m is the mass of a molecule, σ its diameter, ρ the gas density.

In the article in the *Handbuch der Physik* by the same author [24], one reads similar sentences. There the term *Boltzmann gas* is used to describe the system whose dynamics arises from the above-mentioned limit. In more recent literature, the limit when one allows the number of molecules (per unit volume) N to tend towards infinity and the diameter of the molecules to tend to zero, so that $N\sigma^2$ remains finite, is referred to as *the Boltzmann–Grad limit*. This name is extremely appropriate because, although the precise formulation of the limiting procedure is due to Grad, Boltzmann indicated that he had in mind something of this kind, as shown by the passage quoted at the end of the previous section and the passages of his lectures quoted in the previous chapter, to which wc may add thc following onc [6, Vol. 2, Sect. 39]:

We therefore set ourselves the problem of finding the limiting laws of the phenomena for an infinite number of molecules in unit volume, in the presence of external forces, and then assume that the actual phenomena will not noticeably deviate from these limits.

As we have remarked in Chapter 4, these statements occasionally puzzle physicists, who find it strange that Boltzmann talked about infinitesimal sizes and infinite numbers of molecules per unit volume, whereas he knew that the size of the molecules, though extremely small, is finite and the number of molecules per unit volume, though extremely large, is also finite. The statements are, on the other hand, a precise indication that Boltzmann fully appreciated that any statistical theory must leave room for fluctuations, no matter how small; these fluctuations can disappear and give birth to a kind of deterministic theory only if we take appropriate limits.

Let us return to recent developments. Ideas about entropy were changing, under the influence of the developments in physics and information theory. Grad again, in a paper written in 1961 [25] remarks:

One of the reasons of the bewilderment which is sometimes felt at an unheralded appearance of the term entropy is the superabundance of objects which bear this name. On the one hand, there is a large choice of macroscopic quantities (functions of state variables) called entropy, on the other hand, a variety of microscopic quantities, similarly named, associated with the logarithm of a probability or the mean value of the logarithm of a density. Each one of these concepts is suited for a specific purpose. More confusing, however, than the lack of imagination in terminology is the fact that several of these distinct concepts, different in meaning and in numerical value, may be significant in a problem.[...]. A given object of study cannot always be assigned a unique value, its "entropy". It may have many different entropies, each one worthwhile. The proper choice will depend on the interests of the individual, the particular phenomena under study, the degree of precision available or arbitrarily decided upon, or the method of description which is employed; and each of these criteria is largely subject to the discretion of the individual. The fertility of this concept is in large part due to its flexibility and multiple meanings. On the other hand, much of the confusion in the subject is traceable to the ostensibly unifying belief (possibly theological in origin!) that there is only one entropy. Although the necessity of dealing with distinct entropies has become conventional in some areas, in others there is an extraordinary reluctance to do so. The widespread misconception that there exists a paradox in classical statistical mechanics associated with the name Gibbs [see Chapter 7], and the frequent difficulties bound to the classical reversibility–irreversibility dichotomy are directly traceable to this source.

One had arrived at the idea that entropy, in a statistical theory, describes something qualitative about our knowledge of the system, a measure of its disorder. And of course a thing is more or less ordered, according to the level of description (herein lies the discretion of the individual!). This aspect will be taken up again in the next section.

The advancement of science offered something more than just new tools and concepts: on the one hand, experimental technique in fact permitted observation of highly ordered states and the possibility of time-inverting an unusually large number of particles in the system, through the "spin echo" effect [26], in which an unusually large number of degrees of freedom can be actually time-reversed; on the other hand, the use of electronic computers allowed numerical experiments on systems with a large number of degrees of freedom. Among these are found unusual experiments, such as the calculation of the quantity H from molecular dynamics and the study of the effect of the time inversion on that quantity. Boltzmann's ideas were of course confirmed [27, 28]. The quantity H decreases on average (there are fluctuations, because of the finite, and even rather small, number of particles) and the system reaches statistical equilibrium after a few collisions per particle (Fig. 5.1). If however after 50 or 100 collisions we invert the velocities and

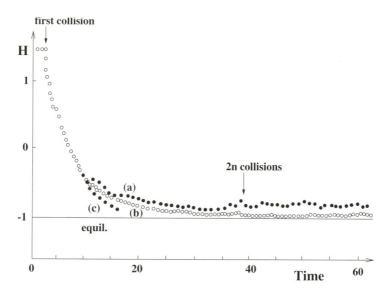

FIG. 5.1. The behaviour of H in a computer experiment [28]. The total number of disks n is 100 in case (a), 484 in case (b), 1225 in case (c).

follow the evolution of H, we find that it initially *increases* until it reaches its initial value (this is true with modern cellullar automata or with numerical schemes designed to preserve time symmetry [29], but only approximately true in the original calculations [27, 28], because of rounding-off errors) and then starts to decrease in a manner similar to that observed in the absence of time inversion (Fig. 5.2). One can thus observe a violation of the Second Law.

This circumstance, far from being in conflict with the statistical interpretation, fully confirms it. Boltzmann had emphasized that his derivation of the evolution equation for the distribution function depended not only on the laws of mechanics but also on the initial conditions. The fact that we usually observe a decreasing H, rather than an increasing one, is related to the small value of the probability of initial data leading to an increase in H; among these initial states however, one finds those which can be obtained by a time inversion of a state reached by the system when evolving from a probable state. The spin echo and numerical experiments are concrete examples of what a hypothetical "Loschmidt's demon", capable of inverting all of the molecular velocities, might do.

At the beginning of the 1970s it was clear what theorem one should try to prove: "If the distribution of the velocities and positions of the molecules is initially factorized, i.e. the probability density of finding the N molecules of the system in a certain state corresponding to the positions and velocities $(x_1, \xi_1; x_2, \xi_2; \ldots; x_N, \xi_N)$ is simply the product of the probability densities of finding the first molecule in state (x_1, ξ_1), the second in state $(x_2, \xi_2), \ldots$, the Nth in state (x_N, ξ_N), then the one-particle distribution function will be, *asymptotically with N tending towards infinity and σ going to zero in such a way that $N\sigma^2$ remains finite*, a solution of the Boltzmann equation; in particular,

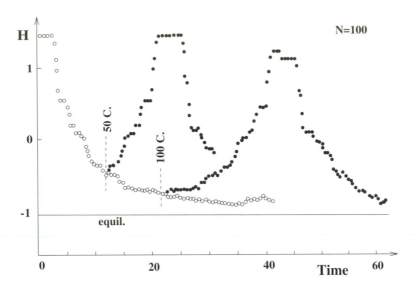

FIG. 5.2.   The effect of time inversion on the behaviour of H in a computer experiment on a system of 100 disks [28]. The white dots correspond to initial data obtained from those of Fig. 5.1 by reversing the velocities, whereas the two sequences of black dots are obtained by reversing the velocities after 50 and 100 collisions in the solution of Fig. 5.1.

the quantity H associated with it will be, in the same limit, a monotonically decreasing function of time".

Anyone who attempts to prove this result with the rigorous techniques of mathematical analysis immediately realizes that his task is at least as difficult as proving a so-called existence theorem for the Boltzmann equation itself, i.e. a theorem that guarantees that given f at $t = 0$, we can find f for $t > 0$. But, the author of this book asked himself in 1972 [30], if the theorem is true, would it not be possible at least to verify its validity, by assuming that all of the limits that are required actually exist in some sense; that they are sufficiently regular for operating on them, in some sense, in the manner required to write down the Boltzmann equation; and finally that a uniqueness theorem is valid for the evolution equations that we obtain for the evolution of the distribution functions when we let N go to infinity. Thus he wrote a paper in which he verified this aspect of the matter. The problem of justifying the Boltzmann equation was essentially reduced to a matter of proving the existence of certain limits and the existence or uniqueness of the solution of certain equations. *Reduce* is of course a rather inappropriate verb, given the complexity of the task that faces anyone who wants to proceed to the next step, a task still uncompleted today.

O.E. Lanford however saw that there was an intermediate step that was within reach. In fact, he proved [31] that the informal derivation becomes rigorous if one limits oneself to a sufficiently short time interval, actually a very short one on a macroscopic scale, since it is of the order of one-fifth of the average time between two subsequent collisions of any given molecule, but quite sufficient from a conceptual viewpoint, since about 20% of the molecules have collided in that time interval. Thanks to Lanford's work, the Boltzmann equation is finally a theorem, although it has been proved for a short time interval only! The physicists were right, but for their statements to have the force of a theorem, it was required to take the limit $N \to \infty$.

What is the situation for times longer than Lanford's critical time? The informal derivation [30] did not show any difficulty related to the length of the time interval; the restriction arose rather from the necessity to prove the *existence* of the mathematical objects about which one is talking, for arbitrarily long time intervals.

The rigorous theory of the Boltzmann equation started in 1933 with a paper by Torsten Carleman, who proved a theorem of global existence and uniqueness for a gas of hard spheres in the so-called space homogeneous case. The theorem was proved under the restrictive assumption that the initial data depend upon the molecular velocity only through its magnitude [2]. This restriction is removed in a posthumous book by the same author [3].

The existence theory was extended by D. Morgenstern [32], who proved a global existence theorem for a gas of Maxwellian molecules in the space homogeneous case. His work was further extended by L. Arkeryd in 1972 [33].

Important methods of analysis developed by H. Grad were brought to completion much later by Japanese authors [34–38].

The problem of proving rigorous theorems on the Boltzmann equation received new impetus when its connection with a rigorous, globally valid justification of the Boltzmann equation became clear.

The results of Carleman, Morgenstern, and Arkeryd referred to the case in which the solution of the Boltzmann equation does not depend on the space coordinates and thus these results do not lend themselves to being applied to the matter of the validity of the Boltzmann equation (because the space correlations that arise for a finite N are present in the space homogeneous case as well). In 1983 however, Reinhard Illner and the late Marvin Shinbrot [39] proved a global existence theorem, i.e. a theorem holding for an arbitrarily long time, for the Boltzmann equation in the case of a sufficiently rarefied gas expanding into a vacuum. When Marvin Shinbrot illustrated this result to the author in October 1983, the latter pointed out to him that the same technique, with suitable adjustments and combinations with Lanford's arguments, should be applicable to the problem of justifying the Boltzmann equation, in the same physical situation, for arbitrarily long times.

The detailed proof was actually obtained by Reinhard Illner and Mario Pulvirenti in two subsequent papers [40, 41].

What is still lacking to bring the theory to completion? The case of a gas in a box, such as the one considered by Lanford, without any restriction on the time interval. Once again our hopes depend upon the existence and uniqueness theorem for the Boltzmann equation. A few years ago the late Ronald DiPerna and Pierre-Louis Lions provided an ingenious proof of existence for the Boltzmann equation [42]. On 4 August 1994 P. L. Lions was awarded the Fields Medal (named after the Canadian mathematician Fields who first conceived this prize as a substitute for a non-existent Nobel prize for mathematics), and his work on the Boltzmann equation is much stressed in the motivation for the choice of the committee. Several technical details of the DiPerna–Lions proof, together with the fact that their existence theorem is not accompanied by a uniqueness theorem, make the application of analogous ideas to the problem of the validity of the Boltzmann equation problematic at the least.

With today's mathematical equipment we are more prepared to discuss the paradoxes described in the previous sections. In fact, we remark that when giving a justification of the Boltzmann equation, one uses the laws of elastic collisions and the continuity of the probability density at impact to express the distribution functions corresponding to an after-collision state in terms of the distribution functions corresponding to the state before the collision, rather than the latter in terms of the former. It is obvious that the first way is the right one to follow if the equations are to be used to predict the future from the past and not the other way around; it is clear however that this choice introduces a connection with the everyday concepts of past and future, extraneous to molecular dynamics and based on our macroscopic experience. When we take the Boltzmann–Grad limit, we obtain equations which describe the statistical behaviour of the gas molecules: a striking consequence of our choice is that the Boltzmann equation describes motions for which the quantity H has a tendency to increase, while the opposite choice would have led to an equation having a negative sign in front of the collision term, and hence describing only motions with increasing H. We must remark that, in order to derive the Boltzmann equation, one takes special (although highly probable) initial data; thus certain special data are excluded. As mentioned before, these excluded data correspond to states in which the molecular velocities of the molecules that are about to collide show an unusual correlation.

In view of the fact that we claim validity for the Boltzmann equation in the Boltzmann–Grad limit only, we do not have to worry about the recurrence paradox either; in fact, Poincaré's theorem can no longer be applied to the set of the possible states of the system when $N \to \infty$ and the recurrence time is expected to go to infinity with N (at a much faster rate than N itself).

A further discussion of this point necessarily leads us to discuss the problem of time's arrow and its connection with the expansion of the Universe. Before doing this in the next section, we should discuss a modern objection to what Boltzmann did.

In fact one might well object that the atoms of today are not the same atoms as considered by Boltzmann. It is not easy to give a complete discussion of this point, in its relevant and irrelevant aspects. But Boltzmann himself was aware of an evolution of the concept of atom and indeed of the entire structure of theoretical physics. His viewpoint is well expressed by this quotation (1899) [43]:

Will the old mechanics, even if divested of metaphysics, continue to exist in its basic features or one day merely live on in history, displaced [...] by some quite different ideas? Of present day molecular theory, notwithstanding any addition and modifications, will the essential features nevertheless survive, or will the future one day bring an atomic theory that is totally different from today's; or, contrary to my demonstration, will it be found one day that the idea of a pure continuum affords the best picture?....Interesting questions indeed! One almost regrets having to die long before they are decided. How immoderate we mortals are! Delight in watching the fluctuations of the contest is our true lot.

This is a very profound remark. To make it sound a bit less sad, I would like to use Piet Hein's verses again [12]:

> *I'd like—*
> I'd like to know
> what this whole show
> is all about
> before it's out.

What does this quotation mean? That, as Boltzmann pointed out, our theories are never a final picture of the world and we presumably discover various layers of reality one after the other, and a profound philosophical view on whether matter is continuous or discrete may be true, false, or even meaningless, depending on the theory that best summarizes our views at a certain depth of our excavation. This undogmatic viewpoint was also adopted by Boltzmann in his lectures on gas theory [6]:

...since the history of science shows how often epistemological generalizations have turned out to be false, may it not turn out that the present "modern" distaste for special representations, as well as the distinction between qualitatively different forms of energy, will have been a retrogression? Who sees the future? Let us have free scope for all directions of research; away with all dogmatism, either atomistic or antiatomistic! In describing the theory of gases as a mechanical *analogy*, we have already indicated, by the choice of this word, how far removed we are from the viewpoint which would see in visible matter the true properties of the smallest particles of the body.

It seems however that Boltzmann captured an important feature which seems to reappear at various depths in one form or another. Irreversibility may appear even if the basic laws are reversible. Initial data and probability are also important. Certainly, the future may have some surprise in store for us, such as a connection between the Second Law and the collapse of the wave-packet in quantum mechanics, as discussed at length by Roger Penrose [44]. It is in the nature of science to speculate and suggest new views, but one should not hasten to subscribe to a brilliant idea before it is carefully tested.

5.5 Time's arrow and the expanding universe

The link between time irreversibility and the Second Law of Thermodynamics was rendered popular by the famous English astronomer Sir Arthur Eddington, in his Messenger Lectures delivered at Cornell University in 1934 [45]. After having asked himself ". . . is there everywhere and everywhen in the physical universe a signpost with one arm marked 'To the Future' and the other arm 'To the Past'?", he concludes that the signpost, the arrow which tells us about the direction of time, is provided by entropy and is based upon the Second Law of Thermodynamics. Eddington (but not Boltzmann, as we saw) does not seem to harbour doubts about the fact that even the statistical treatment would find a time arrow. What can the mathematical treatments that we have just spoken about tell us?

At first sight they would seem to confirm the existence of this arrow, which points towards increasing values of entropy giving a direction to the time axis, but a more careful examination shows us that, from a certain point of view, this is an illusion. In fact, if instead of studying the molecular motion starting from a certain instant with increasing time, we had studied it with decreasing time, we would have obtained that entropy increases with decreasing values of the time variable! In other words, the equations of dynamics, as we already knew, cannot provide a distinction between the two directions along the time axis. If they had in fact done so, we would have a *logical paradox* because the equations themselves are completely reversible. And it is not by treating the problem in a statistical manner, that is, by ignoring part of the information on our gas, that we can make the equations show us the direction of time which they would otherwise not indicate. We therefore conclude that the demonstrated theorems show only that for our gas, given that the initial datum is not in an unlikely set, the entropy increases with increasing or decreasing t. One way of reconciling this result with experience would be to say: "Fine! It is sufficient to *define* as the direction of increasing time that in which entropy increases". As we saw previously, this more or less was Boltzmann's suggestion. This would be satisfactory if there were only one box full of gas in the universe; it would be a very simple but rather boring universe. Our universe is much more complicated: ignoring many other things, we are forced to admit at the very least that there is more than one box full of gas and that some of these could indicate that entropy increases with increasing time, others with decreasing time (with respect to an arbitrarily oriented axis). In the terminology of twentieth century philosophers, these subsystems are called "branch" systems. In this terminology, it would not be possible to change the direction of the time axis so that entropy increases for branch systems. And one of the problems is to explain why these branch systems agree in showing the same time arrow, as experience shows. So what? One must admit that in reality these systems are not isolated but couple with the rest of the universe, so that even when a system is said to be isolated, in reality it must be able to exchange (or to have exchanged in the past) very small quantities of energy, for example in the form of electromagnetic and gravitational radiation. Then the increase in entropy may be applied only to the entire universe (given that it is described by a dynamical system which leads to conclusions similar to those found for a classical gas) and to its parts which are sufficiently isolated (as Boltzmann had indicated in his second answer to Zermelo [22], previously quoted); but the arrow is now the same for all the parts. In reality, since very distant bodies can exchange energy only by radiation and we note that radiation tends to leave rather than arrive (Olbers's paradox [46]), the universe is rather different from a box full of gas, in that it does not seem to reflect radiation. This is related to the fact that the universe is expanding and matter moves away from matter extremely rapidly. But even the phenomenon of expansion (unknown to Boltzmann) is one that throws light upon a privileged direction of time; and it is this, in the final analysis, that is the real arrow of time. This observation seems to have been explicitly made for the first time by Thomas Gold [47].

The expansion of the universe is thus intimately related to all physical processes down to the most minute dimensions, and a deeper comprehension of the laws of physics could allow us one day to deduce the expansion of the universe from the observation of very small-scale phenomena.

The universe appears to be in expansion and not in contraction; does this mean perhaps that between the two possible directions of motion nature has chosen one of them?

Certainly not. This would be true if natural laws were not symmetrical with respect to time reversal: then a universe in expansion would be distinguishable from one in contraction and ours would be of the former type. We could then ask ourselves why we are not in the other type of universe. This however is not the case, and the only time arrow (at a macroscopic level) is provided precisely by this expansion. Naturally, there is nothing new about the idea that the laws of physics are more symmetrical than the universe to which they apply. Although all inertial systems are equivalent, according to the mechanics of Galileo and Newton as well as to that of Einstein's special relativity, one can always find a privileged system defined by the fact that from within it the observer sees an isotropic expansion of the universe about him. And temperature has become the clue to most modern theories of elementary particles.

Therefore the laws of physics would give us two universes, one being symmetrical, with respect to time, to the other; there is no point in seeking why we find ourselves in one instead of the other, because they cannot be distinguished. The same reasoning, as Gold remarks [47], may be applied to the symmetry between matter and antimatter: a world of antimatter is indistinguishable from one of matter. When he wrote this, the violation of symmetry between matter and antimatter [48] had not yet been discovered; only the combination between the symmetry and temporal symmetry is verified. This is even more satisfying from a conceptual point of view, because only two universes are possible [49]: one differs from the other by the distinction between matter and antimatter and by the direction of time, but they are indistinguishable. The symmetries of the laws of physics are precisely those required to describe one single observable universe.

At this point it become inevitable that we ask ourselves why, subjectively, we are so sure that time "really does go" in one direction and not in the other. With physical laws which are symmetrical with respect to time reversal, we can after all predict the present just as well from a knowledge of the future as from a knowledge of the past. Why then are past and future so different for all of us? Why do we not think about evolution in the past, instead of the future, using the laws of physics? Why do we consider information about the past as being completely defined, while, as an anonymous Renaissance poet, probably Lorenzo the Magnificent, put it, *del doman non v'è certezza* ("the future holds no certainty")? Why do we think that the state of a system today is the *cause* of its state tomorrow and not of its state yesterday? We have to conclude from the study of complex systems (such as a gas) that we are dealing only with enormously greater precision in predicting (apart from not so significant statistical fluctuations) the future at a macroscopic level; to predict the past is like predicting in what position a die was at the moment it was cast, by knowing that it lies at a certain point of the table and is showing its six-dot face. And we ourselves are time-oriented: even more complex systems, so complex that for us such precision of a purely statistical nature in the prediction of the future, made in the absence of detailed information on the data, reaches almost absolute certainty as long as we are happy with the same lack of detail in the results of our prediction.

Note also that the term *reversible* (with respect to a time reversal) has different meanings in the mathematical literature; a detailed discussion on this point can be found in a paper by R. Illner and H. Neunzert [50].

It is perhaps not out of place to comment here on a statement that is frequently made, even in very recent books (see [51] and Section 5.6), to the effect that no kind of irreversibility can follow by correct mathematics from the analytical dynamics of a conservative system and hence some assumption of kinetic theory must contradict analytical dynamics. It should be clear that it is not a new assumption that is introduced, but the fact that we study asymptotic properties of a conservative system in the Boltzmann–Grad limit.

Actually, there are models which can be used to illustrate the nature of the paradoxes and the mathematical significance of the growth of entropy, even without using a computer. Perhaps the simplest is a model of an hourglass (Appendix 5.4). An hourglass is rather complicated from a mechanical point of view, but we can give a simplified description of the way it works by ignoring the energy, momentum, etc. of the sand grains and paying attention only to the number of grains in one of the two bulbs of the instrument. The world of an hourglass has a beginning and an end. The beginning is when we put it on a table and the end is when all the grains are in the lower bulb. Usually at the beginning all the sand grains are in the upper bulb, but we can envisage different initial conditions. Somebody who turns the hourglass over, and thus exchanges the positions of the two bulbs so quickly that during the operation no grain passes through the little hole separating them, performs an operation of time reversal, while the "little universe" under consideration can be in any state and thus Loschmidt's paradox occurs before our eyes. It is clear that the Zermelo paradox does not show up in this model, nor do fluctuations. So our model tells only half the story, but is rather instructive anyway; in particular, our conclusions (see Appendix 5.4) will be different from those of the book quoted above [51] and in agreement with Boltzmann's arguments discussed in this and the previous sections.

5.6 Is irreversibility objective or subjective?

Time's arrow is a topic surrounded by an atmosphere of mystery, which has been popularized by science fiction through the possibility of time travel. This concept acquired an aura of scientific respectability thanks to a paper by Kurt Gödel [52]. Not just travel into the future, a consequence of the Theory of Special Relativity, but also travel into the past is theoretically possible. The ground for this provocative view is offered by Gödel's discovery of certain solutions of the field equations of General Relativity that describe worlds in which closed causal chains exist. Gödel himself states: "[...] by making a round trip on a rocket ship it is possible in these worlds to travel into any region of the past, present, and future and back again, exactly as it is possible in other worlds to travel to distant parts of space." [52]

There is also a vast philosophical literature on the past–future asymmetry of time. As we have stressed before, there is no intrinsic difference between the two directions of time in the equations of classical mechanics, but this is not in contradiction with Boltzmann's results.

This section is mainly philosophical in flavour. As such, it should perhaps be put in Chapter 10, devoted to Boltzmann's philosophical views. The fact is that Boltzmann himself never wrote about the remarkable philosophical consequences that one can draw from his H-theorem and, more generally, from the probabilistic interpretation of irreversible processes. But philosophers of the twentieth century have been greatly stimulated by this topic. Although Boltzmann had made a lot of effort to clarify his views on irreversibility, one must admit that he failed in directing most philosophers and scientists towards the real issue of the time asymmetry of thermodynamics. This issue was identified by Boltzmann as one of explaining why the entropy of the universe was so low in the past and still is so low, compared with its equilibrium value. And this is a typically cosmological issue.

As a matter of fact, the H-theorem and the Boltzmann equation itself can be obtained only at the cost of restricting the initial values. Why these initial values are suitable for computing the future and not the past can be explained only at the cost of introducing anthropocentric arguments, which may vary from our conceptions about past and future, which sounds like postulating the irreversibility we wish to prove, to the way we handle macroscopic apparatus.

One of the first and most accurate philosophical treatments of the problem of irreversibility is that of Hans Reichenbach [53]. He starts by giving a clear definition of the problem of time asymmetry. He first considers the case of a symmetric order on a line:

When we say that a line, though serially ordered, does not have a direction, we mean that there is no way of distinguishing structurally between right and left, between the relation and its converse. In order to say which direction we wish to call "left", we have to point to the diagram; or we may give names to points and indicate the selected directions by the use of names. Had we decided to call "right" what we called "left", and vice versa, we would not notice any structural difference; that is the relation *to the left of* has the same structural properties as the relation *to the right of*.

After remarking that this is not the case of the continuum of negative and positive real numbers, where the product introduces a difference (the product of two positive numbers is positive, but the product of two negative numbers is also positive), Reichenbach goes on to say [53]:

Applying these results to the problem of time, we find that time is usually conceived as having not only an order but also a direction. The relation *earlier than* is regarded as being of the same kind as the relation *smaller than*, and as not being undirected like the relation *to the left of*. This means that we believe that the relation *earlier than* differs structurally from its converse, the relation *later than*.

He later defines "the future" as the time direction of entropy increase of the majority of almost isolated ("branch") systems and argues as follows [53]:

Suppose we find in the sand traces of footprints, somewhat smoothed out by wind, but still recognizable as impressions of human feet. We conclude from this "record" that at some earlier time a man walked over the sand, thus causing the footprints. What is the logical schema of this inference?

The different arrangements of grains of sand can be classified into states by using simple rules. [...] it would be incorrect to assume that all the arrangements of sand are equiprobable. [...] the state given by the footprints is a highly ordered state, whereas a smooth surface is an unordered state; [...] and we ask, how can we explain the presence of this ordered state?

[...] The answer is [...] that we assume the observed state to be the product of an interaction, that we prefer an interpretation in which the system was not isolated in the past. And as for thermodynamical statistics, this inference is based on a many-system probability, in contrast to a one-system probability.

Then Reichenbach makes a connection between this problem and one of the basic problems of philosophy, causation. We remark that, although the concepts of cause and effect are very commonly used in everyday life, they become mysterious in science, because of the by now familiar fact that the basic laws are time-symmetric. In physics (at least in classical physics), if a phenomenon B inevitably follows another phenomenon A, we cannot say, strictly speaking, that A causes B; in fact, we could say that B causes A, given the fact that there is no objective difference between past and future. Irreversible processes reintroduce a time order, but this according to Boltzmann is just a macroscopic feature, based on probability arguments.

Let us return to Reichenbach. He says [53]:

[...] the example of the footprints also helps us to analyse the meaning of *causal explanation*. Explanation in terms of causes is required when we meet with an isolated system displaying a state of order which in the history of the system is very improbable. We then assume that the system was not isolated at earlier times: explanation presents order in the present as the consequence of interaction in the past. The otherwise improbable occurrence is thus presented as probable, [...] we arrive at the following explication of terms *cause* and *effect: The cause is the interaction at the lower end of the branch run through by an isolated system which displays order; and the state of order is the effect.*

[...] In fact, the origin of the concept of cause is presumably to be sought in experiences in which man has been confronted by unexpected states of macro-order which have stimulated his imagination to assume past causes. [...]

Thus the explanation of unusually ordered states requires the reduction of order to interaction. And here Reichenbach notices the connection with the concept of entropy:

It is an empirical fact that in all branch systems the entropy increases in the same direction. For this empirical reason, the convention of defining positive time through growing entropy is inseparable from accepting causality as the general method of explanation. [...] If someone argues that it is a matter of convention to select the direction of growing entropy as the direction of time, his conception cannot be called false. But he must not commit the error often connected with other forms of conventionalism: the error of overlooking the empirical content associated with the use of this convention.

Why do we know the past and not the future? Here is Reichenbach's answer: "The statement that although the past can be recorded, the future cannot, is translatable into the statistical statement: *Isolated states of order are always postinteraction states, never preinteraction states.*" [53]

Reichenbach also notes that when correlations between two events at different points in space cannot be explained by a direct causal connection, it turns out that these events are correlated with a third event, their "common cause", in their common past. This obviously asymmetric structure is usually called "the fork asymmetry". In fact the three events are embedded in a V-shaped chain of events determined by physical laws, but need not be embedded in a Λ-shaped pattern (in this graph, time obviously runs upwards); thus there is always a characteristic antecedent event (located at the vertex of the V-shaped diagram), but there need be no characteristic subsequent event (located at the vertex of the Λ-shaped diagram). Thus two bottles of a drink can be indistinguishable even though they have never been in the same place at the same time; their common "cause" is the factory where they were produced or, even more remotely, the designer of their common shapes, labels, and caps.

Reichenbach's extremely ingenious line of thought has been challenged by other authors. Thus Mackie [54] maintains that we extract irreversibility from causation, and not the other way around; Earman [55] claims that the knowledge asymmetry stems from the causal asymmetry and that neither has much to do with entropy; von Wright [56] argues that the time orientation of causation originates from our ability to manipulate the future; Salmon [57] suggests that explanation should be defined as specification of causes.

This striking disagreement about the sources of time-asymmetric phenomena and about their interdependence is, according to Horwich [58], due to the failure to appreciate and accommodate the needs of a comprehensive account. Horwich himself [58] attempts to avoid this shortcoming by considering a broad conceptual network. His theory is that time itself has no intrinsic directionality or asymmetry. His explanation partly agrees with the arguments given in this chapter, as far as entropy is concerned, and with Reichenbach [53]. In particular, though defending Reichenbach's idea against Earman's objection [55], Horwich does not agree with him on the idea that our concept of the future reduces to some entropic relation. Horwich does not share the criticisms of Mackie [54] and von Wright [56] either. He underlines a single physical fact on which all the asymmetries depend: the "fork asymmetry", first introduced by Reichenbach [53] and elucidated by Salmon [57]. He also speculates about what cosmological conditions might account for it (thermodynamics, spontaneous trend from order to disorder, expanding universe); but as he says, this is really the business of physics. His treatment can accommodate any physical theory that gives a unique origin for the fork asymmetry, such as that sketched in the previous section.

We must stress that the work of Reichenbach and the other philosophers we have been quoting is mainly in line with the treatment of this chapter, provided we restrict ourselves to the macroscopic world and do not pretend to apply these arguments to a system of a few molecules.

We must also discuss some basic concepts that have been introduced by philosophers, which maybe have a value if properly understood in a certain context, but are wrong or at least useless if used in scientific explanations. Some of these ideas have unfortunately penetrated certain scientific circles. They are particular dangerous because, thanks to successful books directed to a wide audience and articles in newspapers, many people are led to believe that these wrong ideas are defects of Boltzmann's theory and that new views have been discovered which remedy these defects.

We shall skip here a discussion of the still (after the discovery of DNA!) much quoted [59, 60] views of Bergson [61] on time, which lead to thinking that life and matter are two opposed concepts, since the second can be understood by reason, the first (allegedly) only by intuition. These views may be fine if applied to the personal feelings of a living being, but are utterly out of place in books which purport to explain scientific conceptions.

Let us now turn to the criticism of Popper on Boltzmann's conception of a time arrow defined through entropy increase. He may have been misguided by the statements, correct if correctly interpreted, of scientists such as Heisenberg and Born. The former, ignoring Boltzmann, said [62]:

Gibbs was the first to introduce a physical concept which can only be applied to an object when our knowledge of the object is incomplete. If for instance the motion and the position of each molecule in a gas were known, then it would be pointless to continue speaking of the temperature of the gas.

Born wrote: "Irreversibility is therefore a consequence of the explicit introduction of ignorance into the fundamental laws." [63]

It is thus not surprising to read the following sentence of Popper [64]:

It is clearly absurd to believe that pennies fall or molecules collide in a random fashion *because we do not know* the initial conditions, and that they would do otherwise if some demon were to give their secret away to us: it is not only impossible, it is absurd to explain objective statistical frequencies by subjective ignorance.

Surprisingly, on the subsequent page however he gives what he calls "an objective probabilistic explanation of irreversible processes" not very different from Boltzmann's. In another book [65], after stating that "the strangely law-like behaviour of the statistical sequences remain, for the determinist, *ultimately irreducible and inexplicable*", he invented a rather obscure "propensity" to interpret probabilities. It may be unfair to quote these passages out of context, but this is the impression that a scientist receives when reading philosophical texts. Perhaps it is a problem of language, but then this should be carefully explained.

As we shall see in Chapter 10, Popper admired Boltzmann's philosophy but was very critical about his interpretation of the time arrow and branded it as "idealist". Actually the idea of a large fluctuation can be criticized, but Boltzmann had no other choice. General Relativity was still to come and ideas about cosmology were very rudimentary. However, this is not the criticism raised by Popper. He says, rather [66]:

I think that Boltzmann's idea is staggering in its boldness and beauty. But I also think that it is quite untenable, at least for a realist. It brands unidirectional change as an illusion. This makes the catastrophe of Hiroshima an illusion. Thus it makes our world an illusion, and with it *all our attempts to find out more about the world*. It is therefore self-defeating (like all idealism). Boltzmann's idealistic *ad hoc* hypothesis clashes with his own realistic and almost passionately maintained anti-idealistic philosophy, and with his passionate wish to know.

Here there seems to be a (not uncommon) confusion between idealism and anthropocentrism. In fact this is a very important aspect: we must distinguish between laws of nature and our way of speaking and feeling. The latter is perfectly respectable, acceptable, and justified, provided it is not used in scientific arguments. Thus when an astronomer says: "Tomorrow, the sun will rise at 6.23 a.m.", nobody will think that he is a supporter of the Ptolemaic system. It is a contracted statement of something that could be expressed in strictly rigorous scientific terms. When we come to feelings and historical events, things become even more complicated. One important aim of the philosophy of science should in fact be that of reconciling our intuitive views, arising from everyday life, with the objective findings of scientists. In fact our common views are frequently "illusions" from the viewpoint of scientific laws. Boltzmann made a big step towards clarification of the difference between physical time and the time of everyday life (which is more or less close to the time of thermodynamics, but not to the time of mechanics). Like all great explanations, Boltzmann's contribution shifted the problem to explaining something else, but he certainly provided immense progress in our understanding of different conceptions of time.

Another criticism of Popper's [67] shows that he had a poor understanding of statistical mechanics: in fact he states that Brownian motion is a serious problem for the Second Law. This is a very strange statement if made in 1958, eighty years after Maxwell [68] wrote:

Hence the second law of thermodynamics is continually being violated, and that to a considerable extent, in any sufficiently small group of molecules belonging to a real body. As the number of molecules in the group is increased, the deviations from the mean of the whole become smaller and less frequent; and when the number is increased till the group includes a sensible portion of the body, the probability of a measurable variation from the mean occurring in a finite number of years becomes so small that it may be regarded as practically an impossibility.

But P.K. Feyerabend went even further than Popper. In his popular book *Against method* [69] he invented a *perpetuum mobile* using a single molecule and explained that Brownian motion shows that the Second Law is wrong. This has of course been known since Maxwell and Boltzmann, if the Second Law is applied to atoms and not to macroscopic objects.

The discussion in the previous part of this section has several points of contact with a recent paper by J. Bricmont [70], particularly devoted to clarifying several confusions in the popular literature concerning chaos, determinism, the arrow of time, entropy, and the role of probability in physics. This article is recommended as a detailed study and criticism of some current views on irreversibility, as they are presented in two books quoted above [59, 60]. It has been followed by two short papers, one [71] containing the objections of Ilya Prigogine to [70], the other [72] Bricmont's reply. The main aspects of the issue reduce to matters of terminology. We have already pointed out that there are several possible definitions of irreversibility. In particular, it is true that for the usual, everyday irreversibility, two ingredients are required: an extremely large number of tiny objects (atoms or molecules) and suitable initial data. There might be confusion concerning the role of interaction. Intermolecular forces may enhance or diminish the trend towards irreversible behaviour, depending on whether they are mainly repulsive (as in a gas) or attractive (as in a solid). As pointed out by Bricmont [72], Prigogine, with his brilliant style, writes sentences that may sound appealing to philosophers and laymen (and unfortunately to some scientists as well) but puzzle well-informed scientists. Either his claims are taken literally, and then they are wrong; or they are suitably reinterpreted and then they express standard ideas in a confusing way.

5.7 Concluding remarks

This section is included to summarize the views of the author, which have appeared in this chapter mixed up with other viewpoints.

It seems clear that Boltzmann was able to capture the essence of the Second Law in a mathematical form. Maxwell had foreseen this essence in physical terms, but had missed the possibility of putting it into the form of equations. Perhaps he thought that it was impossible. It is to the everlasting credit of Ludwig Boltzmann that he discovered what Maxwell had left unaccomplished, and devoted his entire life to developing new ideas and interpretations of his discovery. It was a slow process of clarification in an increasingly hostile environment (this applies to Germany and France, not to Great Britain, where even criticism was always friendly).

The impossibility of observing macroscopic phenomena which run backwards with respect to those which are actually observed is, in the last analysis, due to the large number of molecules present even in macroscopically small volumes. The technical trick of taking the Boltzmann–Grad limit is useful in order to eliminate fluctuations which may play a role only for extremely small volumes or for times incredibly long and completely beyond the possibilities of human observation. We have also stressed the fact that, even in the Boltzmann–Grad limit, we cannot prove that after a certain time, initially separated molecules with different properties must necessarily be mixed uniformly, whatever the initial distribution of states. It is only because there are many more uniform distributions than non-uniform ones that the distribution of states will become uniform in the course of time. One therefore cannot prove that, whatever may be the positions and velocities of the molecules at the beginning, the distribution must become uniform after a long time; rather, one can prove only that infinitely many more initial states will lead to a uniform one after a definite length of time than to a non-uniform one. Thus the irreversible behaviour is not a result of the Boltzmann–Grad limit. We must insert suitable initial data. In the words of Boltzmann: "Since in the differential equations themselves there is absolutely nothing analogous to the second law of thermodynamics, the latter can be mechanically represented only by means of assumptions regarding initial conditions." [43, p.170]

The paradoxes only teach us how to construct the highly improbable initial states which actually lead to a very non-uniform distribution after a certain time.

The new facts discovered in the twentieth century are all in favour of Boltzmann's standpoint. We now feel that there must be a connection between the Second Law and the expansion of the Universe; Boltzmann was the first to point out, in his second answer to Zermelo [22], that cosmological arguments were needed to explain the fact that the initial data are a subset of all the conceivable ones. He did it, of course, by using the physics of his time. We cannot blame him for having not known General Relativity!

We have avoided talking about quantum theory, except for mentioning the ideas of Roger Penrose [44] about the most mysterious aspect of quantum mechanics, wave-packet collapse. This very phenomenon occurs during a measurement, i.e. the interaction of a microscopic object with a macroscopic one, the measuring apparatus. It is the only irreversible feature of quantum mechanics and the only one extraneous to the basic equations of this theory, which are perfectly time-reversible. It seems that more investigations are needed on this topic. One possibility is that the apparent irreversibility should be ascribed to the large number of degrees of freedom of the global system (the microscopic system on which we are conducting our measurement and the measuring instrument) and then a typical Boltzmann-type argument should apply (see [73] and references quoted therein). If we accept this kind of conclusion however, the wave function of quantum mechanics just describes our knowledge about the system and not the state of the system, in the same way as the Boltzmann equation tells us nothing about the actual state of the system, but just something about the probable outcomes of a measurement.

We should also quote the remarkable recent book by the philosopher H. Price [74], who summarizes the previous discussions and difficulties with time's arrow and defends the viewpoint that one must take the basic time symmetry much more seriously. Thus the author comes up with a new proposal to avoid the issue of the wave-packet collapse.

It is then surprising to see that there is an abundant literature, even aimed at a wide audience [59, 60, 75], which argues that yes, Boltzmann saw it correctly, but we need modern research to understand it correctly. And this modern research is not the still scanty mathematical proofs of Boltzmann's physical arguments [31, 38, 40, 41], but rather confusing arguments about chaotic phenomena.

6

Boltzmann's relation and the statistical interpretation of entropy

6.1 The probabilistic interpretation of thermodynamics

As we have seen, Boltzmann first used probabilistic arguments in his answer to Loschmidt's objection; up to that moment, though he mentioned probability in his papers, he seemed to think that the distribution function was a way of utilizing the techniques of mathematical analysis in order to count the actual numbers of molecules, and no hidden probabilistic assumption was contained in his arguments. In this connection, we can quote the famous sentence taken from the conclusions of the first part of his basic paper [1], which has been analysed in detail in the two previous chapters: "It has thus been rigorously proved that, whatever may be the initial distribution of kinetic energy, in the course of a very long time it must always necessarily approach the one found by Maxwell".

As we have tried to show in the previous chapters (and their appendices), his statements are actually true only when, by taking a suitable limit, the statistical fluctuations disappear; furthermore, probability plays a heavy role in excluding certain initial data.

None of these problems was apparent in Boltzmann's initial paper [1]. Yet, as remarked by M. Klein [2], Boltzmann, when answering to Loschmidt five years later [3], did not indicate that he had changed his viewpoint, or that he had deepened his understanding of the subject, as a consequence of the reflections caused by the objection that had been raised against him, but acted as if he were simply re-elaborating his old ideas.

Instead, he was conceding that one cannot prove that entropy "necessarily" increases and hence equilibrium is reached, since (as he said in a passage of a paper [3] quoted in the previous chapter) this "is in fact a consequence of probability theory, for any non-uniform distribution of states, no matter how improbable it may be, is still not absolutely impossible." Loschmidt's objection had only made it clearer than ever "how intimately connected are the second law and probability theory". And, remarking that the improbability of certain states, such as those that lead to an entropy decrease, makes them practically impossible, Boltzmann says: "One could even calculate, from the relative numbers of the different state distributions, their probabilities, which might lead to an interesting method for the calculation of thermal equilibrium."

This remark is developed a few months later in another paper, which we shall presently deal with. However, we must first underline the circumstance that, if Boltzmann had already begun to hint at the important role of probability in 1871 [4], the priority in stressing the necessity of a statistical interpretation of the Second Law must certainly be credited to Maxwell because of his invention of the demon now named after him [5, 6], as we mentioned in Chapter 3. In this connection one can also quote from a letter to John William Strutt (later Lord Rayleigh) [7]: "The 2nd law of thermodynamics has the same degree of truth as the statement that if you throw a tumblerful of water into the sea, you cannot get the same tumblerful out of the water again."

6.2 Explicit use of probability for a gas with discrete energies

The remark in paper [3], hinted at above, originated a long memoir [8] that can be considered the culminating point in Boltzmann's studies on the relation between the Second Law and the calculus of probabilities. It is in fact the first paper where he explicitly declares that entropy is a measure of the probability of a state and that the Second Law reduces to stating that the natural evolution is from improbable to more probable states.

The original text of this memoir is clear and suggestive. Boltzmann enunciates the idea that he wants to use in these terms:

The initial state of a system will be, in most cases, a not so probable state and the system will tend always towards more probable states, until it will reach the most probable state, i.e. the state of thermodynamic equilibrium. If we apply this to the second law of thermodynamics, we can identify the quantity that is usually called entropy with the probability of the corresponding state. Let us then consider a system of bodies which is isolated [...]. In such a transformation, because of the second law of thermodynamics, the total entropy of the system cannot but increase. In our present interpretation, this has no other meaning than the fact that the probability of the global state of the bodies of the system must continuously increase: the system cannot but pass from a state to a more probable state.

Boltzmann of course considers the simplest case, that of a gas enclosed in a container with specularly reflecting walls and, in order to avoid the difficulties related to the fact that the position and velocity variables are continuous, he takes as a starting point his favourite model, the discrete one, "an unrealizable fiction", which is useful to allow essential ideas to emerge. Hence he considers a collection of n particles, with discrete energies $(0, \epsilon, 2\epsilon, \ldots, p\epsilon)$ and an assigned total energy, equal to an integer multiple of the same "quantum" ϵ, $\lambda\epsilon$, with λ an integer. In this model, the distribution function is replaced by a collection of integers $n_0, n_1, n_2, \ldots, n_p$ where n_k is the number of molecules with energy $k\epsilon$; this collection is called by Boltzmann the state distribution (*Zustandsverteilung*). Then we have

$$n_0 + n_1 + n_2 + \ldots + n_p = n, \qquad (6.1a)$$

$$n_1 + 2n_2 + \ldots + pn_p = \lambda. \qquad (6.1b)$$

Boltzmann then gives the name *complexion* to a repartition in which each molecule has an assigned energy. It is clear (see Appendix 6.1) that the number P of the complexions compatible with a given distribution is given by the following relation:

$$P = \frac{n!}{n_0! n_1! n_2! \ldots n_p!},\qquad(6.2)$$

where, as usual, we utilize the notation $n!$ to denote the product of the first n positive integers ($n! = 1 \cdot 2 \cdot 3 \cdot \ldots \cdot n$, and, by definition, $0! = 1$); $n!$ is read "n factorial" or "factorial of n".

In order to illustrate this number of complexions and, as a consequence, the probability of realizing a given distribution, Boltzmann examines in detail the example of a system made up of seven molecules (a rather small number) with total energy 7ϵ. It is not difficult then to find the possible distributions, i.e. the possible arrays of eight non-negative integers ($n_0\, n_1\, n_2\, n_3\, n_4\, n_5\, n_6\, n_7$) between 0 and 7, such that:

$$n_0 + n_1 + n_2 + n_3 + n_4 + n_5 + n_6 + n_7 = 7,$$
$$n_1 + 2n_2 + 3n_3 + 4n_4 + 5n_5 + 6n_6 + 7n_7 = 7.$$

One can proceed by trial and error or use some a priori restrictions such as the fact that if one fixes n_0 (between 0 and 6) and n_1 (between 0 and $7 - n_0$), then $n_2 + n_3 + n_4 + n_5 + n_6 + n_7 = 7 - n_0 - n_1$ and $n_2 + 2n_3 + 3n_4 + 4n_5 + 5n_6 + 6n_7 = n_0$ provide rather severe restrictions. Here is a list of the possible states or distributions:

(6 0 0 0 0 0 0 1)	(s_1)	(5 1 0 0 0 0 1 0)	(s_2)	(5 0 1 0 0 1 0 0)	(s_3)
(5 0 0 1 1 0 0 0)	(s_4)	(4 2 0 0 0 1 0 0)	(s_5)	(4 1 1 0 1 0 0 0)	(s_6)
(4 1 0 2 0 0 0 0)	(s_7)	(4 0 2 1 0 0 0 0)	(s_8)	(3 3 0 0 1 0 0 0)	(s_9)
(3 2 1 1 0 0 0 0)	(s_{10})	(3 1 3 0 0 0 0 0)	(s_{11})	(2 4 0 1 0 0 0 0)	(s_{12})
(2 3 2 0 0 0 0 0)	(s_{13})	(1 5 1 0 0 0 0 0)	(s_{14})	(0 7 0 0 0 0 0 0)	(s_{15}).

It is now easy by means of eqn (6.2) to compute the number P_j ($1 \leq j \leq 15$) of complexions of the state s_j. This number is called the *permutability* of the state by Boltzmann. It is sufficient to divide 7! by the product of the factorials of the eight numbers listed above for each of the 15 states, with the following result:

$$P_1 = 7 \quad P_2 = P_3 = P_4 = 42 \quad P_5 = P_7 = P_8 = P_{12} = 105 \quad P_6 = P_{13} = 210.$$

$$P_9 = P_{11} = 140 \quad P_{10} = 420 \quad P_{14} = 42 \quad P_{15} = 1.$$

There are then just 15 possible distributions and $7 + 3 \cdot 42 + 4 \cdot 105 + 2 \cdot 210 + 2 \cdot 140 + 420 + 42 + 1 = 1716$ complexions. The most probable state corresponds to $j = 10$ and is characterized by $n_0 = 3$, $n_1 = 2$, $n_2 = n_3 = 1$ (a rather rough discrete approximation to a decreasing exponential function), whereas the remaining n_k's are zero. The number of the complexions is then given, as we have said, by

$$P_{10} = \frac{7!}{1! 1! 2! 3!} = 2 \cdot 5 \cdot 6 \cdot 7 = 420.$$

It is thus 60 times higher than the number of complexions of the state in which all energy is concentrated in just one molecule (in this case the denominator would be 1!6!) and 420 times higher than the number of complexions of the state in which all energy is equally distributed among the seven molecules (in this case the denominator would be 7!).

Rather than studying the time evolution of the distribution as consequence of the collisions (kinetic method), Boltzmann now proposes to determine the "probability" of a distribution in a way "completely independent of whether or how that distribution has come about". In other words, he suggests determining the most probable distribution (under the constraints of the particle number and total energy conservation, as expressed by eqns (6.1)), on the basis of the intuitive idea that the system's evolution will eventually lead to this distribution. Boltzmann took the number of complexions, given by eqn (6.2), as a measure of the likelihood of a given distribution, underlining that in this way he considered the complexions as a priori equiprobable, in the same way as the probability that the sequence 1, 2, 3, 4, 5 will come out of the hat in a game of lotto is no different from that of any preassigned sequence of five numbers.

In this way, the problem is reduced to the search for a constrained maximum of P. The problem would be complicated if the numbers under consideration were small; but if, as occurs in real life, they are very large, one can make use of Stirling's formula, according to which $n! = n^n e^{-n} u(n)$, where $u(n)$ is a factor that varies very slowly in comparison with the others and hence may be treated as a constant (see Appendix 6.1). The convenience of working with the logarithm of P is then clear. If we use Stirling's formula, the opposite of the logarithm of P is given by:

$$- \log P = n_0 \log n_0 + n_1 \log n_1 + n_2 \log n_2 + \ldots + n_p \log n_p - n \log n + r_n, \quad (6.3)$$

where r_n varies slowly and will be neglected (as a matter of fact, its rate of change tends to zero when n tends to infinity). In eqn (6.3) we have accounted for the constraint given by eqn (6.1a).

Since we consider only very large numbers (and in agreement with the fact that we use Stirling's formula) it is convenient to treat the variables n_k ($k = 0, 1, 2, \ldots, p$) as continuous rather than discrete variables. Then the maximum of P (or the minimum of $- \log P$) can be found by the so-called method of Lagrange multipliers (in order to take the constraints (6.1) into account). This method is based on the fact that if you have a constraint and you search for the maximum (minimum) of a variable quantity, then the extremum will in general be lower (higher) than the one you would obtain if there were no constraints. For instance, if you seek the points on the surface of the earth which are most remote from the equator, and you add no constraints, the answer will be trivially, that there two such points: the North and South Poles. If however you add that the points must lie on a given plane through the centre of the Earth, then unless you chose a very special plane, the answer will be different.

Now Lagrange discovered a very remarkable rule to find the extrema in the presence of constraints, which we shall now recall (those readers who already know the method can skip this paragraph; those who want to avoid calculations with partial derivatives should skip both this and the next one). If we have a function f of many variables and look for its extrema in the presence of some constraints expressed by equating to zero some expressions $\phi_1, \phi_2, \ldots$, containing the independent variables upon which the function depends, then we can look for the extrema of a modified function obtained by adding to f the functions $\phi_1, \phi_2, \ldots$ each multiplied by parameters $\lambda_1, \lambda_2, \ldots$, called the Lagrange multipliers. If we determine the extrema in this way, they will depend upon the multipliers. The value of the latter are now determined by imposing the corresponding constraints.

Let us see how Boltzmann applied Lagrange's method to his function $-\log P$. If we denote the required Lagrange multipliers by α and $\beta\epsilon$ and the linear functions given by the sums on the left-hand sides of eqns (6.1) by $A = A(n_k)$ and $B = B(n_k)$, we must find the zeros of the derivatives of the function $-\log P + \alpha(A - n) + \beta\epsilon(B - \lambda)$ with respect to each variable n_k (the derivatives with respect to α and β yield, as always when applying Lagrange's method, the relations expressing the constraints). Since the derivative of $n_k \log n_k$ with respect to n_k is $1 + \log n_k$ and those of A and B are respectively 1 and k, we obtain:

$$1 + \log n_k + \alpha + \beta k\epsilon = 0 \quad (k = 0, 1, 2, \ldots, p)$$

or

$$n_k = \bar{n}e^{-\beta k\epsilon} \quad (k = 0, 1, 2, \ldots, p), \tag{6.4}$$

where $\bar{n} = e^{-\alpha - 1}$. The two multipliers (or, equivalently, $\bar{n}$ and β) are available to satisfy the two constraints (6.1).

Now for those readers who are not sufficiently familiar with partial differentiation to be able to follow the above passage, we offer an alternative proof that the distribution in eqn (6.4) gives the maximum of $\log P$. This proof is more elegant and complete than the one given above, but, in addition to being a bit longer (if it were not, why should one bother with derivatives?), it has a defect that always puzzles students when they see these elegant proofs. How did people discover them? Or why did they consider this particular expression? The fact is that elegant proofs always come after we already know the result we want to prove, and that is why they look so magic. Behind them lies hard and tedious work which the student does not need to know unless he wants an answer to the questions mentioned above. This at least seems to be the opinion of great mathematicians: no less than Gauss remarked about proofs, "When a proof has constructed a fine building, the scaffolding should no longer be visible". Perhaps with this sentence in mind or knowing that he used to publish only polished proofs, Jacobi called Gauss "the fox of mathematics" because he erased his tracks in the ground with his tail. Nowadays there seems to be more interest in seeing the path followed by the fox.

The starting point is an elementary inequality, which we shall meet again in the next chapter. This inequality reads as follows:

$$x \log x + 1 - x \geq 0 \quad (x > 0). \tag{6.5}$$

Nowadays we can see this with a pocket computer by asking it to plot the function on the left-hand side. We can thus check that this function never becomes negative and that it is zero if and only if $x = 1$. A less empirical proof is easily obtained if we replace x

by $1/x$; then, after an obvious rearrangement, the inequality reads $\log x \leq x - 1$, i.e. it reduces to stating that the logarithmic curve is always below the tangent at the point $(0, 1)$.

How are we going to exploit this result? We first introduce a short notation for $\bar{n}e^{-\beta k\epsilon}$; we simply call it $\overline{n_k}$ ($k = 0, 1, 2, \ldots, p$). Here $\bar{n}$ and β are chosen in such a way that the constraints (6.1) are satisfied when we replace n_k by $\overline{n_k}$ ($k = 0, 1, 2, \ldots, p$) (it can be shown that this choice is always possible and leads to a unique answer). What we want to show is that $-\log P$ has a unique minimum when $n_k = \overline{n_k}$.

To begin the proof, we replace x in eqn (6.5) by $n_k/\overline{n_k}$ ($k = 0, 1, 2, \ldots, p$). After writing eqn (6.5) with x replaced in this way, we multiply each inequality by the corresponding (positive) factor $\overline{n_k}$ and sum the $p+1$ inequalities that we have obtained. The result will be that the sum of $p+1$ terms on the left-hand side will be non-negative and will be zero if and only if each value in the argument of the logarithm is unity. A first simplification occurs in the terms arising from $1 - x$: they will disappear! In fact both the sum of the terms $\overline{n_k}$ and that of the n_k are equal to n, thanks to eqn (6.1a), and cancel each other. Then we are left with

$$n_0 \log(n_0/\overline{n_0}) + n_1 \log(n_1/\overline{n_1}) + \ldots + n_p \log(n_p/\overline{n_p}) \geq 0, \tag{6.6}$$

which, thanks to the properties of the logarithm, can be rewritten in the form

$$n_0 \log n_0 + n_1 \log n_1 + \ldots + n_p \log n_p$$
$$\geq n_0 \log \overline{n_0} + n_1 \log \overline{n_1} + \ldots + n_p \log \overline{n_p}. \tag{6.7}$$

Now $\log \overline{n_k}$ is a constant plus another constant multiplying k. Hence the right-hand side of eqn (6.7) is a constant times the sum in eqn (6.1a) plus another constant times the sum in eqn (6.1b). Since eqns (6.1a) and (6.1b) are satisfied by both n_k and $\overline{n_k}$, we can replace n_k by $\overline{n_k}$ on the right-hand side of eqn (6.7) and obtain

$$n_0 \log n_0 + n_1 \log n_1 + \ldots + n_p \log n_p$$
$$\geq \overline{n_0} \log \overline{n_0} + \overline{n_1} \log \overline{n_1} + \ldots + \overline{n_p} \log \overline{n_p}. \tag{6.8}$$

But the left-hand side is $-\log P$ and the right-hand side the value taken by this function when $n_k = \overline{n_k}$. Thus $-\log P$ has a unique minimum attained when eqn (6.4) is satisfied, as we required to show.

Please note that the second proof, although a bit longer, actually proves more. It proves that the extremum is actually a minimum (and not a maximum), which we did not bother to check in the first proof. Another advantage is that it can be extended to the case when we are dealing with continuous rather than discrete variables, without introducing advanced tools of the calculus of variations, as we shall see in the next chapter.

6.3 Energy is continuous

Boltzmann continued his paper by treating the case of a gas with energies taking continuous values. One must of course find an expression for P, or its logarithm, in this case as well; an obvious method is to discretize the continuous case, use the previous method, and then take the limit when the discrete variables become continuous. This is in fact the method followed by Boltzmann. Essentially, the sums will be replaced by integrals (see Appendix 6.1).

Here a small problem, related to the a priori equiprobability of the states, arises which as underlined by Klein [2], is cleverly exploited by Boltzmann to obtain first a wrong result, perhaps to exhibit that *coup de théatre* that he admired so much in Maxwell's papers. In the continuous case it is not immediately obvious how to define equiprobable states, but, as indicated by Boltzmann himself, there is a guiding principle supplied by the fact that a volume element in phase space is invariant during the time evolution (thanks to Liouville's theorem); hence the equiprobable sets are those which have an equal volume in phase space. The error that one can make is to take the length of the segments on the energy axis. In this way, if E denotes energy, we omit a factor $E^{1/2}$ (we have already met this problem in Chapter 4). Once this point is clarified, it turns out to be clear that the equivalent of $-\log P$ is exactly the quantity H previously discovered by Boltzmann in his basic paper of 1872 [1] and given by the integral of $f \log f$ with respect to the ordinary volume element in phase space. The constraints (6.1) now become that the integrals of f and $|\xi|^2 f$ (here we restrict ourselves, for simplicity, to the usual case of monatomic gases with no body forces, even if the extension is trivial in this case) with respect to the said volume element have assigned values (proportional to the total number of molecules and the total energy). In this case the problem of minimizing H or maximizing $\Omega = \log P$ is an elementary problem in the calculus of variations (see Appendix 7.1) and the result is the Maxwell velocity distribution; the corresponding value of the logarithm of P (which Boltzmann calls permutability) equals, apart from a factor ($2/3$ with Boltzmann's normalizations, since he does not seem to distinguish between internal energy and temperature), the entropy in an equilibrium state.

This relation, written in the form

$$S = k \log W, \tag{6.9}$$

where S denotes entropy, W the probability or likelihood of a state, and k the Boltzmann constant (which, as remarked in Appendix 3.1, Boltzmann never used), is now engraved on Boltzmann's tombstone.

We stop here to mention that according to Ebeling [9], the first mathematician to hit upon the logarithm of a probability was the mathematician De Moivre. We also remark that the same formula appears in modern information theory. It was introduced there by Shannon, who reportedly said that von Neumann suggested the name entropy, since it looked like the statistical formula for entropy, and anyway nobody knows what entropy is.

Boltzmann draws some conclusions from his calculation. He first says [8]:

It is well-known that, when a system of bodies undergoes purely reversible transformations, the total entropy of the system remains constant. If, on the contrary, among the transformations which the system undergoes, some are irreversible, its entropy cannot but increase [...].Because of the previous relation, the same is true of $\Sigma\Omega$, the measure of the permutability for the set of bodies. This measure of the permutability is thus a quantity which, in a state of thermodynamic equilibrium, coincides with entropy, apart from a constant factor, but which has a meaning even during each irreversible process, when it increases continuously.

One can immediately deduce two statements: the first refers to a system of bodies that undergoes several transformations, some of which, at least, are irreversible [...]. If, at the beginning and at the end of the process, the system turns out to be in a state of thermodynamic equilibrium, the total entropy of this system can be immediately calculated; it is in both cases equal to two thirds of the measure of permutability. This first statement then expresses that the total entropy is always greater, after the transformations undergone by the system, than its initial value; this holds, of course, for the measure of permutability as well. The second statement refers to a gas which undergoes a transformation in which the initial and final states are not necessarily thermodynamic equilibria. Hence one cannot calculate the entropy of the gas in the initial and final states, but one can always calculate the quantity that we have called the measure of permutability; and here again its final value is necessarily higher than the initial one. One can equally verify that this last statement can be extended, without any difficulty, to a system made up of several gases, as well as to the case when the molecules are polyatomic and under the action of external forces.

Boltzmann also formulates the following general result:

Let us consider any system which undergoes an arbitrary transformation, the initial and final states being not necessarily equilibrium states; in these conditions, the measure of permutability of the ensemble of the bodies of the system will constantly increase during the process, and will, at most, remain constant in the reversible processes infinitely close to thermodynamic equilibrium.

In formulating this statement, Boltzmann seems to be well aware of having reached a very general principle and considers it likely that this principle should not be restricted to the case of gases, but should extend to solids and liquids, even if an exact mathematical treatment of these more general cases appears to meet with difficulties when a detailed description is attempted.

The next scientist to use Boltzmann's relation was Planck, in his celebrated investigations on black-body radiation (see Chapter 12). In a minor paper by Einstein which we have quoted already in Chapter 3, we read a statement about eqn (6.9): "An expression for the entropy of a system [...] which was found by Boltzmann for ideal gases and assumed by Planck in his theory of radiation..." [10] It seems however that Einstein was not so happy with the way Boltzmann had handled probability. In his basic paper [11] on light quanta in 1905 we read:

The word *probability* is used in a sense that does not conform to its definition as given in the theory of probability. In particular, "cases of equal probability" are often hypothetically defined in instances where the theoretical pictures used are sufficiently definite to give a deduction rather than a hypothetical assertion.

Even if he does not mention it explicitly in this sentence, he seems to be referring to Boltzmann's method of complexions. In the same paper he gives eqn (6.9) the name of Boltzmann's principle. He thought he could dispense with the kind of counting illustrated above and hoped "to eliminate a logical difficulty which still hampers the implementation of Boltzmann's principle" [11].

As a matter of fact, Einstein considered the increase in entropy of a gas undergoing an isothermal process, as provided by standard thermodynamics, and found that, postulating eqn (6.9), the ratio of the probabilities of two states corresponding to volumes

V_a and V_b equals $(V_b/V_a)^N$, where N is the number of molecules in the gas. However, Boltzmann's principle is more powerful and useful than Einstein thought at that time, and another paper that he announced in ref. [11] never appeared. But he did not cease criticizing the concept of complexion, as one can see from the following two quotations taken from two papers of his, appeared in 1909 and 1910:

Neither Herr Boltzmann nor Herr Planck has given a definition of W. [12]

Usually W is put equal to the number of complexions [...] In order to calculate W, one needs a *complete* (molecular-mechanical) theory of the system under consideration. Therefore it is dubious whether the Boltzmann principle has any meaning without a *complete* molecular-mechanical theory or some other theory which describes the elementary processes. [Equation (6.9)] seems without content, from a phenomenological point of view, without providing in addition such a theory of elementary processes. [13]

There are indications however that Einstein, surprisingly enough, had not yet fully understood the subtleties of classical statistical mechanics (this will probably comfort some readers!). In fact, in another paper, written in 1903, he had derived the Second Law in a way that bypassed Loschmidt's reversibility paradox with the following assumption: "We will have to assume that more probable distributions will always follow less probable ones, i.e. that W always increases until the distribution becomes constant and W has reached a maximum." [14]

Clearly the word "always" repeated twice, without any further comment, indicates that Loschmidt's paradox, although mentioned in Boltzmann's lectures [15], which Einstein knew, had escaped his attention. As we shall see in the next section, Boltzmann reached the final stage of his full understanding on this subtle point only in 1895, i.e. almost twenty years after Loschmidt had put forward the reversibility objection.

A few years later, Einstein had become a widely known scientist and a professor at Zurich, where he taught a course on kinetic theory. Yet, in a paper published in 1911, agreeing with a criticism of the sentence just quoted, he says: "Already then [1903] my derivation did not satisfy me, so that shortly after I gave another derivation." [16] This other derivation appears in his paper of 1904 quoted above [10], but again contains the same assumption. This seems to indicate a certain amount of carelessness on Einstein's part, because in the paper of 1910 which has been already quoted, he had written a careful statement of the statistical form of the Second Law: "The irreversibility of physical phenomena is only apparent [...] [A] system probably goes to states of greater probability when it happens to be in a state of relatively small probability." [13] Here the word *probably* has replaced *always*, as required.

A clear indication that Einstein had fully understood Boltzmann's arguments comes from a sentence about Boltzmann written in a paper published in 1915: "His discussion [of the Second Law] is rather lengthy and subtle. But the effort of thinking [required to understand it] is richly rewarded by the importance and the beauty of the subject." [17]

Further comments and information about the relation between Einstein's early papers and Boltzmann's results will be given in Chapter 12.

6.4 The so-called *H*-curve

The connection between entropy and probability certainly clarifies certain aspects of the statistical interpretation of the Second Principle, but many objections can be raised, mainly related to the problem of reconciling the time symmetry of the basic laws with the approach to statistical equilibrium. We shall discuss the ideas of Gibbs and later writers in the next chapter. Here we want to discuss an important part of the historical development, which took place between the answers given by Boltzmann to Loschmidt's and Zermelo's objections. We shall in fact discuss a controversy between Boltzmann and a group of British scientists.

This controversy is not the same as was mentioned in Chapter 1 and related to the problem of specific heat capacities. This new discussion was begun by a paper by E.P. Culverwell [18], who claimed that it is impossible to prove *in general* that a set of particles will tend to the Boltzmann configuration, in which the energy is equally distributed among all the degrees of freedom. Culverwell appealed to the reversibility principle and asserted that for every configuration which tends to an equal distribution of energy, there must be one which departs from it. This is more or less Loschmidt's objection.

The widespread interest in such problems led the British Association to appoint a special committee to investigate "the present state of our knowledge of the Second Law". The committee consisted of just two members, J. Larmor and G.H. Bryan. The latter gave the first report on this subject at the 1891 meeting. The statistical aspects were duly stressed, together with some remarks on the role of instability. No definite conclusion was reached, although qualitative analogies related to mixing were used and the goal of not invoking the role of the ether in the explanation was advocated. The second part of Bryan's report was presented at the Oxford meeting of the British Association in 1894 and can be considered as a fair summary of the state of the subject. The meeting was also attended by Boltzmann, who wrote an appendix.

Culverwell was not so easily convinced and asked the simple question: "Will someone say exactly what the *H*-theorem proves?" [19]. In the discussion following Culverwell's question, S.H. Burbury [20] stated that there is a difference in the way one treats the molecular states before and after collision and that random external disturbances are responsible for that. And he adds: "So there is a general tendency for *H* to diminish although it may conceivably increase in particular cases. Just as in matters political, change for the better is possible, but the tendency is for all change to be from bad to worse."

We remark that Boltzmann quotes Burbury's paper [20] as the origin of his assumption of "molecular chaos" in his lectures on gas theory [15]. We also remark that Burbury was the first to use the letter *H* in place of *E* for what we call the *H*-function, and this choice was adopted by Boltzmann.

The discussion went on in the columns of *Nature* for several months until, in February 1895, a long letter by Boltzmann was received and published [21]. Part of the paper is devoted to the problem of energy equipartition and will be discussed in Chapter 8. Another part is devoted to discussing Culverwell's and Burbury's papers and is of interest here.

Boltzmann first remarks that if one followed Culverwell's reversibility objection, it would mean that we could prove that oxygen and nitrogen do not diffuse, or in other words that they could separate as easily as they mix. Then he reminds readers that

It can never be proved from the equations of motion alone, that the minimum function H must always decrease. It can only be deduced from the laws of probability that if the initial state is not specially arranged for a certain purpose, but haphazard governs freely, the probability that H decreases is always greater than that it increases.

Then Boltzmann considers a vessel with perfectly smooth and elastic walls, containing a given number of gas molecules moving for an indefinitely long time. All *regular* motions shall be excluded; what Boltzmann means by this is explained by an example of excluded motion: one where all the molecules move in one plane. Then for most of the time H will be close to its minimum value H_{min}. Let us construct the so-called H-curve, i.e. let us take time as the abscissa and draw the curve whose ordinates are the corresponding values of H (we saw an example of this curve in the previous chapter, albeit restricted to a rather short time interval). The great majority of the ordinates are very nearly close to H_{min}. But, since larger values of H are not mathematically impossible but only very improbable, the curve has certain, though very few, maxima higher than H_{min}.

We will now consider a certain ordinate $H_1 > H_{min}$. Two cases are possible. H_1 may be very near the top of a summit, so that H decreases if we go either in the positive or the negative direction along the axis representing time. The second case is that H_1 lies in a part of the curve ascending to or descending from a higher summit. Then [in the second case] the ordinates on the one side of H_1 will be greater, and on the other less than H_1. But because higher summits are so extremely improbable, the first case will be the most probable, and if we choose an ordinate of given magnitude H_1 guided by haphazard in the curve, it will not be certain, but very probable, that the ordinate decreases if we go in either direction. [21]

We remark, in order to avoid any possible misunderstanding, that Boltzmann is here talking of conditional probabilities: once H_1 is fixed and has a value significantly larger that H_{min}, then it is extremely improbable that the second case will occur. It is also remarkable that Boltzmann asserts that the value of H must occasionally rise above its minimum value. This is at variance with what he had said in all his previous papers. In particular, we can read in the text of a lecture at the Vienna Academy "On the Second Law of Thermodynamics" [22], delivered on 29 May 1886:

Since a given system can never of its own accord go over into another equally probable state but only into a more probable one, it is likewise impossible to construct a system of bodies that after traversing various states returns periodically to its original state, that is a perpetual motion machine.

In this passage Boltzmann seems to fall into the same mistake which Einstein was to make later, as we have seen in the previous section: evolution towards more probable states is spontaneous. The fact that the presentation is non-technical may be an excuse for his inaccurate statement. Yet we feel that the final stage of his statistical interpretation of the Second Law was reached only after Culverwell's thought-provoking question.

Let us go back to the paper of 1895 [21]. Boltzmann continues with these words:

We will now assume, with Mr Culverwell, a gas in a given state. If in this state H is greater than H_{min} it will be not certain, but very probable, that H decreases and finally reaches not exactly but very nearly the value H_{min} and the same is true at all subsequent instants of time. If in an intermediate state we reverse all velocities, we get an exceptional case, where H increases for a certain time and then decreases again. But the existence of such cases does not disprove our theorem. On the contrary, the theory of probability itself shows that the probability of such cases is not mathematically zero, only extremely small.

Hence Mr Burbury is wrong, if he concedes that H increases in as many cases as it decreases, and Mr Culverwell is also wrong, if he says that all that any proof can show is that taking all values of dH/dt got from taking all the configurations which recede from it, and then striking some average, dH/dt would be negative. On the contrary, we have shown the possibility that H may have a tendency to decrease, whether we pass to the former or to the latter configurations. What I have proved in my papers is as follows: It is extremely probable that H is very near to its minimum value; if it is greater, it may increase or decrease, but the probability that it decreases is always greater. Thus if I obtain a certain value of dH/dt, this result does not hold for every time-element dt, but is only average value. But the greater the number of molecules, the smaller is the time-interval dt for which the results hold good.

Here Boltzmann is again using the fact that we are talking about conditional probabilities, whereas Burbury was looking at the time evolution; only if we fix a value of H significantly larger than H_{min} can we say that the probability of moving towards smaller values is higher than that of moving towards larger ones. But this applies to both the forward and the backward evolution. Thus, if we think of the H-curve, it is true that H increases in as many cases as it decreases. Hence only if the initial data are significantly different from an equilibrium state can we expect H to decrease with a high probability.

In order to explain the point, Boltzmann illustrates with an analogy taken from dice play. If we make an indefinitely long series of throws, we can represent the results in a graph form which has an analogy with the H-curve. Let n be a large integer and let us denote by A_j ($j = 0, 1, 2, \ldots$) the number of times which the face with the number 1 is thrown in the $6n$ throws between the $(j + 1)$th and the $(6n + j)$th inclusive. Let us construct a series of points in a plane, the successive abscissae of which are j/n ($j = 0, 1, 2, \ldots$) and the ordinates are $y_j = (A_j/n - 1)^2$. Boltzmann calls this the P-curve, the analogue of the H-curve for this play. Given the fact that n is large, most of the ordinates of the points of the curve under consideration will be very small, but occasionally there will be points much higher than the average level of the curve. Let us consider all the points of the P-curve whose ordinates are exactly 1, and call these points "the points B"; these points mark the case in which, by chance, we have thrown the number 1 in $2n$ out of $6n$ throws. This is of course extremely improbable, but not impossible. Let m be a number much smaller than n and let us go forward from the abscissa of each point B through a distance $6m/n$. We shall probably meet a point with ordinate less than 1. The probability of meeting an ordinate larger than 1 is extremely small, though not zero. "By reasoning in the same manner as Mr Culverwell" says Boltzmann "we might believe that if we go backward" by the same distance, "it would be probable that we should meet ordinates >1". But this is not correct; the ordinates will probably decrease both in the forward and backward directions.

Boltzmann even ventures to compute a sort of derivative, coming up with the value $-1/3$. He hastens to note that this is not "an ordinary differential coefficient", but only an average ratio of increase. "The P-curve", says Boltzmann, "belongs to the large class of curves which have nowhere a uniquely defined tangent.[...] The same applies to the H-curve in the Theory of Gases."

Boltzmann again took up the subject of the H-curve in another paper, which carries the date Christmas 1897 and is entitled *Über die sogenannte H-curve* ("On the so-called H-curve") [23]. In this paper Boltzmann introduces another analogue of the H-curve by means of a model based on random drawing. N white balls and N black balls are in an urn and one of them is randomly selected and then put back in the urn. We perform $2N + 1$ drawings (where N is very large) designated by

$$Z_{-N}, Z_{-N+1}, \ldots, Z_0, Z_1, \ldots Z_N.$$

Let n be a large positive integer less than $2N + 2$. We denote by a_k the number of white balls drawn during the n drawings $Z_k, Z_{k+1}, \ldots Z_{k+n-1}$, where k is one of the integers (positive, negative, or zero) between $-N$ and $N + 1 - n$. Let us now consider the points B_k having abscissae (k/n) and ordinates $|1 - 2(a_k/n)|$.

Boltzmann calls the H-curve of this lottery the set of points $\{B_k\}$. Since n is large, a_k is close to $n/2$ and most of the curve is close to the x-axis. However, since N is very large, the curve has humps (*Buckel*) and occasionally deviates from the x-axis by a significant distance. If for example $N = 2^n \times 10^3$, there will be 2×10^3 chances that during the aforementioned $2N + 1$ drawings, n black balls are drawn consecutively. If k is the index of the first draw of this sequence, the point with abscissa k/n will have ordinate 1. Similarly there will be 2×10^3 chances of drawing n white balls consecutively and again the corresponding point will have ordinate 1. Thus in a sequence of $2N + 1$ drawings there will be 4×10^3 humps of the greatest possible height, i.e. unity. There will be a much larger number of lower humps, still significantly away from the x-axis.

Again, if we consider an assigned ordinate y_1 and we call Q the points having this ordinate, it will be very unlikely that this point belongs to a hump whose height is much larger than y_1.

Boltzmann considers that he is not wrong in thinking that professional geometers will despise the H-curve (*der H-Curve spotten werden*). He declares that his purpose is to construct a model in order to show that there is no contradiction in assuming that the H-curve in the kinetic theory of gases has certain properties. This H-curve will resemble "the H-curve of the lottery made continuous". In particular this curve will be continuous but without a tangent at any point. One cannot avoid thinking that these papers by Boltzmann are the forerunners of stochastic dynamics and Wiener's theory of Brownian motion, where the set of trajectories having tangent has zero measure.

The second stochastic model certainly inspired the Ehrenfests [24], who introduced a similar model, more widely known than Boltzmann's model. There the balls are numbered from 1 to $2N$ and are distributed in two boxes A and B. An integer from 1 to $2N$ is chosen at random and the ball corresponding to that number is moved from the box in which it lies to the other one. The process is then repeated a desired number of times. It is clear that there will be a tendency towards equalizing the numbers of balls in the two boxes, but there will be fluctuations about the equilibrium value, exactly as in Boltzmann's models illustrated above. This model has been studied and illustrated in detail by M. Kač [25].

It is very important to note the dates of Boltzmann's papers on the H-curve (1895 and 1897). These discussions prepared him for dealing with Zermelo's recurrence objection in 1896. It is also remarkable that in the first of these two papers [21] he discusses a cosmological issue. He says:

I will conclude this paper with an idea of my old assistant, Dr Schuetz. We assume that the whole Universe is, and rests for ever, in thermal equilibrium. The probability that one (and only one) part of the Universe is in a certain state, is the smaller the further this state is from thermal equilibrium; but this probability is greater, the greater the Universe itself is. If we assume the Universe great enough we can make the probability of one relatively small part being in a given state (however far from the state of thermal equilibrium), as great as we please. We can also make the probability great that, though the whole Universe is in thermal equilibrium, our world is in its present state. It may be said that the world is so far from thermal equilibrium that we cannot imagine the improbability of such state. But can we imagine, on the other side, how small a part of the whole Universe this world is? Assuming the Universe great enough, the probability that such small part of it as our world should be in its present state, is no longer small.

If this assumption were correct, our world would return more and more to thermal equilibrum; but because the whole Universe is so great, it might be probable that at some future time some other world might deviate from thermal equilibrium as our world does at present. Then the aforementioned H-curve would form a representation of what takes place in the Universe. The summits of the curve would represent the worlds where visible motion and life exist.

This argument is certainly related to those used by Boltzmann in his second answer to Zermelo [26], published in 1897.

7

Boltzmann, Gibbs, and equilibrium statistical mechanics

7.1 Introduction

In this chapter we shall discuss equilibrium statistical mechanics for systems more complicated than the monatomic gases considered so far, as well as the problem of the trend towards equilibrium of these systems.

Once more must be ascribed to Boltzmann the merit of having begun this branch of statistical mechanics with a basic paper [1], written in 1884 and much less frequently quoted than his other contributions. In this paper he formulated the hypothesis that some among the possible steady distributions can be interpreted as macroscopic equilibrium states. This fundamental work by Boltzmann was taken up again, widened, and expounded in a classical treatise by Gibbs [2], and it is the terminology introduced by Gibbs that is currently used. As a matter of fact a statistical ensemble (in Gibbs's terminology) is called a *monode* by Boltzmann. The question posed in the above-mentioned paper [1] is the following: what statistical families of steady distributions have the property that, when an infinitesimal change is made in their parameters, the infinitesimal changes in the average total energy of the system E, of the pressure p, and of the volume V are such that $(dE + p\,dV)/T$ (where T is the average kinetic energy per particle) is an exact differential (at least in the thermodynamic limit, when $V \to \infty$, $N \to \infty$, whereas N/V remains bounded)? These families are called *orthodes* by Boltzmann. The answer given by Boltzmann to his own question is that there are at least two ensembles of this kind, the *ergode* (Gibbs's *microcanonical ensemble*) and the *holode* (Gibbs's *canonical ensemble*).

Although Boltzmann originated [1] the study of equilibrium states for more general situations than that, already considered by Maxwell, of a dilute gas in the absence of external forces, it is not with his name but that of Gibbs that one usually associates the methods of this area (the most completely developed one) of statistical mechanics. Even the terminology (microcanonical, canonical, grand canonical ensembles) is that due to Gibbs, while the first two ensembles were clearly defined (with different names) and used by Boltzmann. It is then beyond doubt, in the words of Klein [3], that "it was Boltzmann, and not Maxwell or Gibbs, who worked out precisely *how* the second law is related to probability, creating the subject of statistical mechanics".

We may add, for the sake of clarity, that Gibbs invented the name of the subject. Gibbs was of course aware of Boltzmann's priority, as we shall indicate later. Why then, beyond some generic acknowledgements to Boltzmann, it is Gibbs's name that emerges? It is an interesting question, to which one may be tempted to give an easy answer.

Before embarking on this discussion, let us first say something about Gibbs.

7.2 A great American scientist of the nineteenth century: J.W. Gibbs

Let us recall, with the help of two fine papers by M. Klein [4, 5], who was that "Mr. Josiah Willard Gibbs of New Haven" who was appointed Professor of Mathematical Physics by the Yale Corporation on 13 July 1871. He was born on 11 February 1839 (the same month as Boltzmann, five years earlier), the only son and the fourth of the five children of Mary Anna Van Cleve and Josiah Willard Gibbs the elder. His father was a distinguished philologist who had graduated from Yale in 1809 and served as Professor of Sacred Literature in the Divinity School of the same University from 1826 until his death in 1861. Four generations of Gibbs and Willard sons had previously graduated from Harvard College, including Samuel Willards, in 1659, who was acting president of that institution at one time. Gibbs's mother's family background was also impressive: it included a series of Yale graduates, one of whom served as the first president of what is today Princeton University (then College of New Jersey), and several graduates from the latter institution who followed careers in science.

Gibbs graduated from Yale College in 1858, having won a series of prizes and scholarships for excellence in Latin and mathematics. Gibbs continued his studies in Yale's new graduate school and received his PhD in 1863, one of the first scholars to be awarded this degree by an American University. His doctorate was in engineering and his dissertation, entitled "On the forms of the teeth of wheels in spur gearing", was a sort of exercise in geometry and kinematics. He was then appointed a tutor in Yale College, where he gave elementary instruction in Latin and natural philosophy (physics) for three years. In the meantime he continued to work on engineering problems, and in 1866 he obtained a patent on an improved brake for railway cars. In the same year, however, he presented a paper to the Connecticut Academy of Arts and Sciences. The paper deals with the quantities used in mechanics, but contains a remarkably clear discussion of the dual roles played by the concept of mass in classical mechanics—inertial mass and gravitational mass—and of the confusion caused by defining mass as quantity of matter.

In the same year, in the month of August, Gibbs sailed for Europe for what turned out to be his only extended absence from his native city. He travelled with his sisters Anna and Julia, his parents and his two other sisters being no longer alive. He spent a year each at the universities of Paris, Berlin, and Heidelberg, attending a variety of lectures and reading widely in both mathematics and physics. The list of scientists whose lectures he attended is impressive, since it includes Liouville, Darboux, Kronecker, Weierstrass, Helmholtz, and Kirchhoff, but there is no indication that he was a research student anywhere or had yet begun any research of his own or was planning to do so.

Two years after he came back to New Haven, Gibbs had no regular employment and his future activities were not clear either. He was evidently able to manage on the money inherited from his father. He never married, and continued to live in the family home with his unmarried sister Anna, and with Julia, her husband, and their growing family.

His financial independence and his scientific abilities must have been known within the Yale community, since he was appointed to the newly created professorship of mathematical physics in 1871, the official record including the unusual specification "without salary". It is true that he taught only one or two students a year during his first decade as a professor. Two years later he received an offer of $1800 a year from Bowdoin College, which he refused. Only in 1880, when he was tempted to leave because of an offer from the new and appealing Johns Hopkins University, did his own university offer a salary; it was only two-thirds of what Hopkins would have paid, but was enough to convince him to stay.

Gibbs's appointment in 1871 preceded his first published research by two years; this was not unusual at that time in America. We remind the reader that Boltzmann, five years younger than Gibbs, had published papers when he was a student and obtained a chair at the age of 25, in 1869 (see Chapter 1); in 1871 he was about to publish his most famous paper.

Gibbs's first paper [6] was on thermodynamics and immediately demonstrated his mastery of the field. The choice of the subject shows no correlation to the lectures he had attended in Europe, and Gibbs, a very laconic writer, gives no hint as to the reasons for his choice, though his interest in steam engines between 1871 and 1873 might provide a clue.

The title of the paper, "Graphical methods in the thermodynamics of fluids" is not very promising, but its content quietly changed the content of thermodynamics by using entropy as an independent variable, something that not even Clausius had ever done. In particular he analysed the entropy–volume diagram and its "substantial advantages over any other method" because it shows the region of simultaneous coexistence of the vapour, liquid, and solid phases of a substance, a region which reduces to a point in the more usual pressure–temperature plane. We do not know, as Klein [5] points out, whether Gibbs learned of Thomas Andrews's recent (1869) discovery of the continuity of the two fluid states of matter from Andrews's paper itself [7] or from Maxwell's *Theory of heat* [8].

Gibbs's interests are much more apparent in his second paper [9], which appeared a few months later. Although the title might suggest an extension of his previous representation, he had clearly turned from methods to explanations. The problem Gibbs treated is the characterization of the equilibrium state of a material system. This state can be solid, liquid, gaseous, or a combination of these phases according to the circumstances, and was represented as a point of a surface of a space whose coordinate axes are energy, entropy and volume. This surface represents the fundamental thermodynamic equation of the body. Gibbs established the relationships between the geometry of the surface and the conditions for thermodynamic equilibrium and its stability. He showed that for two phases of the same substance to be in equilibrium with each other, not only must they have the same temperature T and pressure p, but also their internal energies E_k ($k = 1, 2$), entropies S_k ($k = 1, 2$), and volumes V_k ($k = 1, 2$) must satisfy the equation

$$E_2 - E_1 = T(S_2 - S_1) - p(V_2 - V_1).$$

This equation answered a question that had puzzled Maxwell when, in July 1871, he was working on Andrews's result on the continuity of the liquid and gaseous states. Maxwell was then writing his *Theory of heat* and corresponded on the subject of Andrews's diagrams with James Thomson (1822–92), the elder brother of William Thomson, whom we have already met more than once. Thomson, a colleague of Andrews in Belfast, suggested that the isothermal curves for a fluid below its critical temperature, for the transition from the gaseous to the fluid state to be possible (in the traditional pressure–volume diagram), should really show a minimum and a maximum [10] rather than a straight line segment parallel to the volume axis, as showed by Andrews's data. Between those extrema, the states pictured on the isotherm would be unstable—since temperature and volume would be increasing together—but such a curve would account for metastable states of supercooling and superheating, and provide what Thomson called "theoretical continuity". In fact these isotherms proposed by Thomson looked very much like those that Johannes Diderick van der Waals (1837–1923) would derive in his dissertation two years later [11]. Maxwell's question was: where must one draw the straight line segment that cuts across Thomson's loop? Or, in physical terms, what is the condition determining the pressure at which gas and liquid can coexist in equilibrium?

In his letter to Thomson, Maxwell proposed a way of answering this question and repeated it in his book [8], arguing that the difference in internal energy between the two phases must be a maximum at the pressure where they coexist. He would change his mind after reading Gibbs's paper.

When he analysed the conditions for the stability of thermodynamic equilibrium states [9], Gibbs arrived at a new understanding of the significance of the critical point. The critical state not only indicated where the two fluid phases became one, but it also marked the limits of instability associated with the two phase system. Gibbs's analysis also led him to a series of new explicit conditions that must be fulfilled at the critical point and that can serve to characterize it.

As remarked by Klein [5], Gibbs could expect his work to be circulated far and wide by the fact that it appeared in the *Transactions of the Connecticut Academy of Arts and Sciences*, a society of which Gibbs had been a member since 1858. Although this academy was of a local character, being centred in New Haven, it had been in existence since 1799 and had developed a regular programme for the exchange of its *Transactions* with similar journals published by some 170 other learned societies, ranging from Quebec and Melbourne to Naples and Moscow. We know however that he did not rely on reaching only those potential readers who might happen to pick up the *Transactions* in their own local academies: he sent copies of his papers directly to some 75 scientists at home and abroad [12]. We cannot tell how many of those actually read his first two papers, but we do know of one, the crucial one, Maxwell. He read Gibbs with enthusiasm and profit. In fact, he had misused the term entropy in the first edition of his book [8], where he had followed his friend Tait. The error was corrected in later editions, after Maxwell had learned the proper definition from Gibbs's papers.

Maxwell talked about Gibbs's work to his colleagues in Cambridge in several fields, and wrote about it to others, recommending it highly. He especially appreciated the geometric approach contained in this work and was fascinated so much by the thermodynamic surface introduced in Gibbs's second paper that he actually constructed such a surface showing the thermodynamic properties of water, and sent a plaster-cast of it to Gibbs. He went so far as to discuss the surface at considerable length in the 1875 edition of his book [13], though that edition appeared in a series described by its publisher as consisting of "text-books of science adapted for the use of artisans and of students in public and science schools".

In his lecture "On the dynamical evidence of the molecular constitution of bodies", delivered at the Chemical Society on 18 February 1875 [14], Maxwell says:

The purely thermodynamical relations of the different states of matter do not belong to our subject, as they are independent of particular theories about bodies. I must not, however, omit to mention a most important American contribution to this part of thermodynamics by Prof. Willard Gibbs, of Yale College, U.S., who has given us a remarkably simple and thoroughly satisfactory method of representing the relations of the different states of matter by means of a model. By means of this model, problems which had long resisted the efforts of myself and others may be solved at once.

In this famous lecture Maxwell introduced his own ingenious method of demonstrating where to draw the horizontal line in the Thomson–van der Waals curve (the so-called *Maxwell rule*, according to which the area above the segment must equal that below) and paid a handsome tribute to van der Waals as well, because the attack of the Dutch scientist on the difficult question of a molecular explanation of the continuity of the liquid and gaseous states "is so able and so brave, that it cannot fail to give a notable impulse to molecular science. It has certainly directed the attention of more than one inquirer to the study of the Low-Dutch language in which it is written."

The Connecticut Academy had some twenty members and had regular meetings. At one of these, in June 1874, Gibbs gave a talk, presumably not too exciting for his audience, on the application of the principles of thermodynamics to the study of thermodynamic equilibrium, which was gradually worked out into a long memoir "On the equilibrium of heterogeneous substances" [15]. This occupies about 300 pages of his collected papers and we can subscribe to the words of M.J. Klein, according to whom it "surely ranks as one of the true masterworks in the history of physical science" [5].

In his memoir, Gibbs greatly enlarged the domain covered by thermodynamics: in fact, he treated chemical, elastic, surface, and electrochemical phenomena by a single, unifying method. He described the basic ideas underlying this work in a lengthy abstract (17 pages) which appeared in the *American Journal of Science* [16], which surely reached a much wider audience than the *Transactions of the Connecticut Academy of Arts and Sciences*. The motivation for his work is clearly stated in the following sentences:

It is an inference naturally suggested by the general increase of entropy which accompanies the changes occurring in any isolated material system that when the entropy of the system has reached a maximum, the system will be in a state of equilibrium. Although this principle has by no means escaped the attention of physicists, its importance does not appear to have been duly appreciated. Little has been done to develop the principle as a foundation for the general theory of thermodynamic equilibrium.

Gibbs's memoir set forth exactly that development. In particular, the general criterion for equilibrium was stated simply and precisely: "For the equilibrium of any isolated system it is necessary and sufficient that in all possible variations of the state of the system which do not alter its energy, the variation of its entropy shall either vanish or be negative."

In order to work out the consequences of this general criterion, and to explore their implications allowing for the variety and complexity that thermodynamic systems can have, Gibbs introduced chemical potentials from the outset. These intensive variables must be constant throughout a heterogeneous system in equilibrium and play a role similar to that of temperature and pressure. Starting from these considerations, Gibbs derived his famous phase rule, which specifies the number of independent variables (degrees of freedom) in a system of a certain number of coexistent phases having a specified number of chemical components [15]. Although this phase rule proved to be a milestone for understanding an incredible amount of experimental material, Gibbs did not underline its importance in any special way.

It was van der Waals who first saw the great power of that simple and basic rule. It took, as Pierre-Maurice Duhem (1861–1916) commented, "a remarkable perspicacity" on van der Waals's part to perceive the phase rule "among the algebraic formulas where Gibbs had to some extent hidden it" [17]. Duhem also wondered how many more such seeds that might have grown into whole programmes of research "had remained sterile because no physicist or chemist had noticed them under the algebraic shell that concealed them?" [17]

This shows that Gibbs's memoir had become widely known in Europe and had received the recognition it merited. It was translated into German by W. Ostwald, into French by Henry-Louis le Chatelier (1850–1936). Duhem wrote a letter to Gibbs on 29 May 1900, stating that this memoir crowned the nineteenth century in much the same way that Lagrange's *Mécanique analytique* crowned the eighteenth. Another opinion of the same memoir is due to Gibbs's student, Edwin Bidwell Wilson, who compared his achievement in it to starting with a knowledge of only the first book of Euclid and developing all the rest for oneself [18].

Gibbs's succinct and abstract style and his unwillingness to include examples and applications to particular experimental situations made his work very difficult to read. Such famous scientists as Helmholtz and Planck developed their own thermodynamic methods in an independent fashion and remained quite unaware of the treasures buried in the third volume of the *Transactions of the Connecticut Academy of Arts and Sciences*.

Gibbs did not write any other major paper on thermodynamics and limited himself to a few short papers elaborating several points in his long memoir [15]. Perhaps he thought that he had said all he needed to say on the subject. He had shifted his attention to other issues of physics and mathematics (vector analysis, calculations of orbits, investigations into the electromagnetic theory of light, a new variational principle for mechanics, the famous Gibbs phenomenon in the Fourier series [12]). Thus he rejected all suggestions to write a treatise on thermodynamics that would expand his ideas and make his work more accessible. Among the people who urged him to do this was Lord Rayleigh, who on 5 June 1892 wrote to Gibbs that the original memoir was "too condensed and too difficult for most, I might say all, readers". Gibbs's answer is rather surprising: he now thought that his memoir seemed "too *long*", and showed a lack of a "sense of the value of time, of my own and others, when I wrote it". These letters, as well as the one by Duhem quoted below, are part of the Gibbs Collection in the Yale University Library.

Gibbs agreed to a republication of his writings of thermodynamics only shortly before his death. He planned to add some new material to the book, in the form of additional chapters. The bare titles of two of these chapters, "On similarity in thermodynamics" and "On entropy as mixed-up-ness" [19] whet our appetite to know more about their content, but we can only guess. The first might have dealt with what is nowadays called the law of corresponding states, the second with the mixing process that Gibbs used, as we shall discuss below, to explain the trend of an isolated system to thermodynamic equilibrium.

7.3 Why is statistical mechanics usually attributed to Gibbs and not to Boltzmann?

At the time when he developed his exposition [2] of statistical mechanics, Gibbs was at the end of a scientific career devoted to the study and application of thermodynamics, in which abstract thoughts were illuminated by several geometric representations, but not by images based on mechanical models typical of atomism. The energeticists, who, as we have seen in Chapter 1 and shall discuss in more detail in Chapter 11, opposed and even scorned the use of molecular ideas, particularly valued Gibbs for having avoided these ideas in his work. In particular, G. Helm praised his thermodynamic writings because they "established the strict consequences of the two laws with no hankering or yearning after mechanics" [20]. Yet Gibbs in his new book was discussing the principles of statistical mechanics.

Although he frequently mocked the procedures typical of mathematicians [21], Gibbs was no less severe than they were in applying stringent logic and preoccupied with avoiding publication of incomplete results. To summarize, he had a character and a way of proceeding that were almost diametrically opposed to Boltzmann's, who had his intuition, his faith in mechanical models, and his enthusiasm as winning tools, with the obvious consequence of a very large number of papers, of preoccupying length. On this aspect, Maxwell had written several years before (1873) to Tait [22]:

By the study of Boltzmann I have been unable to understand him. He could not understand me on account of my shortness, and his length was and is an equal stumbling-block to me. Hence I am very much inclined to join the glorious company of supplanters and to put the whole business in about six lines.

To be sure, Maxwell had died five years before Boltzmann's paper [1], but if this was Maxwell's opinion, we can imagine what opinions people might have had on Boltzmann's style and statements in times and places more hostile to the atomic theory of matter. This is what we have earlier called an easy explanation of the fact that

Boltzmann's paper [1] was practically forgotten and Gibbs's treatise became the standard reference for equilibrium statistical mechanics.

In his paper considering Boltzmann's paper, Gallavotti [23] claims that the reason for its obscurity is due rather to the fact, hinted at in previous chapters, that Boltzmann's work is known only through the popularization of the Ehrenfests' encyclopaedia article [24], which is as good a treatise on the foundations of statistical mechanics as it is in having little to do with many of Boltzmann's key ideas. In fact, people who have had a chance of talking with illustrious physicists who had been students of Paul Ehrenfest may realize, by reading Boltzmann and the encyclopedia article [24], that when such physicists said that they were reporting Boltzmann's views, they were really talking about that article. Gallavotti [23] should be mentioned here, not only because he appears to be the second author after Klein to underline the importance and the basic role of Boltzmann's paper [1], but also because he proposes a new etymology of the words *monode, ergode, holode*. This has some relevance, even if these terms have disappeared from common usage, because the second of them has given rise to the much used adjective *ergodic*, which we shall meet later in connection with the ergodic problem. It is commonly surmised that the second part of these words comes from the Greek term ὁδός (path). Gallavotti argues that it comes from εἶδος (aspect, form, method, way, state; related to the Sanskrit word *vedah*). This is very interesting, because it would suggest something related to a state rather than an evolution, but it is a little puzzling; the present author must confess that he has not the necessary philological competence to discuss this point and in particular why Boltzmann rendered the end of the above German words with *ode* rather than *ide* or *ede*. It is true that the perfect tense of the related verb εἶδον (to see) transforms the ε into an o.   On the other hand, Boltzmann should have written "monhode" if he thought of a relation with ὁδός (as in "hodograph").   Presumably we must accept Gallavotti's interpretation, unless more cogent arguments are suggested against it; in fact he must have discussed it with his father Carlo Gallavotti (1909–92), a distinguished expert on Greek literature, as indicated in the acknowledgements [23].

We can tentatively add a third explanation for the unfamiliarity of Boltzmann's work, by introducing the role played by Niels Bohr. He was probably the most influential scientist of this century (in the words of Max Born, when proposing both Einstein and him as foreign members of the Göttingen Academy of Sciences: "His influence on theoretical and experimental research of our time is greater than that of any other physicist.") Bohr had a great opinion of Gibbs, expressed over and over, because he had introduced statistical ensembles, but did not think highly of Boltzmann, and he may have induced mistrust of Boltzmann's work in a large number of physicists of the twentieth century. It is clear that he had not read Boltzmann, who had been the originator of the first two kinds of ensemble. This is confirmed by his co-worker L. Rosenfeld in an interview on Bohr given to T. Kuhn and J.L. Heilbron, and kept in the Niels Bohr Archive in Copenhagen. These are the relevant sentences of Rosenfeld: "I don't think that he had ever read Boltzmann, or at any rate not Boltzmann's papers. He had read, I think, and studied very carefully Boltzmann's Lectures on gas theory" and "He said that Boltzmann had spoiled things by insisting on the properties of mechanical systems."

Apparently Bohr was not alone, for Einstein once said to one of his students: "Boltzmann's work is not easy to read. There are great physicists who have not understood it." [25]

Bohr's opinions go back to his youth, because he had already expressed reservations about Boltzmann's views in 1912, when lecturing on the statistical foundations of thermodynamics in Copenhagen [26].

7.4 Gibbs's treatise

In his treatise [2], published as one of the volumes in the Yale Bicentennial Series, Gibbs shows himself to be worrying mostly about the degree of generality of his exposition. Thus he abstains from any assumption on the microscopic constitution of matter, and renounces following the time evolution of a particular system. He sets himself the task of determining the way in which an ensemble of systems of the same kind distributes among the possible phases, starting from the distribution at a given instant of time. His theory is thus a branch of rational mechanics, a sort of projection of the latter discipline on to thermodynamics, a projection performed with analogies, the validity of which is discussed by Gibbs himself.

It is to be remarked however that Gibbs, notwithstanding all his above-mentioned preoccupation about the generality of the principles upon which statistical mechanics should be founded and for just this reason not so free with his references, pays a tribute to Boltzmann, underlining the fact that the Austrian scientist was the first, beginning in 1871, to consider explicitly the phase distribution of a large number of systems, as well as to study the evolution of this distribution through Liouville's theorem [2, Introduction and *passim*].

Gibbs's book must have been a real surprise to the scientific community, particularly since it would have seemed to represent not just a change of direction on the author's part, but an actual reversal. In fact, in the introduction he states that only the principles of statistical mechanics could supply the "rational foundation of thermodynamics" and that the laws of the latter discipline were only an "incomplete expression" of these more general principles. Yet it was not the first time that he was referring to molecular behaviour: one may mention his treatment of the reaction of NO_2 to N_2O_4 [15], his mention of "the sphere of molecular action" in the analysis of the thermodynamics of capillarity [15], and the famous Gibbs paradox [15], to which Grad alluded in a passage quoted in Chapter 5.

Actually this paradox is no paradox to those who have understood the true meaning of entropy according to statistical mechanics. It arises when two gases are allowed to mix at constant temperature and pressure; then there is an increase of entropy, which is natural, because as we know, the state in which the molecules of the two gases are intimately mixed is more probable than the state in which the two gases are completely separated. The amount of increase in entropy is "independent of the degree of similarity or dissimilarity" between the two gases, unless they are identical—the same gas—in which case there is no entropy increase at all. Gibbs's description of this situation is explicitly molecular and we can perhaps find here the first seed that would grow into the treatise of 1902. The reason why there is no paradox is that a given system cannot always be assigned a unique value, its "entropy". It may have many different entropies, among which we can choose according to our interests, the particular phenomenon under study, the degree of precision available or arbitrarily decided upon, or the method of description which is employed. For example, an aeronautical engineer studying the motion of air past an aircraft can think of the entropy of the air itself, as if all its molecules were identical, although we know that it is a mixture of gases, mainly nitrogen and oxygen, the molecules of which are decidedly different. If a nuclear engineer wants to separate two isotopes of uranium, he will deal with extremely similar molecules, but he will need an entropy that takes into account the slight difference between them, the only one that can be used for the purpose of separating the two isotopes. Hence there is a hierarchy of entropies which take more and more detail into account. When the molecules are really identical, we can only conceptually think of them (in classical mechanics) as different by applying to each of them a label; then this highly detailed entropy (Gibbs entropy) will remain constant in time, because any state will be equally probable when measured in terms of it. Thus the paradox arises because entropy is a property not of the system but of our description.

The example provided by his paradox must have convinced Gibbs of the importance of studying statistical mechanics. In fact he wrote [15]: "In other words, the impossibility of an uncompensated decrease of entropy seems to be reduced to improbability", the sentence that Boltzmann would choose as a motto for his second volume of his lectures on gas theory [27].

By 1892 Gibbs was devoting much of his energy to writing up the results of his years of work on statistical mechanics [5]. In the above-mentioned letter to Lord Rayleigh, he wrote:

Just now I am trying to get ready for publication something on thermodynamics from the a priori point of view, or rather on 'statistical mechanics', of which the principal interest would be in its application to thermodynamics—in the line therefore of the work of Maxwell and Boltzmann. I do not know that I shall have anything particularly new in substance, but shall be contented if I can so choose my standpoint (as seems to me possible) as to get a simpler view of the subject.

It would not be appropriate here to make a detailed exposition of Gibbs's short treatise. We shall restrict ourselves to briefly examining his ideas on the trend to equilibrium and his thermodynamic analogies. He did not quote Boltzmann's equation or his combinatorial method, or Boltzmann's latest elaborations and clarifications of his views on irreversibility. There are no indications that he read Boltzmann's lectures on gas theory.

In order to make Gibbs's passages quoted below more easily understandable, we recall that he uses the term "index of probability" to mean the logarithm of the probability density in phase space, P, that we considered in Chapter 4. On the problem of trend to equilibrium, we can read [2, Chapter XII]:

It would seem, therefore, that we might find a sort of measure of the deviation of an ensemble from statistical equilibrium in the excess of the average index above the minimum which is consistent with the condition of invariability of the distribution with respect to the constant functions of phase. But we have seen that the index of probability is constant in time for each system of the ensemble. The average index is therefore constant, and we find by this method no approach toward statistical equilibrium in the course of time.

Here Gibbs deals with the fact that at the level of the Liouville equation one cannot talk of a trend to equilibrium, if all the states (the phases) compatible with a given energy can be reached during the time evolution of the system. In order to avoid this deadlock, Gibbs invokes the example of a liquid mass, made of two immiscible components having different colours, white and black, in a container of cylindrical shape: at time zero, the two liquids are clearly separated, but if we spin the cylinder, the white and black parts start tracing narrow ribbons which embrace the axis in a spiral shape, with a thickness that decreases in time: the liquid tends to form perfect mixing of white and black, but after any finite time interval, the total volume will be subdivided into two parts, strictly intertwined, of different colour. He also tries to give a more precise explanation of the trend of a given ensemble towards a state of statistical equilibrium, by imagining subdivision into equal elements ΔV, which are assumed to be not infinitesimal but so small that, at least initially, the probability index is essentially constant in the interior of each element. Then one can see (see Appendix 7.1) that the average index, computed through the subdivision into cells, at subsequent time instants will be lower than its initial value. This does not violate the time reversibility of the equations. It seems appropriate to make two comments here: first, that the analogy with a liquid is justified by the fact that the probability density in phase space behaves like the density of an incompressible fluid, which is a good model for a liquid (see Chapter 4); second, the idea of convergence to equilibrium that Gibbs tries to convey here is akin to the concept of weak convergence in mathematics.

In order to find a basis for thermodynamics in analytic mechanics, Gibbs then studies the coincidence existing between the equations governing the statistical ensembles and the general principles of thermodynamics [2, Chapter XIV]: but, he says,

however interesting and significant this coincidence may be, we are still far from having explained the phenomena of nature with respect to these laws. For, as compared with the case of nature, the systems which we have considered are of an ideal simplicity. Although our only assumption is that we are considering conservative systems with a finite number of degrees of freedom, it would seem that this is assuming too much, so far as the bodies of nature are concerned. The phenomena of radiant heat, which certainly should not be neglected in any complete system of thermodynamics, and the electrical phenomena associated with the combination of atoms, seem to show that the hypothesis of systems of finite number of degrees of freedom is inadequate for the explanation of the properties of bodies [...]. But, although these difficulties, long recognized by physicists (See Boltzmann, Sitzb. der Wiener Akad., Bd. LXIII, S. 418 (1871)), seem to prevent, in the present state of science, any satisfactory explanation of the phenomena of thermodynamics as presented to us in nature, the ideal case of systems of a finite number of degrees of freedom remains as a subject not devoid of a theoretical interest, and which may serve to point the way to the solution of the far more difficult problems presented to us by nature. And if the study of the statistical properties of such systems gives us an exact expression of laws which in the limiting case take the form of the received laws of thermodynamics, its interest is so much the greater.

Gibbs also says that the notion itself of canonical ensemble "may seem to some artificial and hardly germane to a natural exposition of the subject". However, if we take temperature as an independent variable, it is the canonical ensemble that gives the best mathematical representation of a body at a given temperature, whereas it is natural to have recourse to the microcanonical ensemble when energy (a natural notion in mechanics) is considered as an independent variable.

7.5 French scientists on statistical mechanics

It is to be remarked that Hadamard, in 1906 [28], wrote an analysis of Gibbs's treatise, with particular attention to the possibility of explaining the irreversible processes within statistical mechanics.

Hadamard recalls that the equations of analytical mechanics possess, generally speaking, the so-called Poisson stability, according to which the representative point of a system passes again infinitely many times in a neighbourhood as small as we like of any position which it has already occupied, apart from a zero volume set (Poincaré's recurrence theorem, see Chapter 5). He contrasts the analysis of Boltzmann, based on the occurrence of numerous collisions, which may occur in very short time intervals, with Gibbs's, which considers systems without interactions and may be applied only to sufficiently long time intervals $t - t_0$ for two representative points in phase space, close at time t_0, to be very distant at time t. For Hadamard there is no doubt about the fact that the objection based on the reversibility of the equations of motion is not tenable, i.e. it cannot be used to prove the impossibility of systematic irreversible laws, which on the contrary seem to be plausible, as shown by Hadamard's analogy based on card shuffling. Hadamard himself underlines the sensitivity of the solution to a change in the initial data.

According to Hadamard, Gibbs does not exhibit rigorous proofs of the existence of irreversible processes, since a satisfactory definition of randomness of a system is lacking. One may think that the "organized" distributions are the exceptional ones in the sense of Poincaré's recurrence theorem. The conclusion is the following:

If H is a certain function of the distribution of the systems S, the following property: *The quantity H is increasing* can be transformed into a statement of the following kind (in which I no longer see handles for the reversibility objection): Let H_1 and $H_2 < H_1$ be two values of H, T a suitably chosen time interval. Let us denote by M_1 the motions for which, in the interval T, the quantity H takes at least once the value H_1; by M_2 the motions such that (during the same time interval) the quantity H takes at least once the value H_2; by M_3 those that satisfy both of the previous conditions; in other words that are at the same time M_1 and M_2 motions. *The M_3's are exceptional among the M_1's, but not among the M_2's.* I have no doubt, as far as I am concerned, about a statement of this kind after the deductions of Gibbs and Boltzmann; were their conclusion wrong, the error could only concern the expression of the quantity H.

Another great French mathematician, Henri Poincaré, gave his opinions more than once on the kinetic theory of gases, but as we saw in Chapter 5, more with scepticism than with adherence to Boltzmann's and Gibbs's ideas. Concerning Poincaré and Boltzmann, it is ironic that they did not understand each other; they are probably the two scientists who most shaped our ideas about the behaviour of complex classical systems. We have no direct evidence that they met, but they must have done so, since they were both at the meeting in St Louis in 1904 (see Chapter 1).

We recall here the tribute, actually too hasty, by Poincaré before the French Academy of Sciences [29]:

Boltzmann, who died tragically, had been teaching for a long time in Vienna; he had become known especially for his researches on the kinetic theory of gases. If the world obeys the laws of mechanics that allow us to proceed both forward and backward in time, why does it constantly tend toward uniformity without any chance that one may bring it back? This was the problem to be solved which he had devoted himself to, and not without some success.

In an article by Poincaré published in the same year [30], we read:

The kinetic theory of gases still leaves several embarrassing points to be clarified for those who are used to mathematical rigour; several results which have not been made sufficiently precise show up in a paradoxical form and seem to engender contradictions which, on the other hand, are only apparent. Thus the notions of *molar geordnet* (ordered at a molar level) or *molekular geordnet* (ordered at a molecular level) do not seem, in my opinion, to have been defined with sufficient precision. One of the points that were most embarrassing to me was the following: we must prove that entropy decreases [present author's remark: the sign chosen is the same as for Boltzmann's H-function]; but Gibbs's arguments seem to assume that after letting the external conditions vary, one waits for the system to go back to [a steady] regime before letting them vary again. This assumption is essential, or one could arrive at results at variance with Carnot's principle by letting external conditions vary too quickly for the steady regime to prevail. I wanted to clarify this question…

Poincaré continues with his analysis and arrives at the point of discussing a coarse-grained and a fine-grained entropy. The latter is greater than the former, which is the quantity considered by physicists and decreases with time, whereas the other remains constant.

7.6 The problem of trend to equilibrium and ergodic theory

It seems clear that Poincaré's analysis inspired the Ehrenfests in their criticism of Gibbs, who never clarified the difference between the two entropies in a neat way. Although this criticism is relevant, one cannot ignore the fact that Gibbs himself was aware of the difficulties, as shown by the following sentences concerning the motion of a liquid containing colouring matter [2]:

Now the state in which the density of the coloring matter is uniform, i.e. the state of perfect mixture, which is a sort of state of equilibrium in this respect that the distribution of the coloring matter in space is not affected by the internal motions of the liquid, is characterized by a minimum value of the average square of the density of the coloring matter. Let us suppose, however, that the coloring matter is distributed with a variable density. If we give the liquid any motion whatever, subject only to the hydrodynamic law of incompressibility,—it may be a steady flux, or it may vary with time,—the density of the coloring matter at any same point of the liquid will be unchanged, and the average square of this density will therefore be unchanged. Yet no fact is more familiar to us than that stirring tends to bring a liquid to a state of uniform mixture, or uniform densities of the components, which is characterized by minimum values of the average squares of these densities. It is quite true that in the physical experiment the result is hastened by the process of diffusion, but the result is evidently not dependent on that process.

The contradiction is to be traced to the notion of the *density* of the coloring matter, and the process by which this quantity is evaluated. This quantity is the limiting ratio of the quantity of the coloring matter in an element of space to the volume of that element. Now if we should take for our elements of volume, after any amount of stirring, the spaces occupied by the same portions of the liquid which originally occupied any given system of elements of volume, the densities of the coloring matter, thus estimated, would be identical to the original densities as determined by the given system of elements of volume. Moreover, if at the end of any finite amount of stirring we should take our elements of volume in any ordinary form but sufficiently small, the average square of the density of the coloring matter, as determined by such element of volume, would approximate to any required degree to its value before the stirring. But if we take any element of space of fixed position and dimensions, we may continue the stirring so long that the densities of the colored liquid, estimated for these fixed elements, will approach a uniform limit, viz., that of a perfect mixture.

The case is evidently one of those, in which the limit of a limit has different values, according to the order in which we apply the processes of taking the limit.

What does all this lengthy argument mean? Essentially this: if we subdivide phase space (or, better, a hypersurface of constant energy; see Appendix 7.1) in small cells of volume Δ and indicate by $\overline{P}^i(t)$ the average of P over one of these cells (a function of the index i, identifying the cell, and of time t), then $\overline{P}^i(t)$ tends to a constant (independent of i) when $t \to \infty$. The statement is plausible, at least for systems of points which do not exhibit attractive forces, but far from simple to prove. In mathematical terminology we would say that there is a weak convergence to a uniform state.

In fact the result stated above appears to be even stronger than ergodicity, as it is understood today. Here seems to be a good moment to recall the history and the role played by this concept in the development of statistical mechanics.

We have already stated that there is a hint at a property which must be satisfied if the statistical approach to mechanics holds. In particular, since a strict equilibrium is impossible, the analogue of thermodynamic equilibrium is a sort of motion which does not favour any region of phase space, or at least of the hypersurface of constant energy.

The Ehrenfests' article [24] leaves much to be desired from a historical viewpoint. And this is true not only for the role played by Boltzmann in establishing the equilibrium distributions, but also because of certain wrong attributions. Thus S. Brush notices [31] that the authors attribute the expression "continuity of path" to Maxwell, whereas it was first used by J.H. Jeans in 1902 [32], 23 years after Maxwell's death.

Thus it is better to look at what was actually written by the founding fathers, Maxwell and Boltzmann. The latter first published a detailed discussion of ergodic systems in 1871 [33]. He treats the motion of a point mass in a plane under the action of an attractive force which results from the sum of a force $-ax$ along the x-axis and a second force $-by$ along the y-axis. The resulting trajectory will be closed if the ratio between the two constants a and b is rational; in that case one obtains the so-called Lissajous figures well known from elementary textbooks in physics. When the aforementioned ratio is not rational, then exact recurrence may occur for only particular initial data; in general, the motion in the xy-plane will tend to fill a rectangle (*die ganze Fläche*), the size of which is fixed by the initial data (see Fig. 7.1). Similarly in the case of an attractive central force, for values of the total energy which allow only bounded motions (Boltzmann

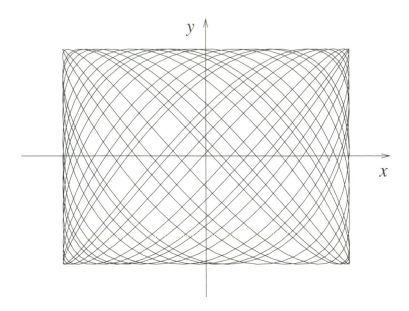

FIG. 7.1.   The trajectory of a point mass subject to a force $-ax$ along the x-axis and a second force $-by$ along the y-axis.

quotes the case of a potential $ar^{-1} + br^{-2}$, for which an exact solution can be obtained), the point mass traverses the entire surface of an annular region (see Fig. 7.2) and the trajectory never becomes closed, in general (notable exceptions are, apart from special initial data, a Newtonian force and a force proportional to the distance from the attracting point). Since Boltzmann studied these simple problems as examples for the property that was required to establish equilibrium statistical mechanics, it is of interest to see what he meant. Did he really mean that the trajectory goes through *every* point in the areas mentioned, or that it covers the area in such a way as to approach as close as one likes to every point? That is the question. Boltzmann then stated that the same property that is in question should hold for a system of n points. Later in the paper he treats the thermal equilibrium of gas molecules on the following assumption [33]:

The great irregularity of the thermal motion, and the variety of forces that act on the body from outside, make it probable that the atoms, thanks to the motion that we call heat, pass through all possible positions and velocities compatible with energy conservation, and that we can accordingly apply the equations previously developed to the coordinates and velocities of the atoms of warm bodies.

The fact that Boltzmann mentions forces acting from outside seems to suggest that he knew that the trajectory was not going through *every* point, but this did not matter if it went as close as we like, since we cannot account for the multiplicity of small external forces.

Perhaps the strongest statement in favour of the ergodic hypothesis, as usually stated, can be read in Maxwell [34]:

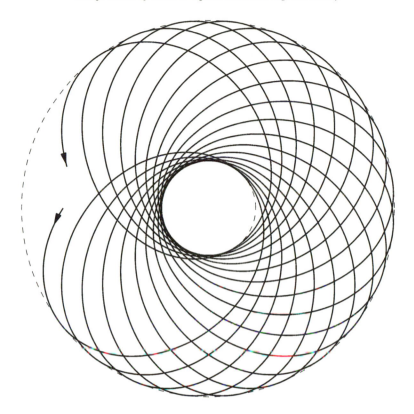

FIG. 7.2.   The trajectory of a point mass subject to an attractive central force slightly different from Newtonian attraction.

The only assumption which is necessary for the direct proof of the expression for the distribution function in equilibrium is that the system, if left to itself in its actual state of motion, will, sooner or later, pass through every phase which is consistent with the equation of energy.

Now it is manifest that there are cases in which this does not take place. The motion of the system not acted on by external forces possesses six equations besides the equation of energy, so that the system cannot pass through those phases which, though they satisfy the equation of energy, do not satisfy these six equations.

Again, there may be particular laws of force, as for instance that according to which the stress between two particles is proportional to the distance between them, for which the whole motion repeats itself after a finite time. In such cases a particular value of one variable corresponds to a particular value of each of the other variables, so that phases formed by sets of values of the variables which do not correspond cannot occur, though they may satisfy the several general equations.

But if we suppose that the material particles, or some of them, occasionally encounter a fixed obstacle such as the sides of the vessel containing the particles, then, except for special forms of the surface of this obstacle, each encounter will introduce a disturbance into the motion of the system, so that it will pass from one undisturbed path to another. The two paths must both satisfy equations of energy, and they must intersect each other in the phase for which the conditions of the encounter with the fixed obstacle are satisfied, but they are not subject to the equations of momentum. It is difficult in a case of such extreme complexity to arrive at a thoroughly satisfactory conclusion, but we may with considerable confidence assert that except for particular forms of the surface of the fixed obstacle, the system will sooner or later, after a sufficient number of encounters, pass through every phase consistent with the equation of energy.

This statement seems inescapable, because the boundary is not introducing any uncontrolled outside influence. The term *every phase* is not identical with *every point*, but using this as the basis of an argument would be really fussy. Thus Maxwell actually stated something similar to the ergodic hypothesis, while Boltzmann was not always very precise (another example is given by Brush [31]), just because he thought that a quasi-ergodic and an ergodic hypothesis, though very different mathematically, were practically equivalent.

We remark that the lengthy and extremely well-documented treatment of the ergodic hypothesis by Brush [31] mentions in passing that Boltzmann had discovered the method of ensembles before Gibbs, but he devotes no discussion to this extremely important point.

To conclude our discussion on the ergodic hypothesis, we must recall that the theory of infinite sets had not been developed in detail in those days. But all it needed was to allow time to pass. Thus it is not surprising, if we call *ergodic systems* those for which the phase space trajectory passes through *every* point, to read the remark of Borel in his supplement to the French translation to the Ehrenfests' article [35]:

It is sufficient to be acquainted with the modern theory of sets to have the certainty that even the definition of ergodic systems is contradictory. [...] this contradiction has been explicitly shown [...] by A. Rosenthal and M. Plancherel [...]; to tell the truth, it does not appear that this abstract hypothesis was ever really considered by physicists; ...

Today, since the work of G.D. Birkhoff [36] and J. von Neumann [37], the term ergodic is equivalent to "metrically undecomposable". This means that a dynamical system has a phase space which does not possess any non-zero measure subsets that are invariant for the motion. In other words, the set of points which cannot be joined by a trajectory has zero measure. It seems likely that many systems of physical interest possess this property, but actual proofs are scant.

7.7 Planck and statistical mechanics

As we have already remarked, Gibbs's treatise was to obscure for most scientists Boltzmann's and Maxwell's kinetic theory, and to develop a tendency to forget the by then old lesson of Boltzmann on the Second Law, albeit that Gibbs himself had at length attracted the attention of his readers to the aleatory nature of his thermodynamic analogies. It is thus interesting to examine what Max Planck was writing in those years. He was a recent convert, if in 1897 he was still writing in the introduction to his treatise on thermodynamics [38] that "obstacles, at present unsurmountable" stood on the way of further progress of kinetic theory, and referred to the "essential difficulties [...] in the mechanical interpretation of the fundamental principles of Thermodynamics".

Planck's conversion was due to the fact that Boltzmann's techniques had played an essential role in the basic idea of the theory of light *quanta* proposed by Planck himself in 1900 (see Chapter 12).

The occasion for taking a position in favour of Boltzmann was offered by the celebration of the latter's sixtieth birthday. Planck [39] remarks that although there was a time, when temperature was the quantity of most interest, since it is directly measurable, whereas entropy seemed to be a complicated concept, the situation had been subverted: one must give a mechanical explanation for entropy, and the definition of temperature will follow. The reason for this inversion of trend is very clear to Planck: temperature is a concept essentially related to the notion of thermal equilibrium, which acquires a meaning only in relation to irreversibility, being the final state toward which irreversible processes tend. Thus the notion of temperature necessarily leads to a study of irreversibility and hence of entropy, a primary, general concept that has a meaning for all kinds of state and hence for all state changes, whereas temperature makes its appearance only in a condition of thermal equilibrium, when entropy reaches its maximum value. He then goes on [39], praising Boltzmann but ignoring, like everybody else, the latter's 1884 paper [1]:

It seems that Clausius and Maxwell have never tried to give a direct and general definition of entropy in mechanical terms. It was left to Boltzmann to accomplish this step, starting from the kinetic theory of gases and defining entropy in a general and univocal fashion as the logarithm of the probability of the mechanical state. In his papers on statistical mechanics, J.W. Gibbs proposes, besides Boltzmann's definition, three new definitions founded upon the calculus of probabilities. The definitions by Gibbs aspire to a wider generality, because they do not imply any particular assumption on the mechanical system under consideration. They allow a successful extension, in principle, to systems that have, or have not, a large number of degrees of freedom and are made up of constituents of the same or different nature. To each of these definitions of entropy there corresponds a definition of temperature through $dQ = T\, dS$. For systems with a large number of degrees of freedom, the three definitions always lead, as shown by Gibbs, to the same result, so that, e.g., for a system formed by very many molecules, there is but one definition, which agrees with the thermodynamic one.

Planck then proceeds to examining the case of equilibrium in a monatomic gas and remarks that both Boltzmann's and Gibbs's definitions are in agreement with thermodynamics:

Whereas Boltzmann defines entropy by the logarithm of probability, entropy, according to the first definition by Gibbs (concerning canonical ensembles), is, with a minus sign, the average value of the logarithm of probability. In the course of irreversible processes, the averaged logarithm of probability *decreases* according to Gibbs, whereas according to Boltzmann the logarithm of probability *increases*. But this contradiction is just apparent, and disappears when one takes into account the fact that the two authors indicate entirely different concepts by the word *probability*. This requires a more detailed consideration. Boltzmann obtains the expression of probability by distinguishing betweeen a given state of the system and a "*complexion*" of the system itself. A state is given by the law of repartition of the positions and velocities, i.e. by the number of particles located in each element of the volume in phase space, by assuming that in each of the elements, considered to be equal, there is always a great number of particles. Hence a given state corresponds to many complexions. In fact when any two particles, belonging to two different elements, interchange their coordinates, one obtains a new complexion, but the state remains the same. If then, with Boltzmann, we consider all the complexions as equally probable, the number of the complexions corresponding to a fixed state gives the probability and hence the entropy of the system in the state under consideration (apart from a constant).

On the contrary, with Gibbs, the question of the uniformity of the particles does not play a more important role for the determination of entropy than for the distribution of positions and velocities. What is being taken into account here is rather the number of the complexions obtained when one considers all the possible values of the coordinates and of velocities within the assigned conditions: the values of the coordinates must be inside the volume v under consideration, whereas the velocities may be constrained by specific conditions.

Although, concerning the calculation of entropy by Boltzmann, Planck refers to the lectures on *Gastheorie*, he devotes more space to the definition given by Gibbs and shows that his three definitions coincide when the number of molecules is large. Then he considers the general case when the particles are not identical, remarking that Boltzmann's definitions provide formulae in agreement with thermodynamics, which Gibbs obtains only by introducing the grand canonical ensemble. He also remarks that Boltzmann's treatment can be extended, at variance with Gibbs's, to non-equilibrium states.

The problem of polyatomic molecules

8.1 The problem of specific heats

Polyatomic molecules created difficulties in kinetic theory from the start, with the problem of thermal capacities (or specific heats). Let us read directly the words of Boltzmann himself [1]:

The distinctively molecular theory of the ratio of the two specific specific heats of gases has likewise resumed an important role today. For the simplest gases, whose molecules behave like elastic spheres, Clausius had worked out the value of this ratio at $1\frac{2}{3}$, which did not fit any gas then known, from which he concluded that there are no gases of such simple constitution. For the case in which the molecules behave like non-spherical elastic bodies Maxwell found the value $1\frac{1}{3}$. Since for the best known gases the ratio was 1.4, Maxwell too rejected his theory. But he had overlooked the case in which the molecules are symmetrical about one axis, in which case theory requires precisely 1.4 as the value of the ratio in question.

Already Kundt and Warburg had found the old Clausius value of $1\frac{2}{3}$ for mercury vapour, but because the experiment was difficult it had never been repeated and all but forgotten. Then the same value turned up again for the ratio of the specific heats in all gases newly discovered by Lord Rayleigh and Ramsay, and all other circumstance pointed to the specially simple structure required by the theory, as had been the case already for mercury vapour. What influence it might have had on the history of the gas theory if Maxwell had not made this slight mistake, or if the new gases had been known as early as the time of Clausius! From the very start one would have found in the simplest gases the values that theory required for the ratio of the specific heats.

We have little to add to this presentation of the early problems of the kinetic theory of specific heats, actually only that it was Boltzmann himself [2] who discovered the "slight mistake" made by Maxwell.

If the problem of specific heats was initially solved with the discovery of noble gases, the reconsideration of the old experiments on mercury vapour for monatomic gases, and Boltzmann's suggestion to consider diatomic molecules as having five degrees of freedom, for more complex molecules the problem was bound to resurface later and to show up as one of the unsolved problems of classical mechanics. Before taking up this subject again, let us first examine another difficulty which kinetic theory had to meet very soon, and which we have already hinted at in Chapter 5.

8.2 The H-theorem for polyatomic molecules

When Boltzmann wrote his equation, he included the case of polyatomic gases as well. He also made a remark on the importance of the extension of the H-theorem to this case in the middle of a section about one-third of the way through his paper [3] (about 100 pages long), but he came back to this point at the end of the same paper. As a matter of fact, though the equation Boltzmann wrote is correct, his original proof of the H-theorem for this case is not completely general, as pointed out for the first time by H.A. Lorentz in 1887 [4–8].

As shown by a letter by Boltzmann dated 11 December 1886, the first paper [4] had been communicated by Lorentz in the form of a manuscript. The basic point is the following: when one proves the H-theorem for monatomic gases, one does not sufficiently emphasize in general (since it is irrelevant for the final result) that the velocities $\boldsymbol{\xi}'$ and $\boldsymbol{\xi}'_*$ are *not* the velocities into which a collision transforms the velocities $\boldsymbol{\xi}$ and $\boldsymbol{\xi}_*$ but the velocities which are transformed into the latter by a collision; in fact the lack of correlation applies to the velocities before a collision and not to those after the same collision (see Chapter 4). If we recall that a collision is parameterized by the unit vector $\boldsymbol{n}$, oriented as $\boldsymbol{x}_* - \boldsymbol{x}$, it turns out that in the gain term $(\boldsymbol{\xi}'_* - \boldsymbol{\xi}') \cdot \boldsymbol{n} < 0$ and $(\boldsymbol{\xi}_* - \boldsymbol{\xi}) \cdot \boldsymbol{n} > 0$, while the signs are exactly opposite in the loss term. In the case considered so far of monatomic gases, one takes advantage of the fact that the expressions of $\boldsymbol{\xi}'_*$ and $\boldsymbol{\xi}'$ are invariant with respect to a change of sign of $\boldsymbol{n}$, which enables one to speak equivalently of two conceptually very different things, i.e.: (a) the pair of velocities which arises in a collision from the pair $(\boldsymbol{\xi}_*, \boldsymbol{\xi})$; (b) the pair of velocities which gives rise in a collision to the pair $(\boldsymbol{\xi}_*, \boldsymbol{\xi})$. The remarkable circumstance that we have just recalled is related to the symmetry of a central collision, so that one may associate with a collision $[\boldsymbol{\xi}_*, \boldsymbol{\xi}] \rightarrow [\boldsymbol{\xi}'_*, \boldsymbol{\xi}']$ an "inverse collision" $[\boldsymbol{\xi}'_*, \boldsymbol{\xi}'] \rightarrow [\boldsymbol{\xi}_*, \boldsymbol{\xi}]$ which differs from the former only because the unit vector $\boldsymbol{n}$, directed along the straight line joining the centres of the molecules, changes to $-\boldsymbol{n}$. In the case of polyatomic molecules, the states before and after a collision will be described by more variables than just the velocities of the centres (e.g. by these velocities and by the angular velocity in the case of a molecule pictured as rigid body with six degrees of freedom). We shall symbolically denote by [A, B] the state of the pair of molecules. Then there is no reason why one may associate with [A, B] $\rightarrow$ [A', B'] an inverse collision [A', B'] $\rightarrow$ [A, B], which differs only for the change of $\boldsymbol{n}$ into $-\boldsymbol{n}$. In the proof of the H-theorem for polyatomic molecules, contained in the original paper by Boltzmann, the assumption was implicitly made that, by analogy with the case of monatomic gases, such inverse collision always exists; but Lorentz remarked that this circumstance had no general validity and proposed a restricted version of the theorem that would apply just to states invariant with respect to a sign change in all the velocities. Boltzmann had no difficulty in admitting his blunder and thanked Lorentz for his interest with the following words: "I am most happy to have found in You a person working to develop my ideas on the kinetic theory of gases. In Germany there is practically nobody who understands these things well." (*Ich freue mich sehr, dass sich in Ihnen jemand gefunden hat, welcher an dem Weiterbau meiner Ideen über Gastheorie arbeitet. In Deutschland ist fast jemand, welcher die Sache ordentlich verstünde.*) In the same year (1887) when the paper by Lorentz [4] appeared, Boltzmann found an "amended" version based on the so-called "closed cycles of collisions" [2, 5, 6]; the initial state [A, B] is reached not through a single collision but through a sequence of collisions. This assumption, though deemed unassailable by Lorentz [4] and Boltzmann [5], did not really satisfy anybody [8, 9]. In fact its original form deals with a cycle of a finite number of collisions, whereas, as Lorentz pointed out in a second letter dated 8 January 1887, infinitely many collisions might be needed to return to the original state. Boltzmann gave a proof for infinite sequences in his 1887 paper [5], but it is certainly much less convincing than his proof for the case of a finite number of collisions.

For a while, this aspect of the matter was forgotten until, with quantum methods, the required property was shown to follow from the property of unitarity of the so-called scattering matrix (*S*-matrix) used to describe collision phenomena in quantum mechanics [10, 11]. A satisfactory proof of the required inequality in the case of polyatomic molecules with purely classical methods was given only in fairly recent times [12] (Appendix 8.1), but it seems to have escaped the attention of historians of kinetic theory [13].

However, both proofs, classical and quantum, incorporate the most characteristic aspect of the argument of Boltzmann's closed cycles, i.e. the fact that one must consider several collisions together (in fact, infinitely many).

8.3 Specific heats again

Let us return now to the problem of specific heats, which should be a problem concerning equilibrium states and hence easily solvable. The difficulties typically arise from the impossibility of applying the theorem of equipartition of energy. If we accept the idea that the gas is in an equilibrium state with a Maxwell–Boltzmann distribution, then every quadratic term—in the velocities or coordinates—appearing in the expression of the total (kinetic + potential) energy of a polyatomic molecule not under the action of external forces contributes with a term $\frac{1}{2}kT$ (where, as usual, k is the Boltzmann constant and T the absolute temperature) to the thermal energy per molecule, which would consequently equal $\frac{\nu}{2}kT$, if ν is the number of terms of the kind indicated above. In this calculation one must consider only the terms coupled with each other in the collision, i.e. actually able to exchange energy; otherwise even in the case of smooth hard spheres (a typical model of a monatomic gas) one should consider the terms of rotational kinetic energy, which however is not convertible into any other form of energy. Thus for diatomic molecules shaped like a hard rod, previously discussed, one neglects the rotational kinetic energy about the axis joining the two atoms and one must take $\nu = 5$, rather than $\nu = 6$. Obviously one would have a discontinuity of behaviour between the case of a perfectly hard rod and a slightly elastic one, because the distance between the two atoms might change in the second case and the molecule would acquire at least one extra vibrational degree of freedom, with the consequence that the thermal energy of the gas would suddenly vary from $\frac{5}{2}kT$ to $3kT$. As in many other problems of classical statistical mechanics, it is commonly thought that this question too is solved by quantum mechanics. In fact, the Planck distribution, which allows the vibration frequency a basic role in the energy distribution, permits gradual unfreezing of the vibrational degree of freedom, thus restoring, in agreement with experience, the continuity of the behaviour of the specific heat when the temperature varies.

However, the problem shows up in another form [14]: why do we not take into account, in the quantum treatment, other degrees of freedom, such as those of the electrons or of the various constituents of the nucleus? The answer is that these fast degrees of freedom are decoupled from the slow ones, and hence their energy cannot be exchanged during a time interval relevant for a macroscopic observation. Now this was exactly the answer given by Boltzmann [15] for the corresponding classical problem of the freezing of the degrees of freedom:

But how can the molecules of a gas behave as rigid bodies? Are they not composed of smaller atoms? Probably they are; but the *vis viva* of their internal vibrations is transformed into progressive and rotational motion so slowly, that, when the gas is brought to a lower temperature the molecules may retain for days, or even for years, the higher *vis viva* of their internal vibrations corresponding to the original temperature. This transference of energy, in fact, takes place so slowly that it cannot be perceived amid the fluctuations of temperature of the surrounding bodies.

In the lines which precede those just quoted, Boltzmann had remarked that the problem shows up in the question of the energy distribution among the various frequencies in the black-body radiation:

The generalised coordinates of the ether, on which these vibrations depend, have not the same *vis viva* as the coordinates which determine the position of a molecule, because the entire ether has not had the time to come into thermal equilibrium with the gas molecules, and has in no respect attained the state which it would have if it were enclosed for an infinitely long time in the same vessel with the molecules of the gas.

In other words, Boltzmann suggested that some of the equilibrium states studied in thermodynamics are really only assumed to be equilibrium states, though they are not, because the times of variation are enormous with respect to our times of observation. Actually, as we saw, this was also his attitude concerning the Zermelo paradox; the recurrence foreseen by Poincaré's theorem will occur after an unimaginably long time. Similar words can be read in section 45 of the second volume of his "Lectures on gas theory" [6]:

If the molecules are spheres filled with mass symmetrically around their midpoints, then of course there is no possibility that they can be set into rotation by collisions, nor that any initial rotation can be lost. Nevertheless it is improbable that such molecules would remain eternally rotationless, or that they would always preserve the same amount of rotation. It seems more likely that they possess this property only to a very close approximation, so that their rotational state does not noticeably change during the time in which the specific heat is determined, even though over a long period of time rotation will be equilibrated with other molecular motions, so slowly that such energy exchanges escape our observation.

Similarly one can assume that in gases for which $\kappa = 1.4$, the constituents of the molecule are by no means connected together as absolutely undeformable bodies, but rather that this connection is so intimate that during the time of observation these constituents do not move noticeably with respect to each other, ...

8.4 Boltzmann's ideas on specific heats, and twentieth century contributions

Boltzmann's point of view on specific heats was taken into serious consideration by Sir James Jeans (1877–1946), and spurred him to discuss in detail the subject of black-body radiation, which, without Planck's hypothesis of light quanta, classically leads to the distribution law named after Rayleigh and Jeans. Here are some quotations from Jeans:

Of course, I am aware that Planck's law is in agreement with experiment if h is given a value different from zero, while my own law, obtained by putting $h = 0$, cannot possibly agree with experiment. This does not alter my belief that the value $h = 0$ is the only value which is possible to take, my view being that the supposition that the energy of the ether is in equilibrium with that of matter is utterly erroneous in the case of ether excitations of short wave-length under experimental conditions. [16]

We may say that the transfer of energy between the $3N$ material degrees of freedom and s (low frequency) degrees of aether freedom (of n degrees of freedom) is comparatively rapid, while that to the remaining $n - s$ degrees is very slow. For an enormous time, these $n - s$ degrees of freedom will not receive their due share of the energy, while the energy will rapidly equalise itself between the remaining $3N + s$ degrees of freedom. During this time, the ratio of the energy of the aether to that of the material system is $s/(3N)$, and this will generally be very small. [17]

A steel ball dropped on to a rigid steel plate will rebound perhaps half a dozen times before its energy is appreciably lessened. If the kinetic theory of gases is true, a system of molecules must rebound from one another and from rigid walls many billions of times before the total energy is appreciably lessened. The aim of the present paper is to show that, in so far as the data available enable us to judge, molecules will possess sufficient elasticity for this to occur. [18]

And, after introducing a model for molecules, in which the internal vibrations are described by harmonic (isochronous) oscillators of frequency p, he finds, with suitable considerations, that after a collision, the vibrational energy acquired by a molecule is proportional to $\exp(-2ap)$, with a certain constant a, and adds [18]:

The appropriate unit of time in this case is of course a. If p in these units has a value 200 we see that the "elasticity" of the molecules has introduced a factor e^{-400} [...]. This means that if the molecules were all moving with average velocity the number of collisions required to dissipate a given fraction of energy would be increased in a ratio of about $e^{400}:1$ by "elasticity". In other words, the "elasticity" could easily make the difference between dissipation of energy in a fraction of a second and dissipation in billions of years.

Considerations of this kind of course disappeared under the impetuous progress of quantum mechanics, but were rediscovered by L. Landau and E. Teller [19]. Whereas up to a few years ago one might have discussed this aspect of the problem from a conjectural or heuristic viewpoint, recently, thanks to the progress of rigorous mathematical methods in analytic mechanics and the availability of powerful equipment for numerical simulation, it is possible to say much more. In fact, nowadays we know that under very general conditions, if two subsystems have an interaction energy of the order of ϵ, the exchange of energy of the two subsystems will be exponentially small in $1/\epsilon$ [20–24], so that the phenomena will appear practically steady if ϵ is small enough. A numerical evaluation, e.g. of the time T required for the vibrational energy of a diatomic molecule to equilibrate with the translational one, gives the formula $T = \tau e^{\alpha v}$ where v is the characteristic frequency of the vibrations, whereas τ and α are constants with the dimensions of time, dependent on the parameters of the interaction potential. Introducing typical values for these parameters, one finds [25] that both of them are of the order of 10^{-13} seconds. This means that the low frequencies will reach an equilibrium state in microscopic time τ of the said order; there will be a frequency $\bar{v}$ (of the order of 10^{-13} seconds) which will relax to equilibrium in one second; but the frequency $2\bar{v}$ will relax in 10^{13} seconds, i.e. in 10^5 years, the frequency $3\bar{v}$ in 10^{26} seconds, i.e. 10^{18} years, etc.

Results of this kind find a place in the modern theory of Hamiltonian dynamic systems, rooted in research carried out in the mid-1950s at a theoretical level by Kolmogorov [26] and at the numerical one by Fermi, Pasta, and Ulam [27]. Kolmogorov [26] found a way to avoid some of the consequences of a famous theorem of Poincaré, which asserts the impossibility of finding, generally speaking, first integrals of the motion for a system close to a completely integrable system (i.e. a system with n degrees of freedom endowed with n first integrals in involution [28]) by perturbation methods. Kolmogorov's theorem, later extended and made more widely applicable by Arnol'd [29] and Moser [30] and thus known as the KAM theorem, asserts that there are sets invariant for the dynamic evolution which have a relative volume arbitrarily close to 1 when we let the perturbation go to zero. What happens to these sets when n tends to infinity is one of the most fascinating open questions of classical statistical mechanics.

If we follow Boltzmann's ideas, however, equally interesting are the properties that do not change (within great accuracy) for very long times. Nekhoroshev's contribution [20, 21] consists in claiming (Appendix 8.2) that, generically, the energy exchange ΔE between two degrees of freedom decoupled for $\epsilon = 0$ is bounded by $|\Delta E| \le A \exp(\epsilon_*/\epsilon)^\alpha$ with suitable positive constants A, ϵ_*, and α. However, this theorem, which is of great interest in itself, does not reveal its practical meaning as long as we abstain from numerical evaluation of the constants under consideration. Particular importance is thus attached to papers by Benettin, Galgani, and Giorgilli [31, 32], who consider molecules made up of two point masses connected by a spring of constant k which tends to infinity. Nekhoroshev's theorem shows that the energy exchanges between the vibrational and the translational degrees of freedom will increase exponentially by a power α of k. In the second of the two papers under consideration [32], the authors show that $\alpha = 1/2$ independently of the number of degrees of freedom (Appendix 8.2). We can thus say that modern developments reveal Boltzmann's deep intuition into the behaviour of complex classical dynamical systems, the qualitative study of which is only now beginning to emerge, with the development of powerful analytic theories and systematic numerical simulations. The latter show in particular that the constants which cannot be easily estimated by theoretical tools are not incompatible with time scales of billions years, as forecast by Jeans.

We are facing here one of those problems which it is easy either to discard as irrelevant or take up with enthusiasm for an unexpected result. Maybe we should take Boltzmann's ideas and their natural development as a guide. Thus, in the same way as in the problem of time's arrow we had to explain why different gas boxes point to "the same future" and we invoked the fact that physical systems are never isolated but interact with the rest of the Universe in many ways (electromagnetic and gravitational forces), we may here consider that when such extraordinarily long time periods are called into play we have either to doubt the validity of our theories or take into account that even weak causes may produce (qualitatively) strong effects. Thus for example the effects of resonance with the frequencies of background radiation should not be neglected. This is only a cursory comment, because to try to develop it is certainly beyond the scope of this book.

9

Boltzmann's contributions to other branches of physics

9.1 Boltzmann's testing of Maxwell's theory of electromagnetism

Boltzmann's scientific activities were mainly devoted to statistical mechanics: in large measure to kinetic theory, but he also founded, as we saw in Chapter 7, the formalism of equilibrium statistical mechanics for systems of a general kind, the paternity of which is usually, in a more or less explicit way, attributed to Gibbs.

We also incidentally remarked in Chapter 1 that Boltzmann was at ease with all the live themes of the physics of his time. Actually, about half his publications deal with contributions to diverse areas, ranging over the fields of physics, chemistry, mathematics, and philosophy. In this chapter we want to examine this secondary but not negligible aspect of Boltzmann's scientific activity. A common aspect of this secondary activity of his seems evident: most of what Boltzmann wrote in science represents some kind of response to an interaction with other scientists or with his students.

We recall from Chapter 1 that, having fathomed the meaning of the theory and implications of Maxwell's equations and in particular the unification between electromagnetism and optics, Boltzmann wrote his first paper on electrodynamics [1] and, at the same time as he was developing his theoretical ideas about the basic integro-differential equation that bears his name, he was also busy with an experimental study on the law that according to the Maxwell picture related the dielectric constant and the refractive index of a given material. As a matter of fact, he rightly thought that this relation between the optical and electrical properties, if ascertained beyond any doubt, would reveal a new route to appreciating the nature of electricity, which appeared to him a rather obscure subject, with absolute certainty.

From a paper published in 1874 [2], one can conclude that the experiments made by Boltzmann on different materials show that the aforementioned law turns out to be correct within the margins of experimental error.

A study of Boltzmann's scientific output reveals a series of papers on the same subject up to 1876. It is notable that these papers deal with both gases and solid bodies. The method used for gases is based on comparison of the capacities of two identical electrical condensers, filled with different gases; for solids, Boltzmann makes use of the attraction exerted by a charged conducting sphere on a non-conducting one and thus shows that for a sulphur crystal the dielectric constant is anisotropic, like the refractive index, and its change with orientation is as predicted by Maxwell's relation.

During the next few years, Boltzmann began an experimental investigation on diamagnetism.

9.2 Boltzmann lays the foundations of hereditary mechanics

Around 1874, Boltzmann's scientific interests turned to the delayed effects (or memory phenomena) in the elasticity of glass, thus laying the foundations for hereditary mechanics [3]. From these papers by Boltzmann one obtains a vivid idea, not so immediate when reading his much more well-known papers on kinetic theory, of the relevance he gave to the experimental foundation of a theory. The phenomenon was not unknown in the literature; by a study of the stretching and torsion of wires, such outstanding scientists as W. Weber [4] and F. Kolrausch [5, 6] had been led to introduce the assumption that the strain at a given instant of time depends not only on the state of stress at the same instant, but on previous states as well. Boltzmann accepts the basic approach of these authors, but he also criticizes the lack of generality both in their papers and in the almost contemporaneous work of O.E. Mayer [7]. After recalling the classical approach of elasticity theory, which he takes as a paradigm of the generality that he had in mind, Boltzmann discusses how one should modify this approach in order to include hereditary phenomena. After considering the classical parallelepiped, with its edges parallel to the coordinate axes, well known from any treatment of the foundations of the mechanics of continua, he says [3]:

The forces acting on the faces of the parallelepiped at a given instant depend not only upon the strain of the body at that instant but also on the previous strains, under the assumption, however, that, for a given strain, the more remote is the instant at which the latter took place, the smaller is the effect produced; in other words the force required to produce a given strain is smaller if a strain of the same kind has already occurred. I want to call *decrease of force caused by the previous strain* the circumstance in which a strain that occurred previously reduces the force required to produce a strain of the same kind.

In this way, Boltzmann introduces the concept of "hereditary phenomenon" (the term would be first used by E. Picard in 1907 [8]) in a form which later became standard after the papers [9, 10] of V. Volterra appeared; and one finds here a clear statement of the concept of "fading memory"—i.e. the circumstance that the delayed effects tend to zero when time t tends to infinity—a property that Volterra would consider as a basic axiom of hereditary elasticity.

Concerning the "constitutive relations" for continua with memory, Boltzmann essentially assumes that they correspond to a sort of time-distributed elasticity. Thus, along with a term which describes the by then classical constitutive relation between stresses and strains of an elastic continuum, containing the Lamé coefficients λ and μ, Boltzmann introduces the contribution of each previous instant τ with infinitesimal coefficients, given by $-\Phi(t-\tau)\,d\tau$ and $-\Psi(t-\tau)\,d\tau$ (where the minus sign accounts for the fact that the required force is decreased and not increased by the presence of the memory effects).

Of course these contributions appear in an integral, where τ varies between $-\infty$ and t; an obvious change of variable from τ to $\omega = t - \tau$ leads to another integral where the integration runs from 0 to ∞.

He then expresses the strain components by the derivatives of the components (u, v, w) of the displacement with respect to three orthogonal Cartesian axes x, y, and z. In this way the problem of the mechanics of a continuum with memory is clearly laid down in a completely general manner. What is left to experiment, of course, is to determine the elasticity coefficients λ and μ, together with the functions $\Phi(\omega)$ and $\Psi(\omega)$ which describe memory effects.

Boltzmann passes on to describe some typical situations, considering as the system to be studied a cylindrical solid body (a deformable wire) subject to a specified torque which varies in time by a defined law, in such a way that one can determine the Lamé coefficient μ and the corresponding memory function $\Psi(\omega)$; he then compares his theoretical analysis with the experimental results obtained by different authors. For his calculations he mainly uses Fourier series series expansions.

A detailed study on the development and historical setting of this paper by Boltzmann has been published by Ianniello and Israel [11].

We end this section by remarking that Boltzmann's paper [3] contains "the first successful theory of rheology" according to an authoritative article by H. Markovitz [12].

9.3 Back to electromagnetism

Certainly Boltzmann had various further interests, but after the paper we have just discussed, his activities both theoretical and experimental appear to be almost exclusively oriented towards statistical mechanics and in particular the kinetic theory of gases. We can infer from a couple of theses of which he was mentor that he had an interest in both ionized gases and irreversible processes in electrolytic conduction, but from his papers we have definite evidence of his interest in the determination of the viscosity coefficient of gases and in diffusion processes in mixtures.

Another subject that interested Boltzmann throughout his scientific activity was Maxwell's electromagnetic theory, the research theme with which he had started out. As we said in Chapter 1, beginning in 1886, deeply impressed by Hertz's experimental verification of the equivalence between electromagnetic waves and light that was predicted by Maxwell's theory, Boltzmann spent considerable effort in redoing Hertz's experiments. These are documented in the last publication he wrote before leaving Graz [13].

Not only did he give a course on Maxwell's electromagnetic theory in Graz in 1890 and in Munich in 1891, but he also published the text of these lectures [14]. This revival of interest in Maxwell's theory was perhaps a result of Hertz's experiments confirming the existence of electromagnetic waves. Also to be ascribed, at least partly, to this interest is his activity in the area of the mechanical models to be used as illustrations of various phenomena. Among these we can recall wave machines, condensers, and a mechanical model called the *Bicykel* which has already been mentioned in Chapter 1 and was built to illustrate the effect of one electric circuit on another; as mentioned in

Chapter 1, the two specimens actually built, one in Graz, the other in Munich, were lost during the Second World War, but we can surmise that they were more or less analogous to the differential gear of a modern car.

In 1895 he also published an annotated German edition of Maxwell's paper "On Faraday's lines of force" in Ostwald's series *Klassiker der exacten Wissenschaften*. It is widely recognized that Boltzmann played a very important role in the eventual acceptance of Maxwell's theory in mainland Europe. It is true that, curiously enough, he did not advance the theory itself as much as Lorentz did, though he had all the necessary tools in his hands. As a consequence, he did not grapple with the difficulties of Maxwell's theory that eventually led to Einstein's theory of relativity.

Boltzmann was the first to prove the property of time reversal in electromagnetic theory: the Maxwell equations are invariant under the joint inversion of the direction of time and of the magnetic field, the electric field being left unaltered [15].

9.4 A true pearl of theoretical physics

Before concluding our consideration of Boltzmann's contributions to physics and giving a short description of his position *vis-à-vis* mathematics, we must recall a basic result of his on the thermodynamics of heat radiation. This new area of investigation had been opened up by Kirchhoff, who in 1859 introduced the concepts of emissive and absorptive powers of a body and showed that their ratio depends only on the temperature of the body, and not on its nature. He also introduced the concept of a *black body*, defined as a body with absorptive power equal to unity. Hence a black body absorbs the whole of the heat radiation falling upon it, and the radiation it emits is a function of temperature alone. It is easy to produce black-body radiation experimentally: if there is a small aperture in a wall of a furnace at uniform temperature, the radiation escaping through it will be black-body radiation. This is due to the fact that the chance of emerging from the furnace is small and thus the radiation inside strikes the walls of the oven repeatedly and is eventually completely absorbed by them.

In 1879 Boltzmann's former mentor J. Stefan [16] had established, or perhaps conjectured, from an analysis of rather rudimentary experimental data in the case of an enclosure which approximates a black body, the proportionality between the density e of the radiation energy and the fourth power of the absolute temperature T. In 1883, Boltzmann was preparing an abstract of a paper by H.T. Eddy on radiant heat as a possible exception to the Second Law of Thermodynamics, for Wiedemann's *Beiblätter*, and learned of a work by the Italian physicist Adolfo Bartoli (1851–96) on radiation pressure. Bartoli's arguments stimulated Boltzmann to work out a theoretical derivation [17] of the same relation on theoretical grounds. Boltzmann's argument combined ideas from two disciplines rather modern at that time—thermodynamics and Maxwell's equations—using in particular a notion now known even by undergraduates but a great novelty in those days, i.e. the fact that electromagnetic waves exert a pressure on the walls of an enclosure filled with radiation. In addition, the Maxwell equations imply the existence of electromagnetic energy distributed within the enclosure.

To put this achievement in its proper historical perspective, it seems appropriate to point out that Boltzmann did this work three years before Hertz demonstrated the existence of electromagnetic waves. Lorentz, then, did not exaggerate when he referred to it as a "a true pearl of theoretical physics" [18], nor did Planck when he wrote that "Maxwell's theory received powerful support through the short but now famous contribution by Boltzmann on the temperature variation of the heat radiation of a black body" [19].

Boltzmann applied what would now be called "the usual thermodynamic relations", using the fact that in equilibrium conditions, and hence those of homogeneity and isotropy, the radiation pressure p equals $e/3$ and depends on the temperature alone. In particular, he was the first to introduce the entropy of radiation. Then it was easy for him, by using the First and Second Laws of Thermodynamics for equilibrium states, to obtain (see Appendix 9.1) that $e = \sigma T^4$.

Although at the time the result obtained by Boltzmann for radiation seemed to be an isolated achievement with no further consequences, it did at least show a possible connection between thermodynamics and electromagnetism that was to lead to quantum theory and to an important aspect of modern astrophysics, i.e. the role of radiation pressure in the study of the equilibrium of stellar atmospheres. Boltzmann's argument was pushed to its widest possible scope by Wien, who, applying thermodynamics to radiation of each single frequency and, taking due account of the Doppler effect, deduced the famous displacement law that carries his name [20] (Appendix 9.2). This is the most advanced result that can be obtained through thermodynamics and the general properties of classical radiation without introducing a detailed model or a more advanced statistical assumption. As is well known, the assumption of classical oscillators subject to an equipartition law leads to the Rayleigh-Jeans law (in complete disagreement with experiment). Hence Planck was led to proposing his law and the assumption of light quanta (see Chapter 12).

9.5 Mathematics and foundations of mechanics

As a final topic for this chapter, we want to deal briefly, as already hinted at, with Boltzmann's position with regard to mathematics and the basic axioms of mechanics.

To start with, we must explode the myth of a Boltzmann not completely at ease with derivatives and integrals. Boltzmann managed very well; otherwise, how would he have dared to be the first to perform complicated calculations in thermodynamics and to introduce integro-differential equations? It is true however that he had a finitist conception of mathematics. Derivatives were for him convenient tools, but the *significant* things were the incremental ratios. There is no lack of passages on this point [21]:

[...] if I tell somebody to sum the series $1 + 1/2 + 1/4 + 1/8 + \ldots$ really to the extent of infinitely many terms, he will be unable to do it; but if I tell him to sum so many terms that a further increase will no longer noticeably influence the result, I have given him a clear and executable prescription, and all proofs that the sum of infinitely many terms equals 2 merely signify that if you add countless thousands of further terms you will never exceed 2, though you will approach it more and more.

And a little further on, in a footnote:

The concepts of differential and integral calculus divorced from any atomist notions are typically metaphysical, if following an apposite definition of Mach we mean by this the kind of notion of which we have forgotten how we obtained it.

And again:

When carrying out the by now customary manipulations with the symbols of integral calculus, one may temporarily forget that in forming these concepts we based ourselves on starting with a finite number of elements, but we cannot really circumvent this assumption. That, too, seems to be the reason why groups of mutually interacting atoms of an elastic body are intuitively much clearer than interacting volume elements. This naturally does not exclude that, once we have become used to the abstraction of volume elements and other symbols of integral calculus and have practised the methods of operating with them, it might be convenient and expedient no longer to remember the peculiar atomistic meaning of these abstractions when we derive certain formulae that Volkmann calls those for coarser phenomena. These abstractions constitute a general schema for all cases where we may imagine the number of elements in a cubic millimetre to be 10^{10} or $10^{10^{10}}$ or milliards of times more still; hence they are indispensable especially in geometry, which must of course be equally applicable to the most varied physical cases where the number of elements can be very different. In using any such schemata it is often expedient to leave aside the basic idea from which they have sprung or even forget it for a while; but I think it would nevertheless be erroneous to believe that one had thus got rid of it.

We can also recall that Boltzmann was concerned with the mathematical problems arising from the atomic models of matter. We have already mentioned (see Chapter 5) a paper in which he gave a mathematical proof that the collision invariants are exactly those that one would expect on physical grounds. Here we mention that an early paper of his, entitled "On the integrals of linear differential equations with periodic coefficients" [22], turns out to be an investigation of the validity of Cauchy's theorem on differential equations with periodic coefficients, which is needed to justify the application of the equations for an elastic continuum to a crystalline solid in which the local properties vary periodically from one atom to the next.

In the last years of his life, Boltzmann became interested in non-Euclidean geometries and Cantor's set theory, on which he even lectured [23], thus originating the verses in a student rag, that we have quoted in Chapter 1.

One of Boltzmann's not negligible interests was of course the axiomatic foundations of mechanics, to which he was being pushed by the criticism of the concept of force and the new exposition of the principles of classical mechanics proposed by Hertz [24]. Although we shall have an opportunity to comment on the philosophical aspects of this interest in the next chapter, we cannot refrain here from remarking that Boltzmann felt the need to write a book, in two volumes, on classical mechanics [25] as a sort of interlude between the two volumes of his lectures on gas theory. For the benefit of the reader we mention that some important parts of this book have been translated into English [26].

FIG. 9.1.   Boltzmann applying the principles and basic equations of mechanics, as pictured in a cartoon by K. Przibram.

In the first volume we find a conceptual contribution by Boltzmann concerning the definition of distinguishability of particles. What does it mean that two particles are identical but distinguishable? It seems that Boltzmann was the first to feel the need to specify what we mean when we say "the same mass point". To make the issue clear, suppose that we show somebody two identical balls lying on a table and then ask this person to close his/her eyes and a little later to open them again. We then ask whether or not the two balls have been switched around in the meanwhile. He/she cannot tell, because the balls are perfectly identical. Yet we know the answer. If we have switched the balls, then we have been able to follow the continuous motion that took them from their initial to their final positions. This elementary example illustrates the first axiom of classical mechanics laid down by Boltzmann, which essentially states that identical material particles which cannot occupy the same point of space at the same time can be distinguished by their initial conditions and by the continuity of their motion. This assumption alone, Boltzmann underlines, "enables us to recognize the same material point at different times" [25, Vol.I, p.9; 26, p.230]. This property acquires a particularly pioneering aspect, because it shows what we understand by classical identical particles, as opposed to the indistinguishable particles of modern quantum mechanics.

Boltzmann is also willing to consider rather unusual ideas concerning atoms. Thus we read [25, Vol.I, p.4; 26, p.227]:

Nor must one ever seek metaphysical reasons for the picture nor draw hasty inferences from it, for example that chemical atoms are material points. Nor should we lose from sight the possibility that it might one day be displaced by quite different pictures, let us say, to avoid appearing small-hearted, ones taken from manifolds that lack even the properties of our three-dimensional space, so that for example simple geometrical constructions of atomism would have to be replaced by manipulations with numbers forming a complicated manifold.

More pioneering aspects are contained in the discussion of the possibility of changing the axioms of mechanics. Hans Motz [27] suggested that these passages were known to Einstein and paved the way to General Relativity. Here we shall restrict ourselves to bare facts.

When discussing inertial systems and the possibility of identifying them, and having said that "we are going to think that the world is finite", Boltzmann continues [25, Vol.II, p.334; 26, p.264]:

Quite independently of this there is the question whether the mechanical equations here developed and therefore also the law of inertia might perhaps be only approximately correct and whether, by formulating them more correctly, the improbability or rather inhomogeneity of having to adopt into the picture a co-ordinate system as well as material points would disappear of itself.

After this remarkable sentence admitting that the principles of Newton's mechanics, and in particular one of its pillars, might be only approximately correct, Boltzmann goes on to discuss what we call Mach's principle [25, Vol.II, p.334; 26, p.264]:

Here Mach pointed to the possibility of a more correct picture, obtained by assuming that only the acceleration of the change of distance between any two material particles is determined mainly by the neighbouring masses, its velocity being determined by a formula in which very distant masses are decisive. This naturally avoids the adopting of any co-ordinate system into the picture, since now it is only a question of distances.

Here perhaps "relative position" would be a better term than distance.

Boltzmann notes that, with this suggestion, Mach introduces other difficulties which he, at variance with some other physicists, does not consider "so particularly great", though, he remarks, they seem "to exclude all empirical test forever". This is another remarkable statement if read nowadays, when astrophysicists are looking for the "missing mass" needed to confirm Mach's principle.

But this is not the end of it. After a few further comments, Boltzmann says [25, Vol.II, p.334; 26, p.264]:

At all events I think that such an extension of our vision, by pointing out that what we regard as most certain and obvious may perhaps be only approximately correct, is most valuable. It is in line with the suggestion that the distances of fixed stars may perhaps be constructed only in a non-Euclidean space of very small curvature, which is of course connected with the law of gravity in that a moving body not acted on by forces would then after aeons have to return to its previous position if the curvature is positive.

This is the most extraordinary passage, which according to Motz [27] seems to contain the germ the General Relativity. Unfortunately the English translation, as in other places, is unfaithful, as stressed by Motz [27]. The most remarkable deviation from the German text concerns the statement that the space curvature is *of course* connected with the law of *gravity*. The German text says that the curvature is connected with the law of *inertia (in einem nichteuklidischen Raume von enorm geringer Krümmung [...], was ja insofern auch mit den Trägheitsgesetze zusammenhängt*). Even if we ignore this incredible "discovery," introduced only by the hindsight of the translator, Boltzmann makes a number of explicit statements: space may be non-Euclidean with (very small!) curvature which is connected with the law of inertia. The possibility that space might be non-Euclidean had been considered by both Gauss and Riemann, and the latter had suggested that this assumption might be useful to describe physical phenomena (see Chapter 2). What appears to be new with Boltzmann is the connection between Mach's principle and curvature and the possibility that in a finite space the problem of inertial frames would be automatically solved, the laws being the same for all observers: two of the basic ideas of Einstein's General Relativity. What Boltzmann of course lacks is the idea that space must be replaced by space–time, a concept which needed the ideas of Special Relativity to start with.

What about Special Relativity? Had Boltzmann any connection with its birth? This point has been investigated by Siegfried Wagner [28]. One point that he stresses is that Boltzmann was the first to discuss in a textbook a displacement of the reference system, although a few authors (Ludwig Lange, Streintz, Neumann) had used related terminology. Boltzmann is well aware of the basic problems concerning space and time; he explicitly assumes the need for rigid rods and a universal clock and discusses in detail the change of a reference frame. The next author to do this was Einstein when he laid down the foundations of Special Relativity [29]; the only change was of course that there was no absolute clock but the speed of light was absolute.

Boltzmann discusses in detail the use of a coordinate system. Here we just quote a few sentences [25, Vol.I, p.7–8; 26, pp.229–30]:

To define the position at any time we imagine that at all times there is in space a definite rectangular system. [...] The co-ordinate system is of course nothing real, but this offers no difficulty according to the views we base ourselves on here, since we are at present concerned only with construction and mental pictures. [...] the cause why the above pictures are clear is obvious: they are prescriptions for thinking spatial circumstances that everybody can palpably represent for himself in approximation, by means of ruler and pencil or wooden sticks and knitting needles...

More straightforwardly, Einstein defines a system of coordinates as "three rigid material lines, perpendicular to one another, and issuing from a point" [29].

In the second volume of Boltzmann's lectures, published one year before Einstein's paper, at the end of the discussion on the law of inertia quoted above in connection with Mach's principle, we read an interesting sentence [25, Vol.II, p.335; 26, p.265]:

However, the law of inertia does not hold for the particles of aether itself; Maxwell's equations would have to be formulated in such a way as that they determine only the mutual actions of adjacent volume elements so that we need no absolute space to formulate them. A working out of this as yet quite undeveloped theory is no concern of ours here.

Again some remarks on the English translation are in order. First, let us remark that the word *Lichtäther* (luminiferous aether) is simply translated as "aether". This is of course reasonable, but this comment is required for what we are going to say. Second, the last part of the passage is rather unfaithful to the original, which ends with *liegt uns hier ferne* ("lies far from us now"). As remarked by Wagner [28], the development foreseen by Boltzmann occurred just one year later. This indicates that Boltzmann was a very good physicist, but no prophet. We remark that at about that time Lorentz and Poincaré, who are rightly considered as forerunners of Special Relativity, were still using either an ether or an absolute space.

Now we wish to compare, as Wagner [28] suggests, the above passage by Boltzmann with a sentence from the second page of Einstein's epoch-making paper [29]:

The introduction of a "luminiferous ether" will prove to be superfluous inasmuch as the view here to be developed will not require an "absolutely stationary space" provided with special properties, nor assign a velocity-vector to a point of the empty space in which electromagnetic processes take place.

The similarity is striking; in particular the word *Lichtäther* is used by both Boltzmann and Einstein, whereas as remarked by Wagner [28], usually *Äther* would suffice; thus we cannot blame the translator in this case.

Are these coincidences just accidental? According to Wagner [28] they are not. Although Einstein does not explicitly mention Boltzmann as an author that he had studied when he was a student, his authorized biography by Philipp Frank [30] confirms that he had in fact done so. In particular, Frank says that from the works of Hertz and Boltzmann Einstein "learned how one builds up the mathematical framework and then with its help constructs the edifice of physics".

After what we have just said here (and in Chapter 6) we shall return to the connection between Einstein and Boltzmann in Chapter 12.

10

Boltzmann as a philosopher

10.1 A realist, but not a naïve one

As we have already remarked in Chapter 1, in the last few years of his life Boltzmann devoted a significant part of his time to the philosophical aspects of science and, more generally, of knowledge. He left written documents of his engagement in this area in the volume *Populäre Schriften* [1], partly translated into English [2]. In 1903 his philosophical work was sufficiently recognized to warrant his appointment to the chair of philosophy of science at the University of Vienna, as successor to Ernst Mach. The notes of some of his more scientific lectures, written in a special shorthand, have recently been interpreted and published (in German) [3].

Professional philosophers may find Boltzmann's ideas somewhat naïve, but this is not the case. It is true that Boltzmann is a realist, but not a naïve realist. A philosopher reading the following pages should always take into consideration that Boltzmann's vision of the world is the picture that every physicist has in mind when investigating his problems, even when he does not admit it. In addition, as we shall see, Boltzmann originated ideas about scientific revolutions, later spread by and attributed to Kuhn.

The central point of philosophy is, according to Boltzmann, the problem of the relationship between existence and knowledge, gnoseology (this term is usually rendered as epistemology in the English translation [2], although this somehow restricts its scope). To characterize his attitude in this respect, nothing serves better than a sentence thrown almost casually into the least philosophical and most entertaining of his essays, describing his trip to "Eldorado" in 1905. This sentence is frequently quoted and follows his description of the campus of the University of California at Berkeley [1]: "The name of Berkeley is that of a highly esteemed English philosopher, who is even credited with being the inventor of the greatest folly ever hatched by a human brain, of philosophical idealism that denies the existence of the material world."

As is well known, Berkeley's ideas were among the starting points of Kant's analysis, when he stated that any specific knowledge we claim to have of external objects is obtained through our senses (which is hardly debatable) and *hence* is at best only indirect and questionable, with the consequence that what we know directly and with certainty is only the set of our ideas (including the sensory data).

Before beginning a more detailed analysis of Boltzmann's views on philosophy, we remark that it is always difficult to make statements about philosophical themes, because

FIG. 10.1. Boltzmann pondering on the principles of philosophy, as pictured in a cartoon by K. Przibram.

the meaning of words changes in time and even from one philosopher to another in the same period. We shall have to say something more about the term "realist" later. For the moment we shall be content with the following meaning: a realist is somebody who believes that the world outside us exists independently of our sensations, observations, and consciousness, and the human mind can construct an image of this world with the help of sensations and more or less accurate experiments in such way that the objective image does not explicitly contain our sensations, but can explain all of them. This detailed definition is required because otherwise even Berkeley might be called a realist, since he says that the objects of sense are nothing else but ideas which cannot exist unperceived, yet their existence may be guaranteed by some spirit (such as God) that perceives them, though we do not.

Another term that needs definition is "metaphysics". Although some philosophers may give a wider meaning to this term, scientists in general and Boltzmann in particular have in mind Kant's definition in his tract "*Prolegomena zu einer jeden künftigen Metaphysik die als Wissenschaft auftreten können*" [4]:

First, as concerns the sources of metaphysical cognition, its very concept implies they cannot be empirical. Its principles (including not only its maxims but its basic notions) must never be derived from experience. It must not be physical but metaphysical knowledge, viz., knowledge beyond experience.

This will be the meaning of metaphysics we shall stick to.

How can we then convince ourselves that existence does not reduce to the perceptions of a subject, according to Boltzmann? We find his answer in one of his essays contained in the aforementioned book collecting his writings for a wide audience [1, 2]. The essay is entitled "On the question of the objective existence of processes in inanimate nature" (1897). There we can read:

We said that the purpose of thinking was the foundation of rules for our ideas such that future sensations are thereby announced in advance. This aim is attained in large measure if we apply the experience gained from perceptual complexes concerning our own bodies also to the interaction of those very similar complexes that relate to the bodies of others.

As typical examples of the way in which we more or less unconsciously proceed, Boltzmann indicates that somebody else's hand behaves just as though on touching a fire a feeling of pain occurred, his lips as though volitions acted on them. Of these sensations and volitions of other people we have not the least knowledge. In fact, we know only our own ideas of them, with which we operate as with those of our own sensations and volitions. In this way, we obtain useful rules for constructing and predicting the course of our own sensations concerning the bodies of other people. What do we mean then when we assert that these alien sensations and volitions exist as much as our own? It appears that this adds some hypothetical and unprovable element to the facts, contravening the task of our ideas as merely to describe the facts. This however, Boltzmann notices, leads to solipsism (which he calls ideologism) [1, 2]:

If, however, someone were to assert that only his sensations existed, whereas those of all the others were merely the expression in his mind of certain equations between certain of his own sensations (let us call him an ideologist), we should first have to ask what sense he gives to this and whether he expresses that sense in an appropriate way.

What ought the ideologist rather to assert, according to Boltzmann? He should say: "I use the term 'sensation' or 'act of will' as a thought symbol in three ways: firstly, to represent those sensations and volitions immediately given to me; secondly, if I find it useful to link the same terms according to the same laws in order to describe certain regularities between my perceptual complexes". Before indicating the third way of using the aforementioned terms, Boltzmann notes that one should distinguish the second mode of employment by saying that the terms are signs for sensations and volitions of others. The third way applies either if previously a person wrongly thought that the terms under consideration would be useful for representing the perception regularities mentioned above, or, without ever believing this and for quite different reasons (such as practising, playing), he combines terms that are quite analogous: these could be called terms for sensations and volitions of non-existing people that are merely imagined.

Restated in this form, however, the ideologist's assertion no longer differs, according to Boltzmann, from the ordinary way of expressing matters. The second way expresses the enormous subjective difference that, for the person describing his own sensations, exists between him and others. How can we attach an objective existence to somebody else's sensations? Our common language has long since found the solution: when we say that they are not ours but those of others, we recognize that we implicitly admit that there must be somebody, because this is by far the simplest way of explaining our complexes of sensations.

Analogously to the sensations of others, processes in inanimate nature likewise exist for us merely in imagination. In other words, we mark them by certain thoughts and verbal signs, because this facilitates our construction of a world picture capable of foretelling our future sensations. As Boltzmann says [1, 2]:

Processes in inanimate nature in this respect are thus just like the sensations of others, and inanimate objects themselves like those of others, except that the signs and laws of their conjunction are rather more different from those used in representing our own sensations.

The sentence "an inanimate object exists or does not exist" thus has the same significance as "a person exists or does not exist". It would therefore be a mistake, says Boltzmann, to believe that in this way one had established that matter is more of a mental entity than another person is. Thus to say that atoms exist, we may add as a corollary, means also that they existed before their existence was experimentally proved, or even before anybody used the word "atom", in the same way as a child exists before we saw it, or before it was given a name.

To recognize the existence of our fellow humans is of course basic. In fact, one of the most important ways in which our world picture develops further is through interaction with others. It goes without saying that, to start with, everyone will naturally distinguish himself as speaker (the subject) from those spoken to (objects) and at first adopt the (subjective) point of view, the language of the ideologist, corrected in the way proposed by Boltzmann.

In scientific matters, however, language must use some other terminology that is equally appropriate for all persons, or, as we usually say, we must adopt the objective point of view. It then turns out that the concepts previously linked with "existing" and "not existing" largely remain applicable without change. Those people or inanimate things that we merely imagine or conceive without being forced to do so by regularities in complexes of sensations (the third way) do not exist for others either; they are "objectively" non-existent.

There is an important point that must be examined, however. And here we see how Boltzmann was ahead of his time; this theme is as important today as it was for him. Do processes in inanimate nature have many analogies with psychological ones, or can one draw such a sharp line between the two that the former can be described as objectively not existing?

The argument used by Boltzmann is a continuity argument. To start with, the sensations of higher animals are so perfectly analogous to human ones, he says, that we must of necessity ascribe objective existence also to them. Where then is the boundary? Here is Boltzmann's answer [1, 2]:

One does indeed hear occasional doubts whether insects or divisible animals like certain worms have sensations, but a sharp boundary where sensing stops cannot be given. In the end we reach organisms that are so simple that their world pictures and thoughts are zero. If we are not suddenly to deny existence to the sensation of animals below a certain level, which would be quite inappropriate, then we must ascribe existence also to this unthinking organised matter, in which sensation can hardly be discovered, but this in turn runs through continuous gradations as far as the level of plants. But then it would seem to me an unjustified and inappropriate jump to deny existence to unorganised matter.

Boltzmann also remarks that a series of sensations and volitions that we call a single individual always soon come to break off again, that individual people die, whereas the matter to which those mental phenomena were tied remains. A subjective world picture that construes matter as merely expressing that certain complexes of human sensations persist in time thus starts by trying to imitate the transient and complicated features by means of symbols and only later using these pictures to represent simple and more permanent features of matter. In Boltzmann's words: "It construes the pyramids of Egypt, the Acropolis of Athens as mere equations existing between the sensations of generations through thousands of years." [1, 2]

Common usage and subsequently science (which ought to be the continuation of the former with subtler methods) construes a simpler (objective) world picture that starts from the simple and represents the transient by means of laws that govern the more permanent features. In Boltzmann's words [1, 2]:

Pursuing our mental picture consistently, that is according to the rules that have always led to confirmation by experience, we reach the conclusion that the planet Mars is of similar size to the Earth, that it has continents, oceans, snowcaps and so on, indeed it seems not at all impossible to us that on planets of other suns there are the most splendid landscapes without their producing sense impressions on any animate being.

In contrast, the idealist compares the assertion that matter has the same degree of existence as our sensations with the opinion of the child that a stone feels pain when struck. A similar, and more well founded, contention is attributed by Boltzmann to the realist [1, 2]:

The realist compares the assertion that he could never imagine how the mental could be represented by the material let alone by the interaction of atoms with the opinion of an uneducated person who says that the Sun could not be 93 million miles from the Earth, since he cannot imagine it. Just as ideology is a world picture only for some but not for humanity as a whole, so I think that if we include animals and even the Universe the realist mode of expression is more appropriate than the idealist one.

Boltzmann's opinions, expressed in such a vivid manner, differed of course from those of his colleague Ernst Mach, who also was both a physicist and a philosopher. Here are two examples of claims made by Mach [5]: "The assertion, then, is correct that the world consists only of our sensations. In which case we have knowledge only of our sensations..."; "...there is no rift between the psychical and physical, no inside and outside, no 'sensation' to which an external 'thing' different from sensation, corresponds."

To use the very words used by Boltzmann in his polemic against Ostwald (1904) [1]:

Mach pointed out that we are given only the law-like course of our impressions and ideas, whereas all physical magnitudes, atoms, molecules, forces, energies and so on are mere concepts for the economical representation and illustration of these law-like relations of our impressions and ideas.

Now whereas Boltzmann undoubtedly appreciated the importance of economy of thought in the classification and ordering of the results of science, he did not identify this economy with the final aim of scientific research. In his paper of 1892 "On the methods of theoretical physics" [1, 2] we can in fact read:

In mathematics and geometry the return from purely analytic to constructive methods and illustration by means of models was at first occasioned by a need for economy of effort. Although this need seems to be purely practical and obvious, it is just here that we are in an area where a whole new kind of methodological speculations has grown up which were given the most precise and ingenious expression by Mach, who states straight out that the aim of all science is only economy of effort. With almost equal justice one might declare that since in business the greatest economy is desirable, the latter is simply the aim of shops and money, which in a sense would be true. However, when the distances, motions, size and physico-chemical properties of the fixed stars are investigated, or when microscopes are invented and with their help the causes of diseases are discovered, we would hardly wish to call this mere economy.

The intimate connection of the mental with the physical is, according to Boltzmann, in the end given to us by experience. This connection makes it very likely that to every mental process there corresponds a physical process in the brain, in a one-to-one correspondence, and that the latter processes are all genuinely material, that is, are representable by the same pictures and laws as processes in inanimate nature. As a consequence, all mental processes must, at least in principle, be predictable from the pictures used to represent brain processes.

This leads us to another theme present in Boltzmann's philosophical writings, that of Nature as a grand mechanism, whose culminating point is to be identified with Darwin's theory of evolution and natural selection. Even the success of theoretical physics is explained on evolutionistic grounds, as one can read in the previously quoted paper "On the question of the objective existence of processes in inanimate nature" [1, 2]:

The brain we view as the apparatus or organ for producing world pictures, an organ which because of the pictures' great utility for the preservation of the species has, conformably with Darwin's theory, developed in man to a degree of particular perfection, just as the neck in the giraffe and the bill in the stork have developed to an unusual length. [...] The moment we subscribe to the view [under consideration], we must suppose that the pictures and the laws that serve the representation of processes in inanimate nature will suffice unambiguously to represent mental processes too; we say, in brief, that mental processes are identical with certain material processes in the brain (realism).

In 1886 Boltzmann had already written on these themes, in particular on the relation between the origin of life and the phenomena of thought and sensation [1, 2]:

However, in scientific questions I want to deprive such feelings of authority: the contemporaries of Copernicus were equally directly conscious and felt that the earth did not revolve. Still, the most direct path would be to start from our immediate sensations and to show how by means of them we attained knowledge of the Universe. However since this does not seem to lead to our goal, let us follow the inverse path of natural science. We frame the hypothesis that complexes of atoms had developed that were able to multiply by forming similar ones round them. Of the larger masses so arising the most viable were those that could multiply by division, and those that had a tendency to move towards places where favourable conditions for life prevailed.

This was greatly furthered by the receptivity for external impressions, chemical constitution and the motion of the circumambient medium, light and shade and so on. Sensitivity led to the development of sensory nerves, mobility to motor nerves; sensations that through inheritance led to constant compelling messages to the central agency to escape from them we call pain. Quite rough signs for external objects were left behind in the individual, they developed into complicated signs for complex situations and, if required, even to quite rough genuine imitations of the external, just as the algebraist can use arbitrary letters for magnitudes but usually prefers to choose the first letters of the corresponding words. If there is such a developed memory sign for the individual himself, we define it as consciousness. In this there is a continuous path from the closely connected clear conscious ideas to those stored in memory and unconscious reflex movements. Does not our feeling tell us once more that consciousness is something quite different? However, I have silenced feeling: if the hypothesis explains all the phenomena concerned, feeling will have to give way as in the question of earth's rotation.

In January 1905 Boltzmann delivered a speech at the Vienna Philosophical Society originally bearing the title "Proof that Schopenhauer is a stupid, ignorant philosophaster, scribbling nonsense and dispensing hollow verbiage that fundamentally and forever rots people's brains". This title sounds a bit harsh and as such it was refused, to become simply "On a thesis of Schopenhauer's" [1, 2], but in the written text of his conference Boltzmann indicates that he had originally given the former title and explains that he had taken it from a paper by Schopenhauer himself, just changing the name of the philosopher to whom it referred (and who, though Boltzmann does not say this, was Hegel). In this conference, the thoughts of all philosophers, Kant included (although Boltzmann somehow respected him), are declared to be fundamentally unsound. His aim, he claims, is the liberation of mankind from that mental headache which is called metaphysics. In this text Boltzmann explicitly calls his own philosophy "materialism", where he says: "Idealism asserts that only the ego exists, the various ideas, and seeks to explain matter from them. Materialism starts from the existence of matter and seeks to explain sensations from it." [1, 2]

How then are ideas, or anything that Kant would call a priori, conceived of in Boltzmann's philosophy? Man owes his own ideas to evolution [1, 2]:

In my view all salvation for philosophy comes from Darwin's theory. As long as people believe in a special spirit that can cognize objects without mechanical means, or in a special will that likewise is apt to will that which is beneficial to us, the simplest psychological phenomena will defy explanation.

Only when one admits that spirit and will are not something over and above the body but rather the complicated actions of material parts whose ability so to act becomes increasingly perfected by development, only when one admits that intuition, will and self-consciousness are merely the highest stages of development of those physico-chemical forces of matter by which primeval protoplasmatic bubbles were enabled to seek regions that were more and avoid those that were less favourable to them, only then does everything become clear in psychology.

Later in the same speech, Boltzmann gives an entertaining example of the role of evolution in human habits [1, 2]:

Consider another example that is quite simple and banal. Of our original ancestors countless numbers must have died of drinking bad water. Those who preferred the juice of fruit had an advantage. But unfermented fruit juice too could easily contain bacteria, so that those who preferred fermented juices had an advantage in the struggle for existence, and by hereditary development of this predilection it has become a habit which of course often overshoots the mark. I must confess that if I were antialcoholic I might not have come back from America, so severe was the dysentery that I caught there as a result of bad water; even bottles carrying labels of mineral water probably contain mostly river water, and it was only through alcoholic beverages that I was saved.

"What then will be the position of the so-called laws of thought in logic?" Boltzmann asks himself. His answer is immediate [1, 2]:

Well, in the light of Darwin's theory, they will be nothing else but inherited habits of thought. Men have gradually become accustomed to fix and combine the words through which they communicate and which they rehearse in silence when they think, [...] in such a manner as to enable them always to intervene in the world of phenomena in the way intended, and inducing others to do likewise, that is to communicate with them. These interventions are greatly promoted by storing and suitable ordering of memory pictures and by learning and practising speech, and this promotion is the criterion of truth.

This method for putting together and silently rehearsing mental images as well as spoken words became increasingly perfect and has passed into heredity in such a way that fixed laws of thought have developed. It is quite correct that we should not bring these laws with us if all cognition were to cease and perception were totally disconnected.

Endeavour towards acting in an advantageous way has perfected these ideas to produce a world of will and representation. "Even Schopenhauer could not wish better", adds Boltzmann, ironically.

In what sense can we can call these laws of thought a priori? Here is Boltzmann's answer [1, 2]:

[...] because through many thousands of years of our species' experience they have become innate to the individual, but it seems to be no more than a logical howler of Kant's to infer their infallibility in all cases.

According to Darwin's theory this howler is perfectly explicable. Only what is certain has become hereditary; what was incorrect has been dropped. In this way these laws of thought acquired such a semblance of infallibility that even experience was believed to be answerable to their judgement. [...] Just so it was at one time assumed that the ear and the eye were absolutely perfect because they have indeed developed an amazing degree of perfection. Today we know that this is an error and that they are not perfect.

Similarly I would deny that our laws of thought are absolutely perfect. On the contrary, they have become such firmly established habits that they overshoot the mark and will not let go even when they are out of place. In this they behave no differently from all inherited habits.

We have other wrong convictions, Boltzmann remarks in the final part of his address, referring to mathematical advances, not so widely known in his days [1, 2]:

The prejudice against non-Euclidean geometry and four-dimensional space is likewise in the course of disappearing. Most people still believe that Euclid's geometry alone is possible, that the sum of the angles of a triangle must be 180°, but some have already come to admit that these are mental ideas become habitual, from which we can and must free ourselves.

10.2 Laws of thought and scientific concepts

At this point one can understand the criticism that Boltzmann addresses to Hertz, who in his famous book on the principles of mechanics [6] had proposed that our mental pictures must conform to the laws of thought. Boltzmann had already written in his lectures at Clark University in 1899 [1, 2]:

Against this I should like to urge certain reservations or at least to explain the demand a little further. Certainly we must contribute an ample store of laws of thought; without them experience would be quite useless, since we could not fix it by means of internal pictures. These laws of thought are almost without exception innate, but nevertheless they suffer modifications through upbringing, education and our own experience. They are not quite the same in a child, a simple uneducated person, or a scholar.

After indicating a similar evolution in philosophy, Boltzmann recognizes that there are certain common assumptions which we regard as basic, not only in everyday life but in science as well [1, 2]:

Certainly there are laws of thought that have proved so sound that we place unconditional confidence in them, regarding them as unalterable a priori principles of thought. However, I think nevertheless that they developed gradually. Their first source was the primitive experience of mankind in its primeval state, and gradually they grew stronger and clearer through complicated experience until finally they assumed their precise present formulation; but I do not wish to recognize the laws of thought as supreme arbiters. We cannot know whether they might not suffer this or that modification in future.

Then Boltzmann recalls how certain children or uneducated people have the conviction that our feeling alone must be able to decide what is up and down at all points of space, from which they imagine they can even deduce that antipodes are impossible. This, as we have seen in Chapter 3, was also the view of Epicurus, illustrated in detail by Lucretius. What is the main danger? That if such people wrote on logic, they would very likely regard the existence of an absolute vertical as an a priori evident law of thought. Boltzmann continues by saying [1, 2]:

Just so people raised many a priori objections to the Copernican theory and the history of science contains many cases where propositions were either founded or refuted by arguments that at the time were regarded as self-evident laws of thought, whereas we are now convinced that they are futile. I therefore wish to modify Hertz's demand and say that insofar as we possess laws of thought that we have recognized as indubitably correct through constant confirmation by experience, we can start by testing the correctness of our pictures against these laws; but the sole and final decision as to whether the pictures are appropriate lies in the circumstance that they represent experience simply and appropriately throughout so that this in turn provides precisely the test of the correctness of those laws.

Another occasion when Boltzmann had a chance of talking about this subject was the meeting in St Louis, where he presented an updated survey of statistical mechanics [1, 2], preceded by a discussion on why physicists are interested in questions that were once left to philosophers. Here is the part of his address we are alluding to, where he deals with the usual way in which philosophers proceed [1, 2]:

To call this logic seems to me as if somebody for the purpose of a mountain hike were to put on a garment with so many long folds that his feet become constantly entangled in them and he would fall as soon as he took his first steps in the plains. The source of this kind of logic lies in excessive confidence in the so-called laws of thought. It is indeed true that we could have no experience if certain forms of linking perceptions, that is forms of thought, were not innate to us. If we wish to call these forms laws of thought, they are of course a priori in the sense that they are in our minds, or if we prefer in our heads, prior to any experience. However nothing seems less founded than an inference from the a priori in this sense to absolute certainty and infallibility. These laws of thought have evolved according to the same laws of evolution as the optical apparatus of the eye, the acoustic machinery of the ear and the pumping device of the heart. In the course of mankind's development everything inappropriate was shed and thus arose the unity and perfection that can give the illusion of infallibility.

After remarking, in a way analogous to that already quoted from his lecture on Schopenhauer, that the perfection of the eye, the ear and the arrangement of the heart are such that we cannot claim that these organs are absolutely perfect, Boltzmann draws the conclusion that "just as little must the laws of thought be taken as absolutely infallible". They have just evolved to the point of grasping what is necessary for life and practically useful. The results of experimental research are much closer to life than our thinking equipment. It is thus not surprising if the forms of thought that have become habitual are not "quite adapted to the abstract problems of philosophy which are so far removed from what is applicable in practice, nor have yet become so from Thales till now" [1, 2].

This explains, according to Boltzmann, why the simplest things of life and science are to the philosopher the most puzzling, and why he finds contradictions everywhere. But what is given cannot contain contradictions; these are found in the "inappropriate and mistaken mental re-shapings" of philosophers. What should we do when contradictions show up and seem unavoidable? Here is the answer [1, 2]:

[...] we must immediately seek to test, extend and alter what we call our laws of thought which actually are nothing but inherited and acquired ideas, confirmed throughout the ages, for denoting practical requirements. Just as the inherited inventions such as the roller, cart, plough, have long since been joined by countless artificial ones that have been created with full consciousness, so here too we must rearrange ideas artificially and consciously into a better order. Our task cannot be to summon data to the judgement-throne of our laws of thought but rather to adapt our thoughts, ideas and concepts to what is given.

How can we proceed, given the fact that we cannot clearly express such complicated conditions except by words (even if written or thought in silence)? We must collocate these words in such a way as to apply the most fitting expressions to the given; then the connections we create between words will be everywhere as adequate as possible to the connection in reality. This of course sounds too easy; the most appropriate solution to a problem can still present the greatest difficulties, but, says Boltzmann, "at least one knows the goal being aimed at and will not stumble over obstacles of one's own making" [1, 2].

Boltzmann also remarks that many inappropriate features in the behaviour of living beings are provoked by the fact that a mode of action that is appropriate in most cases becomes a habit and, so to speak, second nature. Thus "adaptation overshoots the mark". And this happens especially often with mental habits and becomes a source of those apparent contradictions previously mentioned. The law-like character of the processes of nature is a basic prerequisite for all cognition and has the consequence that we always ask the cause. This has become an irresistible urge and we even ask the cause for everything's having a cause.

Some philosophers, says Boltzmann, "racked their brains over the question whether cause and effect represent a necessary link or merely an adventitious sequence" [1, 2]. One can sensibly ask only whether a specific phenomenon is always linked with a definite group of others, thus appearing their necessary consequence, or whether this grouping may at times be absent. Another example of a useless question, according to Boltzmann, is to ask what is the value of life, since something is called useful or valuable if it furthers the living conditions of the individual or of mankind, but we overshoot the mark if we ask the value of life itself.

A similar pointless attempt occurs when we take the simplest concepts out of which everything is built, and try in vain to build these in turn out of simpler ones or try to explain the simplest fundamental laws. Boltzmann addresses this topic by saying [1, 2]:

> We must not aspire to derive nature from our concepts, but must adapt the latter to the former. We must not think that everything can be arranged according to our categories or that there is such a thing as a most perfect arrangement: it will only ever be a variable one, merely adapted to current needs. Even the splitting of physics into theoretical and experimental is only a consequence of the two-fold division of methods currently being used, and it will not remain so forever.

The view that certain questions fall outside the scope of human cognition immediately leads to the idea that this is a defect in man's cognitive capacity. But, says Boltzmann, the existence of these questions and problems themselves is an illusion. Not without irony, Boltzmann notes that it may be surprising that after the illusion has been recognized, the drive to answer these questions does not cease. Why is this? Again, because the mental habit is much too powerful to loosen its hold on us.

There is an analogy with ordinary sense illusions, which continue to exist even after their cause has been recognized. Thus a man of science who tries to philosophize experiences a feeling of insecurity, a lack of satisfaction. Then what is the purpose of philosophy, when all these illusions will slowly and gradually recede? Boltzmann [1, 2] regards it as a central task of philosophy

> to give a clear account of the inappropriateness of this overshooting the mark on the part of our thinking habits; and further, in choosing and linking concepts and words, to aim only at the most appropriate expression of the given, irrespective of our inherited habits. Then, gradually, these tangles and contradictions must disappear. What is brick and what is mortar in the intellectual edifice must be made to stand out clearly and we should soon be freed by the oppressive feeling that the simplest feeling is the most inexplicable and the most trivial the most puzzling. Unjustified modes of thought can in time recede, witness the fact that any educated persons now understand the theory of the antipodes and many that of non-Euclidean geometry.

This is the price that philosophy, according to Boltzmann, must pay to become the queen of sciences: to create a system, in which certain questions are unjustified and therefore not asked.

Boltzmann recognizes that our innate laws of thought are indeed the prerequisite for complex experience, but they were not so for the simplest living beings. Evolution started there, however; the laws of thought developed slowly, but simple experiences were enough to generate them. They were then bequeathed to more highly organized beings. Boltzmann concludes this argument by saying: "This explains why they contain synthetic judgements that were acquired by our ancestors but are for us innate and therefore a priori, from which it follows that these laws are powerfully compelling, but not that they are infallible." [1, 2]

10.3 Ethics, aesthetics, religion

What is the place occupied, in the philosophy proposed by Boltzmann, by one of the practical problems that in the eyes of many people motivate metaphysics, or at least a philosophy founded upon spirit rather than matter: the problem of ethics? Obviously the foundation is again evolutionism. In the previously quoted address on Schopenhauer, Boltzmann in fact states [1, 2]:

Ethics must therefore ask when may the individual insist on his will and when must he subordinate it to that of the others, in order that the existence of family, tribe, or humanity as a whole and thereby of each individual is best promoted. This innate love of asking, however, overshoots the mark if we ask whether life as such should be promoted or inhibited. If some ethics were to cause the decline of a tribe adhering to it, that would refuse it. In the last instance it is not logic, philosophy nor metaphysics that decide whether something is true or false, but deeds. 'In the beginning was the deed', as Faust says. What leads us to correct deeds, is true.

In 1904, Boltzmann gave a speech criticizing a lecture, in which Ostwald had given a mathematical formula for happiness (according to this formula, happiness would be given by $E^2 - W^2$, where E indicates energy spent intentionally and successfully, W that spent with dislike). Leaving aside the discussion devoted to the formula itself (to refute it), it is of interest to remark that Boltzmann did not let slip the opportunity to speak of the role of Darwin's evolution in ethics [1, 2]:

As regards the concept of happiness, I derive it from Darwin's theory. Whether in the course of aeons the first protoplasm developed "by chance" in the damp mud of the vast waters on the Earth, whether egg cells, spores or some other germs in the form of dust or embedded in meteorites once reached Earth from the outer space, is here matter of indifference. More highly developed organisms will hardly have fallen from the sky. To begin with there were thus only very simple individuals, simple cells or particles of protoplasm. Constant motion, the so-called Brownian molecular motion, happens with all small particles as is well known; growth by absorption of similar constituents and subsequent multiplication by division is likewise explicable by purely mechanical means. It is equally understandable that these rapid motions were influenced and modified by the surroundings. Particles in which the change occurred in such a way that on average (by preference) they moved to regions where there were better materials to absorb (food), were better able to grow and propagate so as soon to overrun the others.

In this simple process that is readily understood mechanically, we have heredity, natural selection, sense perception, reason, will, pleasure and pain all together in a nutshell. It merely requires a change of quantitative degree with constant application of the same principle, to proceed via the whole world of plants and animals to mankind with all his thinking, feeling, willing and acting, its pleasure and pain, its artistic creation and scientific research, its nobility and vices.

Cells that had come together into rather large collections in which there occurred division of labour and hived off, by division, cells with similar tendencies, had greater opportunities in the struggle for existence, especially if certain cells when subjected to harmful influences did not rest until the working cell had removed such inroads as much as possible (pain). The action of these cells was particularly effective, if, so long as the harmful influences were not completely eliminated, it continued and left behind a tension that diminished only very slowly, thus stressing the memory cells and inciting the motor cells to even more vigorous and circumspect collaboration when similar harmful circumstances recurred. This state is called lasting displeasure, a feeling of unhappiness. The opposite, complete freedom from such vexatious after-effects and a warning to the memory cells that the motor cells are to act in the same way again if similar circumstances supervene, is called lasting pleasure, a feeling of happiness.

This does not of course exhaust all the gradation of these feelings in highly organized beings. Not even a beginning has been made for a physiology of happiness; but the point of view has been fixed from which relevant phenomena have to be regarded if one is to explain them scientifically, rather than merely turn beautiful, uplifting, poetic and inspiring phrases about them.

The same considerations hold for aesthetics, as discussed by Boltzmann in his inaugural lecture in Leipzig (1900) [1, 2]:

We must mention also that most splendid mechanical theory in the field of biology, namely the doctrine of Darwin. This undertakes to explain the whole multiplicity of plants and the animal kingdom from the purely mechanical principle of heredity, which like all mechanical principles of genesis remains of course obscure.

The explanation of the exquisite beauty of flowers, the great wealth of forms in the world of insects, the appropriate construction of organs in human and animal bodies, all this thereby becomes a domain of mechanics. We understand why it was useful and important for our species that certain sense impressions were flattering and therefore sought after, while others were repellent; we see how advantageous it was to construct the most accurate pictures possible of our surroundings and strictly to keep apart that which coincided with experience as true from what did not as false. Thus we can explain the genesis of the concept of beauty, just as that of the concept of truth, in terms of mechanics.

Moreover, we understand why only those individuals could continue to exist which abhorred certain very noxious influences with all the nervous energy at their command and sought to keep them in the background, while with equal vigour aiming at other influences that were important for the preservation of themselves or their kind. In this way we grasp how intensity and power of our whole affective life developed, pleasure and pain, hate and love, happiness and dispair. Just as we cannot rid ourselves of bodily disease, so likewise with the whole gamut of passions, but again we learn how to understand and bear them.

Boltzmann was conscious of the difficulties that these ideas encountered, as one can surmise from the following passage in his commemoration of Kirchhof in Graz (1887) [1]:

Nay, the most distant posterity will not spare admiration for the great men whom our century gave birth to. If something could equal this admiration, it would be the fact that the same century did not succeed in getting rid of so much ridiculous pedantry, of the heritage of so much nonsense and silly superstition [...] Does there not sound louder than ever the whimper of all obscurantists, the enemies of freethinkers and of enquiry, against the new Pythagoras theorem, Darwin's teachings? [...] But, lucky us! It is the thunderstorm that forecasts the arrival of spring. Until then, however, light-hearted pleasantries are premature; arm yourselves for a bitter, bloody struggle.

Nor did Boltzmann neglect the spontaneous repulsion that one might feel when seeing the sensations of beautiful and good reduced to a mere mechanism. A poet might lament the incrustation built up by civilization on the simplest feelings about nature and exclaim with Nobel laureate Odhysseas Elytis: "...όταν ανεβαίνει ο έλιος τα πυροβόλα όλων των μεγάλων κοσμοθεωριών παθαίνουν αφλογιστία ..." (...as the sun rises, the guns of all the great world theories are silenced...). But a scientist can react by saying that he is no less moved by a blue sky or a red sunset because he knows the laws of light scattering that are responsible for those beautiful colours. In the same way, a physician does not appreciate good food less because he knows the mechanisms of digestion and the origin of body pleasures. We must distinguish between things and our ideas about them ("The concept of dog does not bark", as Spinoza put it). But let us consult Boltzmann again on this topic. In the same inaugural lecture at Leipzig [1, 2] one can in fact read:

I foresee how the enthusiast will be horrified by these last remarks of mine, how he fears that everything great and noble is degraded to a dead and unfeeling mechanism and all poetry falls away. However, it seems to me that all this apprehension is based on a complete misunderstanding of what I said.

Indeed our ideas of things are never identical with the nature of things. Ideas are only pictures or rather signs that necessarily represent one-sidedly and indeed can do no more than imitate certain modes of connection of the things signified without in the least touching their essential character.

We need therefore take back none of the sharpness and definiteness of our previous expressions. We have in any case used them only for asserting a certain analogy between mental phenomena and the simple mechanisms in nature. We have merely constructed a one-sided picture in order to illustrate certain connections between phenomena and to predict new ones unknown to us. Alongside this one picture there can and because of its one-sidedness must be others that represent the inward and ethical side of the matter; these latter will no longer hinder the exaltation of the soul, as soon as we take the right view of the mechanical picture which should be applied only where it belongs. Yet we shall not deny its usefulness but reflect that even the noblest ideas and conceptions are again only pictures or external signs for the ways in which phenomena are linked.

This does away with the objection that might perhaps have been raised against my remarks, that they ran counter to religion. Nothing is more perverse than linking religious concepts which rest on a quite different and immeasurably firmer basis with the vacillating subjective picture that we form of external things. I should be the last to put forward the views here mentioned, if they harboured any danger for religion. Yet I know that the time will come when all will own that they are as irrelevant to religion as the question whether the earth is at rest or moving round the Sun.

It is not inappropriate to recall here another passage in which Boltzmann hints at religion. This passage can be interpreted as an explicit declaration of pantheism. It can be found in the previously quoted lecture "On the question of the objective existence of processes in inanimate nature" (1897) [1, 2]:

It is certainly true that only a madman will deny God's existence, but is equally the case that all our ideas of God are mere inadequate anthropomorphisms, so that what we thus imagine as God does not exist in the way we imagine it. If therefore one person says that he is convinced that God exists and another that he does not believe in God, in so saying both may well think the same thoughts without even suspecting it. We must not ask whether God exists unless we can imagine something definite in saying so; rather we must ask by what ideas we can come closer to the highest concept which encompasses everything.

10.4 Philosophy of science

Boltzmann of course did not restrict his philosophical interests to knowledge in general, with *excurses* in ethics, aesthetics, and theology. He was deeply interested in the philosophy of science itself, as should be clear indirectly from the passages already quoted. His contributions to this branch of knowledge certainly show the influence of Maxwell and of the change of attitude towards science in general, and the foundations of mechanics in particular, which took place in Germany in the second half of the nineteenth century. But Boltzmann's viewpoint appears to be clear and pragmatic, in the sense that he accepts the new and the criticism of the old, in so far as they serve the purpose of improving our understanding of the foundations of mechanics; certain passages are surprising for their freshness and modernity. The problem that preoccupies him most is that of the atomic hypothesis.

Here it seems better not to follow a chronological order. Rather, we shall start by perusing the first pages of the inaugural lecture of the course of philosophy of science (1903) [1, 2], which also give an idea of Boltzmann's style:

I have hitherto written only one single dissertation of philosophic content and what moved me to do so was pure chance. On one occasion in the assembly hall of the academy, I was involved in the liveliest debate with a group of academics, Professor Mach amongst them, on the newly revived controversy about the value of atomistic theories.

I here mention in passing that in the assignment that begins with today's lecture I am in a certain sense Mach's successor and I should really have started my talk by paying homage to him. However, I think that to express his special praise would amount to carrying owls to Athens, as far as you are concerned, and indeed not only you but any Austrian and even any educated person throughout the world.

Then Boltzmann underlines the fact that Mach himself has ingeniously discussed the fact that no theory is absolutely true, and equally, hardly any absolutely false either. Each of them must be gradually perfected, as organisms must according to Darwin's theory. By being strongly attacked, a theory can gradually shed inappropriate elements while the appropriate residue remains. Boltzmann therefore states that the best way to honour Mach is in this sense to contribute as much as possible to the further development of his ideas. He then goes back to the debate with Mach [1, 2]:

In that group of academics during the debate on atomism Mach suddenly said laconically: 'I do not believe that atoms exist'. This utterance ran in my mind.

It was clear to me that we unite groups of perceptions into ideas of objects, for example of a table, a dog, a man and so on. Moreover we have memory pictures of these groups of ideas. When we form new groups of ideas which are quite similar to these memory pictures, there is sense in the question whether the corresponding objects have existence or not. We here have as it were an accurate measure for the concept of existence. We know exactly what is meant by the question whether the griffin, the unicorn or a brother of mine exists. However, when we form quite novel ideas, such as those of space, time, atoms, the soul, or even God, does one know, so I asked myself, what is meant by asking whether these things exist? Is not the only correct thing to do here to try to clarify what concepts one is linking with the question as to the existence of these things?

Discussions of this kind formed the topic of my one and only dissertation in the field of philosophy. As you see, it was genuinely philosophical; abstruse enough, at least, to deserve the name. Apart from this I have published nothing in this field. This much might of course pass; if one wanted to be malicious one might say that here and there people have taught at universities who had written even one publishable piece less about their fields.

In any case, however, it must fill me with utter modesty. It is said that if God gives one a task, he will give one the wit to do it. Not so the ministry; it can of course make appointments and fix salaries, but it can never furnish the wit; for that I alone must bear the responsibility.

Let us go back now to the lecture on the Second Law of Thermodynamics (or, as Boltzmann says, of the mechanical theory of heat) (1886) [1, 2]. In it we read:

Perhaps the atomic hypothesis will one day be displaced by some other but it is unlikely.

This is not the place for naming all the reasons that might be advanced for this. There will be no need to recall the ingenious inferences of Thomson who used the most varied methods to work out with quite satisfactory agreement how many of these individual things make up a cubic millimetre of water. I need not mention that, besides many facts of chemistry, it was by means of the atomistic hypothesis that science succeeded in calculating in advance the temperature dependence of the frictional constant for gases, the absolute and relative constants of diffusion and thermal conduction, which can surely be put alongside Leverrier's calculation establishing the existence of the planet Neptune or Hamilton's prediction of conical refraction.

Another paper which plays an essential role in grasping how Boltzmann estimated the successes and the difficulties of the atomic hypothesis is entitled "On the indispensability of atomism in natural science" (1897) [1, 2]. There we can read the following lines:

While phenomenology requires separate and mutually rather unconnected pictures for the mechanical motion of centres of gravity and rigid bodies, for elasticity, hydrodynamics and so on, present day atomism is a perfectly apt picture of all mechanical phenomena, and given the closed nature of this domain we can hardly expect it to throw up further phenomena that would fail to fit into that framework. Indeed, the picture includes thermal phenomena: that this is not so readily proved is due merely to the difficulty of computing molecular motions. At all events all essential facts are found in the features of our picture. Further, it proved itself extremely useful for representing crystallographic facts, the constancy of proportions of mass in chemical compounds, chemical isomerisms, the relations between the rotation of the plane of polarisation and chemical constitution, and so on.

For the rest, atomism remains capable of being developed much further. One may conceive of atoms as more complicated individuals endowed with arbitrary properties, as for example the vector atoms...

And, in a long footnote (note 4) which precedes the last passage:

Thus the view of atoms as material points and of forces as functions of their distance is no doubt provisional but must at present be retained failing a better one.

Of course, elementary reflection and experience alike will tell us that it would be hopelessly difficult to hit at once upon appropriate world pictures merely by aimless guesswork; on the contrary, they always emerge only slowly from adaptation of a few lucky ideas. Rightly, therefore, epistemology is against the doings of those framers of hypotheses who hope to find without effort a hypothesis that would explain the whole of nature, as well as against metaphysical and dogmatic foundation of atomism.

Another interesting remark against dogmatism in the matter of scientific models can be found in another footnote in the same paper:

Such a feature arbitrarily ascribed to the picture of atoms is their invariability. The objection, that this was an unjustified generalisation of invariability of solid bodies observed for only a finite time span, would certainly be justified as soon as one tried to prove atomic invariability on a priori grounds as used to be the custom. However, we merely take this feature into the picture in order that the latter should be able to represent the essential concept of the greatest number of individual phenomena, just as one takes the first time derivative and the second space derivatives into the equation of thermal conduction in order that it should fit the facts. We are prepared to drop invariability in those cases where some other assumption would represent the facts better. For example, the vector atoms of the aether, mentioned in Note (4) above, would not be invariable with time.

Thus, atomic invariability belongs to those notions that show themselves very serviceable although the metaphysical considerations that led to it will not stand up to unprejudiced criticism. However, just because of this many-sided usefulness one must allow a certain likelihood that so-called radiant energy may be represented by pictures similar to those for matter (that is, luminous aether is a substance).

In the address at St Louis (1904) "On statistical mechanics" [1, 2], one can read similar considerations: "The word atom must not mislead us here, it has been taken over from ancient times, no physicist ascribes indivisibility to atoms today."

If, as we know, atomic theory was not the only interest of Boltzmann as a scientist, it was even less so when he dealt with the philosophy of science. His essential point was the usefulness of the models in producing new ideas and new conceptions. His viewpoint is clearly expressed in the paper that opens the *Populäre Schriften*, entitled "On the methods of theoretical physics" (1892) [1, 2]. There one can read among other remarks: "Faraday's ideas were much less clear than the earlier hypotheses that had mathematical precision, and many a mathematician of the old school placed little value on Faraday's theories, without however reaching equally great discoveries by means of his own clearer notions."

In another lecture, "On the development of the methods of theoretical physics in recent times" (1899) [1, 2], one can read an attack to positivism:

Phenomenology believed that it could represent nature without in any way going beyond experience, but I think this is an illusion. No equation represents any process with absolute accuracy, but always idealizes them, emphasizing common features and neglecting what is different and thus going beyond experience. That this is necessary if we are to have any ideas at all that allow us to predict something in the future, follows from the nature of the intellectual process itself, consisting as it does in adding something to experience and creating a mental picture that is not experience and therefore can represent many experiences.

Only half of our experience is ever experience, as Goethe says. The more boldly one goes beyond experience, the more general the overview one can win, the more surprising facts one can discover but the more easily too one can fall into error. Phenomenology therefore ought not to boast that it does not go beyond experience, but merely warn against doing this in excess.

From these words it should be clear what is a physical theory in Boltzmann's sense, but we need not make any effort at interpretation: in his address "On the significance of theories" (1890) [1, 2] he gives us his viewpoint in very clear terms; after the introductory sentences which have been quoted in Chapter 1, he says:

I should not be a genuine theoretician if I were not first to ask: what is a theory? The layman observes in the first place that theory is difficult to understand and surrounded by a tangle of formulae that to the uninitiated speak no language at all. However they are not its essence: the true theoretician uses them as sparingly as he can; what can be said in words he expresses in words, while it is precisely in books by practical men that formulae figure too often as mere ornament.

A friend of mine has defined the practical man as one who understands nothing of theory and the theoretician as an enthusiast who understands nothing at all. The rather pointed view contained in this we will likewise oppose.

Boltzmann is of the opinion that the task of theory consists in constructing a picture of the external world that exists purely internally and must be our "guiding star": in all thought and experiment; that is, in completing, as it were, the thinking process and carrying out globally what on a small scale occurs whenever we form an idea.

He remarks that it is a peculiar drive of the human spirit to construct such a picture and increasingly to adapt it to the external world. Then he says [1, 2]:

If therefore we may often have to use intricate formulae to represent a part of the picture that has become complicated, they nevertheless remain inessential if most serviceable forms of expression, and in our sense Columbus, Robert Mayer and Faraday are genuine theoreticians. For their guiding star was not practical gain but the picture of nature within their intellect.

The immediate elaboration and constant perfection of this picture is then the chief task of theory. Imagination is always its cradle, and observant understanding its tutor.

The first theories of the Universe from Pythagoras and Plato until Hegel and Schelling were, according to Boltzmann, childlike, because the imagination was over-productive, test by experiment was lacking. His comment however is favourable [1, 2]:

No wonder that these theories became the laughing stock of empiricists and practical men, and yet they already contained the seeds of all the great theories of later times: those of Copernicus, atomism, the mechanical theory of weightless media, Darwinism and so on.

In spite of all mockery the drive to form a theoretical view of external things remained unconquerable in the human breast and it constantly gave rise to new flowers. As Columbus set course always to the west, so this drive always unswervingly directed us towards this great goal.

What was lacking was sober experimental understanding and the dexterity needed for handling the many devices and machines invented; when these came, the old and variegated imaginative structures were sifted and refined and, with amazing speed, became good and important models of nature. After this remark, Boltzmann says:

Today one can aver that theory has conquered the world.

Who can see without admiration how the eternal stars slavishly obey all laws that the human spirit has not indeed given to but learnt from them. And the more abstract the theoretical investigation, the more powerful it becomes. If, still somewhat mistrusting the path on which, being led by the formulae rather than leading them, we have reached a theorem of arithmetic, we test it on numerical examples, we are even more strongly haunted by the feeling that numbers without exception must inevitably bow to our formulae.

But even those who value theory only as a milk cow, can no longer doubt of its power. Are practical disciplines all of them by now not penetrated by theory and do they not follow their reliable guiding star? The forms of Kepler and of Laplace not only show the stars their celestial courses, but along with Gauss's and Thomson's calculations on the earth's magnetism they show ships their way on the high seas. The gigantic structure of the Brooklyn bridge that stretches beyond sight and the Eiffel tower that soars without end rest not only on the solid framework of wrought iron, but on the solider one of elasticity theory. Theoretical chemists have become rich through the practical application of their syntheses, not to mention the electrical engineer! Does he not pay constant homage to theory by the fact that next to pound and penny the names that are most familiar to him are Ohm, Ampère and so on, all of them great theoreticians, none of whom, alas, were blessed by the lucky fate of the chemists just mentioned; for their formulae did not become fruitful in practice until after they had died. Indeed, it may well not be long before these great electric theorists will be glorified in every domestic bill and in the next century every cook may know how many "Volt-Ampères" one fries meat and how many "Ohms" her lamp has got.

Boltzmann also has something to say about the dexterity of technicians working with recently invented equipment [1, 2]:

It is precisely the practical technician who as a rule treats the complicated formulae of electricity theory with a surer hand than many a tiro scientist, because he has to pay for his errors not only by way of reproof from his teacher, but in hard cash. Indeed almost any carpenter or metal worker knows today how much a grasp of descriptive geometry, the theory of machines and so on make him more competitive. I must mention also the splendid field of medical sciences, where theory gradually seems to gain a footing too.

This passage serves as an introduction to a favourite theme of his, theory as a practical tool:

One is almost tempted to assert that quite apart from its intellectual mission, theory is the most practical thing conceivable, the quintessence of practice as it were, since the precision of its conclusions cannot be reached by any routine of estimating or trial and error; although given the hidden ways of theory, this will hold only for those who walk them with complete confidence. A single mistake in a drawing can multiply a result thousandfold, whilst an empirical worker never errs so far; for that reason there will no doubt always remain some cases where the thinker who is immersed in his ideas and always bent on what is general will be outdone by the clever and self-interested practical man; witness Archimedes who fell victim of the attacking Roman, or another Greek philosopher who, while looking at the stars, stumbled upon a stone. Let silence overtake the question "what is the use of it?" which is customarily thrown at any more abstract endeavours. One would like to ask the counter-question: "what is the use of furthering by gaining mere practical advantages at the expense of that which alone makes life worth living, namely the tendance of the ideal?"

However, theory keeps all away from overrating itself; its very defects are grounded in its own nature and is theory itself that uncovers its own errors; indeed, already Socrates placed the main emphasis on the recognition of the gaps in his own knowledge. All our ideas are purely subjective. That this is so even as regards our views on being or not-being is shown by Buddhism which reveres nothingness as the really existing. I called theory a purely mental inner picture, and we saw to what a high degree of perfection this may be brought. How then, as we become more and more immersed in theory, could we fail to take the picture for what really exists? It is in this sense that Hegel is said to have regretted that nature was unable to realize his philosophic system in its full perfection.

Thus it may happen to the mathematician, who is constantly occupied with his formulae and blinded by their inner perfection, that he takes the mutual relations as the really existing and turns away from the real world. What the poet laments then holds for the mathematician, that his works are written with the blood of his heart and highest wisdom borders on supreme folly.

10.5 Boltzmann's views on scientific revolutions

An important aspect of Boltzmann's thought, which shows that Boltzmann anticipated the views of Thomas Kuhn (1922–96) on scientific revolutions, can be grasped from the following passage in an obituary for Joseph Stefan (1895) [1]:

The layman may imagine that new notions and causes of phenomena are gradually added to the existing basic ones and that in this way our knowledge of nature undergoes a continuous development. This view, however, is erroneous, and the development of theoretical physics has always been by leaps. In many cases it took decades or even more than a century to fully articulate a theory such that a clear picture of a certain class of phenomena was accomplished. But eventually new phenomena were discovered which were incompatible with the theory; in vain was the attempt to assimilate the former to the latter. A struggle developed between the followers of the theory and the advocates of an entirely new conception until, eventually, the latter was generally accepted. Formerly one used to say that the old view had been recognized as false. This sounds as if the new ideas were absolutely true and, on the other hand, the old (being false) had been entirely useless. Nowadays, to avoid confusion in this respect, one just says: the new way of ideas is a better, a more complete and adequate description of the facts. Thus the following are clearly expressed: (1) the earlier theory, too, had been useful because it gave a true, though partial, picture of the facts; (2) the possibility is not excluded that the new theory in turn will be superseded by a more fitting one.

The same theme is also discussed in the previously quoted lecture "On the development of the methods of theoretical physics in recent times" (1899) [1, 2]:

A closer look at the course followed by developing theory reveals for a start that is by no means as continuous as one might expect, but full of breaks and at least apparently not along the shortest logical path. Certain methods often afforded the most handsome results only the other day, and many might well have thought that the development of science to infinity would consist in no more than their constant application. Instead, on the contrary, they suddenly reveal themselves as exhausted and the attempt is made to find other quite disparate methods. In that event there may develop a struggle between the followers of the old methods and those of the newer ones. The former's point of view will be termed by their opponents as out-dated and outworn, while its holders in turn belittle the innovators as corruptors of true classical science.

This process incidentally is by no means confined to theoretical physics but seems to recur in the developmental history of all branches of man's intellectual activity. Thus many may have thought at the time of Lessing, Schiller and Goethe, that by constant further development of the ideal modes of poetry practised by these masters dramatic literature would be provided for in perpetuity, whereas today one seeks quite different methods of dramatic poetry and the proper one may well not have been found yet.

Just so, the old school of painting is confronted with impressionism, secessionism, plein-airism, and classical music with music of the future. Is not this last already out-of-date in turn? We therefore cease to be amazed that theoretical physics is no exception to this general law of development.

And further on, in the same lecture:

From this it follows that it cannot be our task to find an absolutely correct theory but rather a picture that is as simple as possible and that represents phenomena as accurately as possible. One might even conceive of two quite different theories both equally simple and equally congruent with phenomena, which therefore in spite of their difference are equally correct. The assertion that a given theory is the only correct one can only express our subjective conviction that there could not be another equally simple and equally fitting image.

Thus Boltzmann appears to have been the first to suggest that a scientific discipline develops in two alternating phases, one marked by a fairly continuous development (*normal science* in Kuhn's terminology), the other related to the discovery of new phenomena incompatible with the accepted theory, recalcitrant to any attempt made by scientists to assimilate them (the *crisis* in Kuhn's terminology). And when Boltzmann talks of the "struggle developed between the followers of the theory and the advocates of an entirely new conception until, eventually, the latter was generally accepted", is he not talking about what Kuhn calls a *scientific revolution*?

In spite of the vital interest in these questions characterizing the philosophy of science in the last three decades, this circumstance is not reported either in Kuhn's work [7, 8] or in I.B. Cohen's fairly comprehensive treatment [9]. The first to remark this curious circumstance appears to have been E. Scheibe [10], who also discusses the development of Boltzmann's conception in the writings of many physicists (notably W. Nernst, see Chapter 11).

As a matter of fact, a surprising but undeniable aspect of the contributions of Boltzmann to philosophy lies in the fact that they seem to have remained unknown to most philosophers of the twentieth century. On this point, we can quote an illuminating passage of Karl Popper (1902–95), himself an Austrian by birth [11]:

Boltzmann is little known as a philosopher; until quite recently I too knew next to nothing about his philosophy, and I still know much less about it than I should. Yet with what I know I agree; more closely perhaps than with any other philosophy. Thus I greatly prefer Boltzmann to Mach—not only as a physicist and a philosopher but also, I admit, as a person. [...] To this day Boltzmann's realism and objectivism have been vindicated neither by him nor by history. (The worse for history.)

On the other hand, Popper criticizes Boltzmann's physics, going back to the so-called "Zermelo recurrence paradox", which we discussed in Chapter 5, and in particular to the answer by Boltzmann to Zermelo. Let us recall that, according a computation by Boltzmann, even for just one cubic centimetre of gas, the time required to "go back in time" is unimaginably long, of the order of a number of years expressed by 10^{18} figures (as opposed to the 17 figures required to express the age of the Universe, according to modern estimates). Probably Popper perceived an example of a theory which can be conceptually but not practically falsified in Boltzmann's idea ("staggering in its boldness and beauty" [11]) (who will be here to falsify a theory after such a long time?). Rather than taking note of this fact, however, he accuses Boltzmann of idealism. Nevertheless we are facing not idealism but, as we have occasionally pointed out before, the fact that a macroscopic instrument, in particular an observer of the size of a man, cannot but observe, in a world made up of extremely tiny molecules, statistically averaged events.

10.6 Boltzmann's education in philosophy

It is clear from our quotations that Boltzmann was an undogmatic realist. It is sometimes claimed, perhaps following Popper, that scientific difficulties led Boltzmann to abandon realism. This is clearly due to a misunderstanding (only people expressing dogmatic views cannot be misunderstood).

Before we discuss this point further in the next section, we must ask ourselves the question: what was Boltzmann's education in philosophy?

A biographer who has assiduously studied Boltzmann's thought, E. Broda, says without hesitation [12]: "We have seen that in philosophy Boltzmann was essentially an autodidact. Probably he discussed his philosophical views with fellow physicists, but no important school of philosophy existed in Austria during most of his life." This opinion is vigorously opposed by A.D. Wilson in a well-documented paper [13], where one can read:

But no one, on the other hand, can reasonably accept Broda's account of Boltzmann's early lack of formal exposure to and interest in philosophy. Documents deposited at both the *Akademisches Gymnasium* in Linz and at the University of Vienna show that Broda is simply wrong. As I have already suggested, and as we shall see, Boltzmann was by no means a philosophical autodidact; for, he did in fact study philosophy systematically and formally during his student's years.

Wilson then proceeds to illustrate Boltzmann's curriculum at both the *Akademisches Gymnasium* in Linz (with detailed quotations from the textbook for the course in philosophy [14]) and the University of Vienna. As for the latter, if one reflects on the fact that the students in mathematics and physics at the University of Vienna had to enrol in the Philosophical Faculty, it is not surprising to discover that Boltzmann enrolled for ten courses of philosophy; as for his high school education, we can infer that Boltzmann was able to receive very good preparation, as good as a high school student was able to get in Europe until about thirty years ago in a secondary school specializing in the humanities. Indeed, his reactions to idealistic philosophers and, more generally, to the German philosophy of his days were not so different from those of a student of the kind just described, if endowed with a scientific attitude, as we can read in "An inaugural lecture of natural philosophy" (1903) [1, 2]:

Not only while writing my only dissertation but at other times too, I have often speculated about the enormous field of philosophy. It seems infinite and my powers slight. A whole life would be but little time to struggle through to some results in it; the tireless activity of a teacher from youth to old age would not suffice to transmit philosophy to the next generation, am I then to pursue it as a subsidiary occupation along with another subject that by itself requires all my powers?

Schiller has said that a man grows with his purposes. Dear old Schiller! I fear that man does *not* grow with his higher purposes.

When I had qualms about taking this heavy burden, I was told that another would do not better than I. How paltry this consolation seems at the moment when I must shoulder the load.

And yet, what bows me down now could surely raise me up again? If I, who have busied myself so little with philosophy, was found to be the most deserving person to lecture on it, is this not a twofold honour?

If it is desirable for a professor of medicine or engineering that he should continue in practice alongside his teaching, lest he become ossified, indeed if Moltke was made a member of the historical division of the Berlin Academy because he had not written but made history, perhaps I too was chosen not because I had written about logic, but because I belong to a science that offers the best opportunities for daily practice in strict logic.

If it was only with hesitation that Boltzmann followed the call to meddle in philosophy (though, as he notes, philosophers have the more often meddled in natural science) [1, 2],

They have now for many years invaded my preserves and I could not even understand what their views were and therefore wanted to improve my knowledge of the fundamental theories of all philosophy.

To go straight to the deepest depths, I went for Hegel; what unclear thoughtless flow of words I was to find there! My unlucky star led me from Hegel to Schopenhauer. In the preface to his first work I found the following passage which I will report verbatim here: "German philosophy stands laden with contempt, derided by other countries, banished from honest science like a"...I suppress the next bit since there are ladies present. "...The heads of the present generation of learned men are disorganized by Hegelian nonsense. Incapable of thought, uncouth and stupefied, they fall prey to shallow materialism that has crawled forth from the basilisk's egg." With that I was of course in agreement, except that I found that Schopenhauer, too, really deserved the blows of his own cudgel.

Indeed, we have already seen with what title Boltzmann wanted to cudgel Schopenhauer!
    In the same inaugural dissertation we read:

Metaphysics seems to cast an irresistible spell on the human mind and all the abortive attempts at lifting its veil have not impaired its power. The drive toward philosophizing seems ineradicably innate. Not only Robert Mayer, who was a philosopher through and through, but Maxwell, Helmholtz, Kirchhof, Ostwald and many others too made willing sacrifice to it and recognized its questions as the very highest, so that today metaphysics figures again as the queen of sciences.

We can perhaps agree, in a Solomonic way, with both Broda and Wilson. Boltzmann had an exceptional preparation in philosophy, compared with a physicist of today (especially if young and/or American), but certainly at a level that was not impossible to find in the *Mitteleuropa* of his days, and as such he is to be considered an autodidact. We recall, by way of analogy, that Faraday was essentially an autodidact in physics!

10.7 Did Boltzmann abandon realism?

Before concluding this chapter, it seems appropriate to deal here with the question whether the opposition to kinetic theory towards the end of the nineteenth century was primarily due to scientific difficulties encountered by the theory or to philosophical objections. A related question is whether Boltzmann abandoned realism as a consequence of these philosophical difficulties.

As for the former question, it appears that for some time the view generally accepted by historians and philosophers of science was in favour of the second interpretation, as summarized by S.G. Brush [15]:

In retrospect it seems clear that the criticism of the kinetic theory in this period was motivated not primarily by technical problems, such as specific heats of polyatomic molecules, but rather by a general philosophical reaction against mechanistic or 'materialistic' science and a preference for empirical or phenomenological theories, as opposed to atomic models.

In the 1970s however, a different view was proposed, claiming that the opposition to kinetic theory must be attributed to scientific difficulties it met with. In fact, P. Clark [16] states that it "was the degeneration of the kinetic programme compared with the empirical success of thermodynamics which accounts for the rise of scientific positivism". On the other hand, M.R. Gardner [17] maintains that there was a gradual transition from an instrumentalist to a realistic acceptance of the atomic theory. For the reader's benefit we recall that, according to instrumentalism, a physical theory is not an explanation but a system of mathematical propositions, aiming to represent a set of experimental laws in a simple, complete, and exact fashion. A few years ago, J. Nyhof thoroughly examined this matter [18] and called the attitude summarized in this definition *philosophical instrumentalism*, as opposed to *methodological instrumentalism*, the practice of a scientist who is a realist, yet makes use of a theory merely as an instrument. Thus, Nyhof argues not without reason, Gardner's claim makes no sense if one talks about *philosophical instrumentalism*, but it can be understood and turns out to be helpful if we refer it to *methodological instrumentalism*, i.e. if we remain within *realism*.

Mach's philosophical influence on scientists grew when he published his book on the development of mechanics [19]. Together with several most valuable observations, Mach also stated that there should be no mechanical explanations and went so far as to say [19]: "But now, [...] after our judgement has grown more sober, the world conception of the Encyclopaedists appears to us as a *mechanical mythology* in contrast to the *animistic* one of the old religions." This was an open blow to kinetic theory and may explain the bitter sentences of Boltzmann in the preface to the second volume of his lectures on gas theory [20]:

It was just at this time that attacks on the kinetic theory began to increase. I am convinced that these attacks are merely based on a misunderstanding, and that the role of gas theory in science has not yet been played out. [...] I am conscious of being only an individual struggling weakly against the stream of time. But still remains in my power to contribute in such a way that, when the theory of gases is again revived, not too much will have to be rediscovered.

It seems appropriate to recall here that the most widely known scientist who was influenced by Mach was Einstein. His witness and judgement are worth quoting here [21]:

It was Ernst Mach who, in his *History of Mechanics* shook this dogmatic faith: this book exercised a profound influence upon me in this regard while I was a student. I see Mach's greatness in his incorruptible scepticism and independence; in my younger years, however Mach's epistemological position also influenced me greatly [...] a position which today appears to me to be essentially untenable.

This passage follows one in which Einstein recalled how nearly all physicists of the nineteenth century saw in mechanics a firm foundation for all physics; after the passage just quoted, he mentions kinetic theory as an example of a theory to which Mach's philosophy cannot do justice.

We might be tempted to conclude, with Nyhof [18], that the difficulties with the specific heats of polyatomic gases which we discussed in Chapter 8 have little if anything to do with the opposition to kinetic theory. Positivism and similar philosophies clashed with the principles on which kinetic theory based its explanations and thus the theory was considered to be philosophically objectionable.

However, it is difficult, as indicated by H.W. De Regt [22] in a recent paper, to distinguish between the cause and the effect when two events (in our case, the decreasing popularity of the kinetic theory and the rise of positivist philosophy) occur at the same time: we can only argue that they are correlated. The main point in De Regt's paper however is that both Clark and Nyhof [16, 18] agree in seeing philosophy as playing an *external role*. With reference to Maxwell and Boltzmann however, he argues that in this episode their scientific methodologies and the philosophical standpoints were interconnected and influenced each other. He first gives a detailed analysis of the philosophical views of Maxwell, who never gave a systematic presentation of these views. Thus his ideas are reconstructed from observations and remarks scattered throughout Maxwell's work. Then De Regt examines Boltzmann's philosophical views, with which we have dealt in detail above. He finds that both men's views are realist, mechanicist, and materialist, but indicates that the fundamental difference between them lies in the fact that theory came first and empirical reality was only secondary for Boltzmann, whereas the opposite order must be applied to Maxwell's scientific view. As a matter of fact there is a considerable amount of confusion in the papers of historians and philosophers of science on whether Boltzmann rejected or weakened his realist attitude. According to Broda [12] and Nyhof [18] he never gave up realism, whereas Clark [16], Elkana [23], and Hiebert [24] hold the opposite view in a more or less strong form. According to De Regt, the issue can be resolved by carefully distinguishing between three levels (ontological, epistemological, methodological) at which one can talk of realism. There is no doubt that Boltzmann is a realist at the *ontological* level: it is enough to recall the quotation about Berkeley at the beginning of this chapter and his rejection of Mach's phenomenalism. In this respect Boltzmann's and Maxwell's views are very similar.

The question of the interpretation of theories is of *epistemological* nature. What can be known about unobservable reality? We have already quoted a passage about the task of theory according to Boltzmann [1, 2]. He made some concession to Hertz's views: theories are mentally constructed pictures, related to phenomena, as "sign to designatum" [1, 2]. De Regt proposes to call his position *constructive realism*. We have used the term *undogmatic realism* before, starting essentially from the same idea, i.e. that this realism should not prevent us from accepting freely created models for reality, even more than one at the same time.

In fact, when we come to the *methodological level*, Boltzmann, contrary to Maxwell, argues that the empirical world is so complex that one would not go very far if one made it the starting point of scientific work. Several theories can exist at the same time and can be appropriate for different purposes. Only their fertility is the touchstone. This theoretical pluralism is also underlined by Hiebert [24]: theories, like species in Darwin's theory, must struggle to survive, but they can also find a way of surviving together. This view is also supported by a quotation given above. In this sense Boltzmann's philosophy is close to Popper's, as Popper himself was ready to admit.

De Regt [22] also addresses the question of how the philosophies of Maxwell and Boltzmann influenced their scientific work. He does this by examining their contributions to the kinetic theory of gases, with particular reference to the specific heat anomaly, discussed in Chapter 8. The main argument is based on the rejection of Boltzmann's explanation of the ratio of specific heats of diatomic molecules; indeed, we remarked that this was a surprising feature of Maxwell's analysis. Obviously, Maxwell was guided by the circumstance that spectroscopic data deny that a rigid molecule makes sense. Probably Boltzmann might have answered with the sentence that he used in 1895 to justify his even more venturesome explanation of why one can neglect the vibrational degrees of freedom (see Chapter 8) [25]:

It may be objected that the above is nothing more than a series of imperfectly proved hypotheses. But granting its improbability, it suffices that this explanation is not impossible. For then I have shown that the problem is not insoluble, and nature will have found a better solution than mine.

We have so far avoided talking about an important, but debatable, contribution by Mach's biographer, J. Blackmore [26]. The main theme of this paper is the concessions to Mach in Boltzmann's papers.

Blackmore gives an extensive appraisal of Boltzmann's and Mach's viewpoints. In particular he recognizes that Mach and his followers, by discouraging the use of atomic theory, held back the progress of science, but he maintains that Boltzmann adopted Mach's *philosophy* of science to prevail over Mach's *methodology of science*. The main features of Blackmore's analysis are the stress on linguistic philosophy in Boltzmann's work and the claim that Boltzmann's philosophy of science becomes more phenomenalistic than Mach's. Blackmore essentially claims that Boltzmann eventually abandoned realism at both the epistemological and the ontological level. He quotes from his inaugural lecture on natural philosophy, from which we have already taken some passages. He concentrates on the following part [1, 2]:

Here I always had a nightmarish feeling that it was an unresolvable puzzle how I could exist at all, or that a world could, and why it should precisely as it was rather than otherwise. The science that would succeed in resolving this puzzle seemed to me the greatest and true queen of sciences, and this I called philosophy.

I gained more and more knowledge of nature, I absorbed Darwin's theory and saw from it that it was really a mistake to ask in such a way that the question cannot be answered, but the question always returned with the same compelling violence. If it is unjustified, why can it not be dismissed? Following on from this there are countless other questions: if there is something else behind perceptions, how can we even suspect that there is? If there is nothing behind them, would then a landscape on Mars or on a planet of Sirius really not exist if no living being is ever able to perceive them? If all these questions are senseless, why can we not dismiss them or what must we do in order finally to silence them?

Boltzmann, argues Blackmore [26], seems to be looking for ways not to solve ontological or metaphysical problems but to dissolve them like a "Machist" philosopher. He further says that Boltzmann failed to distinguish between truth and certainty.

We come now to the most interesting part of Blackmore's essay, which follows his comment that Boltzmann redefined philosophy as a linguistic approach which could prevent us from even asking metaphysical questions. Blackmore goes on to remark that, in this respect, the basic paper by Boltzmann is his lecture on statistical mechanics in St Louis. As is clear from our previous quotations, this lecture pays only lip-service to the topic mentioned in the title, being rather a call for philosophy to abandon metaphysics. We have already commented upon these passages, but here are Boltzmann's exact words [1, 2]:

My present theory is totally different from the view that certain questions fall outside the boundaries of human cognition. For according to that latter theory this is a defect or imperfection of man's cognitive capacity, whereas I regard the existence of these questions and problems themselves as an illusion. [...] If therefore philosophy were to succeed in creating a system such that in all cases mentioned it stood out clearly when a question is not justified so that the drive towards asking it would gradually die away, we should at one stroke have resolved the most obscure riddles and philosophy would become worthy of the name of queen of the sciences.

Blackmore calls this drastic surgery and compares Boltzmann's ideal philosophy to Bloody Mary or the Red Queen in *Alice in Wonderland*, who would shout "Off with his head!". But Boltzmann thought that this was the way to heal human understanding.

Blackmore asks himself: now, was this a tactic that Boltzmann chose to protect his scientific work or did he seriously believe in what he was saying? Then he admits that he had thought for a long time that Boltzmann was a realist, but he also says that he might be in error. In fact he claims that Boltzmann wants to reduce the difference between idealists and realists to a difference in language. After examining the unfriendly remarks on Berkeley and Schopenhauer, Blackmore concludes that Boltzmann, like Mach, had an unreasonable, exaggerated fear of metaphysics (perhaps here the word is used in a meaning different from that defined in the first section of this chapter) and speculation, and as a consequence he was an idealist (without realizing it).

Now this appears unconvincing to the author of this book in the case of Boltzmann. It is true that Boltzmann indicated that many questions reduce to a matter of definitions. But this is of course necessary: either we take it for granted that in a discussion two persons mean the same thing when they use a certain word or expressions, or we must clarify the meaning of words and expressions. This analysis may show that the difference is purely verbal or substantive. The more science departs from everyday concepts, the more this analysis is needed. We construct certain pictures or models to explain phenomena and we have to make a correspondence between the model and our perceptions and experiments. Boltzmann was a passionate advocate of the objective existence of the real world, to the point of considering the possibility of changing the rules of our logic if they do not conform to our experimental findings.

It is also true that Boltzmann shifted his philosophical position a little without informing his listeners or readers (we have seen that he did so even in scientific matters). But his passionate reference to Darwin's theory of evolution as a foundation for philosophy is the touchstone for his realism.

Blackmore subdivides Boltzmann's philosophy into three periods (realism until the 1890s, phenomenalism or Machian idealism until about 1904, linguistic philosophy until his death). As is clear from what has been said, it is hard to share this view. An interesting remark by Blackmore is that Boltzmann was never a real believer in any type of philosophy; this of course we shall never know.

We have of course pointed out the necessity of placing an adjective next to the noun "realist", because that word is used in many different meanings (Blackmore gives four conditions that realism must satisfy, and the fourth has ten sub-conditions). We feel that our characterization of Boltzmann as an "undogmatic realist" is reasonable.

We must of course distinguish between Boltzmann and the influence that he exerted on certain philosophers. In fact, it appears that the most valuable part of Blackmore's paper resides in the elucidation of the role of Boltzmann as father or godfather of a great deal of modern linguistic philosophy, and especially of the ideas of Wittgenstein. But it would be a complete exaggeration to say that Boltzmann tried to reduce everything to mere matters of language, or denied that we can refer to or think about a physical world beyond the appearances and consciousness, much less understand such a world. This of course may be true of the Vienna Circle, logical positivism, and their followers.

Let us end this chapter by remarking that some of Boltzmann's concessions to his opponents confused even his fellow-physicists, beginning in the early 1890s. But we can stick to the opinion of George Hartley Bryan, who says [27] that Boltzmann's views "appear to have been interpreted in a different light to what he doubtless intended" and that he was sure "that Prof. Boltzmann will be much astonished to learn that his statements are now widely circulated and quoted as being an authoritative admission that the Kinetic Theory of Gases is nothing more than a purely mathematical investigation, the results of which are not in accord with physical phenomena; in short a mere useless mathematical plaything." It is clear that the undogmatic views of Boltzmann and his very modern conception of theories as models of reality lent themselves to misinterpretation. We can thus agree with Bryan when he says [27] "I gather that his views are not nearly as pessimistic as the opponents of kinetic theory would wish to maintain."

11

Boltzmann and his contemporaries

11.1 The contacts between Boltzmann and his colleagues

Some important aspects of the relations between Boltzmann and several great scientists who were his contemporaries have already emerged in the previous chapters. In Chapter 1 we recalled his master Josef Stefan, who introduced him to Maxwell's papers, the leaves of absence that Boltzmann took to work with Bunsen and Könisberger in Heidelberg and with Kirchhoff and Helmholtz in Berlin, the friendship with Josef Loschmidt, the contacts with H.A. Lorentz, H. Helmholtz, J. Rayleigh, W. Ostwald, and his colleagues in Munich (van Dyck, Pringsheim, Lommel, Sohnke, Nayer, Seeliger, and Linde). More than once we have met the name of Ernst Mach, and a couple of times that of Helm. We also mentioned his students Paul Ehrenfest, Fritz Hasenöhrl, Stefan Mayer, Lise Meitner, Svante Arrhenius, and Walter Nernst, and his contacts with Brentano. Other names which have occasionally appeared are those of Gibbs, Poincaré, Planck, Hertz, Zermelo.

There is more than enough to belie the idea of a Boltzmann isolated from the science of his days. Yet it is a fact that his scientific position appears to be rather singular and almost isolated in the framework of the scientific research of his century. This chapter is devoted to the problem of singling out the objective data and motivation of this circumstance.

11.2 Maxwell

Let us begin with the greatest of the above-mentioned physicists, one with whom, so far as is known, Boltzmann had no direct contact, either in person or by correspondence. We allude of course to J.C. Maxwell, who exerted, through his outstanding work in electromagnetism and in kinetic theory, as well as through his conception of physical models, a lasting influence on Boltzmann's work. We shall restrict ourselves, for obvious reasons, to the development of the kinetic theory of gases, which in its modern form owes everything to these two great scientists, whose relations will occupy us at present. We shall initially take up again some aspects already touched on in Chapters 3 and 4. Maxwell made his first appearance in kinetic theory [1] in 1860, making use of the concept of mean free path, which had been introduced by Clausius [2] two years earlier. And we ought to call his contribution a masterly début if we were talking about a scientist of lesser capacity. In fact he not only developed a systematic theory of transport processes, but he also posed for the first time the problem of calculating the distribution function and solved it with an original argument which, for all its limitations from a logical standpoint, shows what results can be reached by powerful intuition. It is in this paper that we first meet what we are now used to calling Maxwell's distribution, or Maxwellian.

Maxwell immediately realized, however, that the concept of mean free path was inadequate as a basic tool of kinetic theory, in spite of its qualitative and heuristic importance. In fact he wrote a couple of manuscripts based on this method that he never published. Instead he developed a completely new method, which he published in 1866 in his most important work in kinetic theory [3]. We refer to the method of transfer equations which bear his name and constituted the source of inspiration for the basic work written by Boltzmann in 1872 [4]. The aim of Maxwell's paper is to write the equations describing the time evolution of the average values of any function of the molecular velocity. The average is performed with the help of a distribution function, which however can no longer be generally speaking a Maxwellian, since the latter describes equilibrium states only. In the same paper Maxwell discovered the particularly simple properties of a molecular model, according to which the molecules are point masses (thus not hard spheres) which interact with a repulsive force inversely proportional to the fifth power of their distance (these fictitious particles are usually called Maxwell molecules). He also gave a better justification of his formula for the distribution function of the velocities in a gas in a state of statistical equilibrium.

After considering repulsive forces which vary with the inverse nth power of their distance and remarking that the term describing the effect of collisions contains the power $V^{\frac{n-5}{n-1}}$ of the relative speed V of two colliding molecules, Maxwell introduces the molecules which bear his name with these simple sentences: "It will be shewn that we have reason from experiments on the viscosity of gases to believe that $n = 5$. In this case V will disappear from the expressions [...] and they will be capable of immediate integration [...]. If we assume $n = 5...$" [3]. And he immediately proceeds to perform the integration, which can be done—this is the essential point—without knowing the distribution function. This circumstance allows him to compute the transport coefficients in terms of the constant giving the intensity of the intermolecular force (as a matter of fact, the computation is not so easy and requires the use of elliptic functions). The experiments which Maxwell alludes to had been described in a previous paper [5].

With his transfer equations, Maxwell nearly obtained an evolution equation for the distribution function, but this last step must be attributed beyond any doubt to Boltzmann [4], as we saw in Chapter 4. Before writing his basic paper [4], Boltzmann had learned how to manage Maxwell's techniques and in 1868 he had already extended Maxwell's distribution to the case when the molecules are in equilibrium in a conservative force field [6], including the case of polyatomic molecules as well [7]. This result was independently rediscovered by Maxwell after some years [8]. Maxwell was informed about the activity of the young Austrian scientist, because in one of his last papers [9] he unambiguously ascribes to Boltzmann both this result and the evolution equation for the distribution function f ("Boltzmann has shewn that the function f_1 must satisfy the equation [...]"), remarking that one can solve it only in particular cases. He later says: "We shall suppose, however, with Boltzmann, that" f is, approximately, a Maxwellian multiplied by a polynomial. He thus succeeds in discovering the stresses arising from differences of temperature. This paper by Maxwell [9] is also of great importance because, upon the request of a referee, he added an appendix where the boundary conditions for the Boltzmann equation are discussed for the first time [10–12].

There is a singular circumstance in the relation between Maxwell and Boltzmann. The former never mentions the H-theorem, i.e. the most important result that can be obtained from the Boltzmann equation. What is the reason for this omission? We cannot believe that Maxwell was not sagacious enough to be able to estimate the importance of Boltzmann's result. Did the quotation of the theorem not fit in his 1879 paper [9]? Was he annoyed by Boltzmann's statement that he had obtained a "purely mechanical proof" of the Second Law, while Maxwell was well aware of the role that probability must play in this law? Probably the answer is simply that he read only some parts, if any, of the 1872 paper [4]; the quotation of Boltzmann by Maxwell in 1879 should then be either ascribed to a partial reading or understood as a reference to a by then well-known fact. We can even lean towards the second hypothesis without hesitation, if we note that the two aforementioned passages of Maxwell's 1879 paper [9], where Boltzmann's contributions are quoted, are not accompanied by any precise citation of the corresponding papers, as is the case for quite a number of other authors who are also mentioned. Further support for this interpretation comes from a passage of a letter written by Maxwell to Tait in 1873 [13], i.e. one year after ref. [4], which has been already quoted in Chapter 7:

By the study of Boltzmann I have been unable to understand him. He could not understand me on account of my shortness, and his length was and is an equal stumbling-block to me. Hence I am very much inclined to join the glorious company of supplanters and to put the whole business in about six lines.

Maxwell wrote another paper referring to Boltzmann's work, actually containing Boltzmann's name in the title [14]. This paper is remarkable in many respects; in particular, we find here the first direct computation of the Maxwell–Boltzmann distribution as the limit—when the number of degrees of freedom tends to infinity—of a uniform distribution on the surface of a sphere of constant energy.

11.3 Lorentz

The scientist who, being a contemporary of Boltzmann, read his papers and understood them so deeply as to correct some errors—as we have already seen in Chapter 8—and to extend his methods to the then unexplored field of electron theory was Hendrik Antoon Lorentz (pronounced as "Lawrence" [15]).

Lorentz (1853–1928) was nine years younger than Boltzmann and had similar interests but a completely different life. At the age of 24, he was offered a mathematics chair at the University of Utrecht; he turned it down and instead applied for a position as a teacher at a secondary school in Leiden. One year later he was offered the newly created chair of theoretical physics at the university of the latter city, which he accepted. He did not travel outside his native Netherlands for scientific purposes until the age of about 45, when he attended a conference in Düsseldorf, to which he was invited by Boltzmann. Despite this, Lorentz was fluent in German, French, and English. He wrote flawlessly in these languages and also corresponded in Italian.

Let us pause for a moment to comment on the invitation by Boltzmann, which introduced Lorentz to the world of German physicists. Lorentz was asked to give a lecture on any subject of his choice at the *70. Versammlung deutscher Naturforscher und Ärzte*. Boltzmann's letter is dated 13 October 1897 and the meeting was planned for September of the subsequent year. The invitation changed Lorentz's life; afterwards, hardly a year passed in which he did not travel to a foreign country, attending conferences or giving lectures. The correlation between Boltzmann's invitation and Lorentz's change is supported by a reminiscence by the latter's daughter: "Seldom have I seen my father in such good spirits as after his return from this congress" [16, p. 89].

Lorentz's theory of electrons so extended the domain of electrical science that eventually it became necessary to modify our concepts of space and time to make its results compatible with experience. The first memoir in which the new concepts were disclosed was published by Lorentz in 1892. All electrodynamic phenomena were ascribed to their source, i.e. moving electric charges. This idea was not completely new, because it had been advanced by Weber, Riemann, Clausius, and Lorentz himself in his earlier work. However, in the previous theories, the electrons were assumed to be able to act on each other at distance, with forces depending on their charges, mutual distances, and velocities, whereas in Lorentz's 1892 memoir the charge carriers were supposed to interact not directly with each other but with the medium (the ether) in which they were embedded. His ether however was far from having the properties of a ponderable body; in particular, it looked more like empty space endowed with certain dynamic properties, described by Maxwell's equations. When Paul Drude (1863–1906) began a kinetic theory of electrons in solid conductors, Lorentz, with his deep knowledge of kinetic theory, produced an approach that at the time represented a great advance on everything that had preceded it. Only the discovery of the new statistics produced a better theory, more than twenty years later.

Lorentz's place among his contemporaries of the twentieth century is perhaps described best by Albert Einstein (1879–1955), who wrote three times on Lorentz [17]: the first piece (1927) concerns Lorentz's action in favour of international cooperation, the second is the allocution on Lorentz's grave (1928), and the third is a speech given in Leiden on the occasion of the centenary of Lorentz's birth (1953). It is in the last of these writings that we read:

At the turn of the century the theoretical physicists of all nations considered H.A. Lorentz as the leading mind among them, and rightly so. The physicists of our time are mostly not fully aware of the decisive part which H.A. Lorentz played in shaping the fundamental ideas in theoretical physics. The reason for this strange fact is that Lorentz's basic ideas have become so much part of them that they are hardly able to realize quite how daring these ideas have been and to what extent they have simplified the foundations of physics.

As is made clear by the subsequent sentences, Einstein is here referring to the theory of electrons. The outcome of this theory are terms well known to physicists, such as Lorentz transformations, Lorentz–Fitzgerald contraction, Lorentz force, Lorentz invariance, Lorentz gauge.

We must also mention another area in which Lorentz worked with success, i.e. fluid mechanics. He wrote papers on slow viscous flows and turbulent flows and in 1918 (at the age of 65) he became a mathematical engineer and oceanographer. In fact the Dutch government had asked him to take up the presidency of a committee whose task was to investigate the effect of a proposed giant dam for the biggest engineering project ever to be undertaken in his country: closing off the Zuiderzee. Eight years later the committee produced their final report [18], which contained predictions which later proved of great value: it was Lorentz who carried the modelling burden! We refer to the paper by Kuiken [15] for more detail.

Let us now deal with the relations between Lorentz and Boltzmann, not without remarking that the last sentence of Einstein's passage quoted above could apply to Boltzmann too.

After receiving the manuscript in which Lorentz pointed out his blunder which we mentioned in Chapter 8, Boltzmann thanked him (on 11 December 1886) and admitted his mistake (on 1 January 1887), while also expressing (in his first letter) the pleasure which Lorentz's interest in his work afforded him: "I am very happy to have found in You a person who works to develop my ideas on the kinetic theory of gases. In Germany there is nobody who correctly understands these things." [19]

A few years later, while Lorentz was writing a paper on the molecular theory of dilute solutions [20], Boltzmann wrote a paper [21] in which he gave a derivation of van't Hoff's law of osmotic pressure, a subject which was also treated by Lorentz [20]. Boltzmann's proof was very simple but contained an assumption that was not justified, as Lorentz informed him in a letter dated 16 December 1890. Boltzmann started to write another paper [22] in which he corrected his mistake and sent an answering letter dated 21 December, without any grudge, which contains the following remarkable sentence: "From the stamp and the handwriting, I recognized that the letter came from You and I had a moment of joy. True, every letter of Yours means that I made a mistake; but I learn so much on these occasions that I would even like to make more mistakes, in order to receive more letters from You." [22]

Let us also recall that Boltzmann was commemorated by Lorentz (and not by a scientist whose mother tongue was German) in 1907 before the German Physical Society [22]. Of this address we quote just one sentence:

The old doctrines about which Boltzmann speaks [see the passage quoted at the beginning of Chapter 2] have blossomed, thanks only to his efforts, into a new and strong life and, even if their aspect has changed and will certainly often change again in the course of time, we can however hope that they will never be lost to science.

11.4 Boltzmann and the energetists

After reviewing the relations between Boltzmann and the two great scientists who were the closest to him as far as methodology and conception of physical theories go, we must now discuss those who, on the contrary, were in bitter disagreement with him. We have already seen in some of the previous chapters the vehemence with which Boltzmann opposed the supporters of energetics, real pontiffs of German physics at the end of the nineteenth century, ready to excommunicate all those who opposed their axioms.

The supporters of so-called "energetics" considered Leibniz as their founding father; in fact, as we saw in Chapter 2, one can trace back to him, probably influenced by the researches of Huygens, the concept of energy (although, in one of the passages quoted in the aforementioned chapter, Leibniz clearly indicated that the conservation of energy could be correctly understood only with the help of the atomic hypothesis). But again as we saw in Chapter 2, the concept under consideration was to ripen fully in the nineteenth century. Ostwald, the great supporter of energetics, puts Robert Mayer in first place among the modern founders of the theory: "an aspect which contributes to making him a true energetist, an energetist with the modern spirit, is his dislike for hypotheses" [24]. Actually, Mayer was the first, after Leibniz, to talk of energy as of a substance. Ostwald goes even beyond this [24]:

If nowadays a physicist or a chemist wants to appear up-to-date, he declares that matter and energy are two entities similar or parallel and defines the physical sciences as the sciences of the transformation of these two indestructible things, matter and energy, without knowing in most cases that he is but reproducing Mayer's concepts. It will be seen later that one must not consider this concept as something definitive and that even the matter–energy dualism can be suppressed, given the circumstance that the notion of matter is not even a well-chosen one. Clearly, the spirit–matter dualism disappears at the same time, and the question arises of knowing what is the relation of energy to spirit. Well—and it is here that the most considerable progress in this set of ideas has occurred—as far as science is concerned, these two notions are of the same kind, and the notion of spirit merges with that of energy.

What is Ostwald's opinion about Joule, Helmholtz, who played a very important role in the clarification and the diffusion of the concept of energy, and Rankine, who was the first to speak of *energetics*? The sentence delivered is a final one: excessive adherence to the model of point masses interacting with central forces. The only merit of Rankine [24] is that he

observed that the most different energies can *all be represented as the product of two factors*, markedly different in their nature, which is the same in all kinds of energy. If we call *intensity* the factors of one kind and *extension* those of the other kind, it will be said that one of the two factors of any type of energy has the nature of intensity, the other of extension. Rankine presented his remark in a rather peculiar form, but in spite of this, it has played a very important role in building up a general theory of energetics.

Before considering the practical consequences of the concepts of energetics, we remark that its main leader, Ostwald, was a chemist. Now, although chemists talked freely about atoms and molecules, it was a widespread idea that these were just schematic representations which had no existence whatsoever. They seemed to think that science would gain by entirely discarding these shameful concepts.

We may start with John Dalton (1766–1844), whose *New system of chemical philosophy* [25] may be considered to mark the birth of modern chemistry, since it reduced all kinds of matter to a finite number of elements (only eighteen in those days). He called atoms what we call molecules, and this confusion in terminology was rather common in the first half of the nineteenth century and beyond, because Dalton did not accept the hypothesis put forward in 1811 by Amedeo Avogadro, whom we already mentioned in Chapter 2. This hypothesis fixes the concept of the molecule in a very precise way, by stating that at a fixed pressure and temperature, equal volumes of different gases contain equal numbers of molecules. At the top of the agenda of the first international conference of chemists, held in Karlsruhe in 1860, which was also mentioned in Chapter 2, we find the discussion: Should a difference be made between the expressions "molecule" and "atom"...? Even more interesting is the fact that no consensus was reached. The still young but quite well-known August Kekulé von Stradonitz (1829–96) maintained that physical and chemical molecules are not the same thing. In the discussion that followed his talk, Stanislao Cannizzaro, whom we also met in Chapter 2, commented that this difference had no experimental basis and hence was unnecessary. In his own talk, Cannizzaro emphasized the importance of Avogadro's principle for *chemistry*. Thirty years later, Dmitri Ivanovich Mendeleev (1834–1907) recalled that "the law of Avogadro received by means of the congress a wide development, and soon afterwards conquered all minds" [26].

We may also recall that even William Prout (1785–1850), who was the first to put forward the bold hypothesis that essentially all the elements are made of the same stuff, since the atomic weights can be expressed as integral multiples of a fundamental unit, considered the atomic theory a conventional artifice, exceedingly convenient for many purposes but which does not represent nature.

It was only natural that Ostwald, who had mastered the concepts of thermodynamics, highly useful for chemistry, would join the battle against atomic theory. His main argument was essentially the Loschmidt paradox. Yet he should have known better, because Jacobus Henricus van't Hoff and Joseph Achille Le Bel had independently explained, in 1874, the isomerism of certain compounds in terms of molecular stereochemistry.

To claim that energy is the essence of reality may be an acceptable credo, but then one must explain everything, in particular the principles of mechanics, the mass of bodies, etc. in terms of energy. Nowadays these things are easier, but certainly they were impossible with the conceptual tools available in the nineteenth century and Boltzmann seized this opportunity. In particular, he censured Helm and Ostwald for their confused and erroneous derivations in dynamics and thermodynamics.

Boltzmann's first answer, published under the title *A word from mathematics on energetics* [27], started by pointing out that the trend of his time was to construct a description of phenomena as clear and free from assumptions as possible, but capable of yielding precise laws: this was the goal that Kirchhoff, Helmholtz, Clausius (in his general thermodynamics), Hertz, Thomson (Lord Kelvin), Gibbs, and so on, had set themselves.

A parallel trend, whose perpetrators did not want to deny the interest of the other, says Boltzmann, was to continue with great success to picture models of the phenomena of elasticity, of fluid mechanics, of light, of heat, of electromagnetism. By their nature, these models were incomplete, but always led to general theorems which were capable, given sufficient effort, of yielding a proof or a formulation, such as the Hamilton principle, the Second Law of thermodynamics, the Maxwell–Hertz equations of electromagnetism.

However, a new trend was developing [27]:

Rather recently, some scientists have thought that they could free themselves from these complications and express the basic laws in a much simpler manner. Since, in the final analysis, they arrived at giving a true existence to energy only, they called themselves energetists. We ignore whether our present conception of nature is the most adequate; the mere fact of striving for a more general and higher viewpoint than that of today's theoretical physics is thus perfectly justified. But today's energetists do not restrict themselves to such striving: they are convinced that they have already reached a higher viewpoint and think that the methods of expression prevailing in theoretical physics must henceforth be completely abandoned, or at least modified in their essential principles, an opinion which I think I can refute in what follows.

After stating that he counts many friends among the energetists, that he admires their papers having a different nature and does not want to give a personal slant to the discussion, Boltzmann declares that the goal that Helm had set him, of deriving the equations of mechanics from the mere conservation of energy, is impossible even in the case of only one point mass (Helm [28] wrote the balance of energy in differential form, and treated the coordinates of the moving point as independent variables). Ostwald, setting the same problem, restricted himself to a hint. Boltzmann tries to explain Ostwald's ideas in a synthetic fashion, summarizing them in three principles which seem to be devoid of any contradiction. The first principle states that mechanical systems are endowed with a total energy, the sum of a kinetic energy (function of velocity alone) and a potential one (function of position alone); the second principle is that if all the point masses of a system are at rest at some time instant, in a subsequent infinitesimal time they move in such a way that as much as possible of potential is transformed into kinetic energy (this is admittedly rather unclear, but this is the way it was stated); the third principle is that if the point masses are already moving at a certain instant, they will subsequently move in such a way as to add to their velocity the same amount that they would acquire according to the previous principle. Boltzmann also remarks that one can admit with Planck that, for the motion of a point mass, the energy principle holds for the projection of the motion on each of the coordinate axes. However, he adds as if he wanted to become holier than the Pope [27]:

But none of these constructions satisfies the promises of energetics. In fact, to start with, they are grounded on the assumption that bodies are made up of point masses, something that, in the spirit of energetics, constitutes an unwarranted restriction of the freedom of our thought. I do not know of any direct proof of Euler's equations for the motion of solid bodies, of the equations of elasticity and hydrodynamics starting from the principles of energetics alone, without making use of the atomic assumption. Further, these constructions do not produce progress in our knowledge of the principles of mechanics; they do not introduce general theorems which hold equally in other areas, but annex the foundations of mechanics by restricting themselves, without much success, to partially translating these foundations into the language of energetics. The principle of maximum transfer [from potential to kinetic energy] turns out to have a reduced meaning—whereas one wanted it to be universal—since it applies but to short motions starting from rest. The elevation of the principle of superposition to the role of a general law of nature seems premature to me [...]. Lastly, when several point masses come into play, the separation of the work supplied by the point masses through their motions appears to be arbitrary to me.

If, following the more advanced picture of Ostwald, one admits that there is just energy and not necessarily any material support for it, this seems to make the question easier in the case of thermal energy, but it is not the same for kinetic energy. In fact, we ought to fix some laws for its space distribution, to recover from these the concept of mass, and to define the speed as $(2T/m)^{1/2}$. Boltzmann then enters into a detailed discussion of energetics in thermodynamics, showing that the energetists (Helm in particular) have taken up again results found by Gibbs in the framework of the usual ideas and generalized them in an arbitrary fashion. At most, says Boltzmann, the principles of energetics might be applied to very slow motions; Ostwald's formulation of the Second Law would in fact be more restrictive than the usual one.

Boltzmann then recalls the address by Ostwald at Lübeck in 1895, by saying [27] that there

he proceeds through images and general considerations, about which we ought to avoid to express an opinion, because, albeit brilliant, they neither prove nor confirm anything. But the illusion of rigorous logic exerts such an extraordinary power of conviction on many minds that it seems useful to examine more closely at least certain aspects of this address. I think that this is so much more necessary, in that many young people turn towards the easy harvests promised by the different areas of energetics, without possessing the apparatus of mathematical criticism indispensable for operating usefully in theoretical physics.

The criticism by Ostwald of the Laplace picture of determinism based on atoms and forces, thought of as the final reality, is addressed to something that nobody wants to support any longer. In fact Boltzmann states [27]:

There is no longer, so to speak, anybody that considers the force to be a reality, or thinks that one may exhibit a proof that the set of natural phenomena is liable to a mechanical explanation [...]. I also once pleaded the cause of a universal mechanism, but I did this just because I wanted to show its immense superiority over the mystic explanations which were accepted before. The opinion according to which the only way to explain nature would reside in considering point masses between which forces act at a distance, had been almost unanimously abandoned well before Ostwald's statement. Today we have become decidedly more cautious: this representation is for us a mere picture which we do not worship, which is perhaps capable of being perfected, but which, maybe, we shall be obliged to renounce altogether. This picture is of the utmost interest still, since it is the only one which we possess and can be developed in certain important aspects in agreement with experience.

Scientists have thus reached unanimity in paving the way for descriptions freed as much as possible from arbitrary assumptions, says Boltzmann. This trend can be found in Maxwell [29], whose position on the principles of optics is analogous to that of Ostwald, of Hertz [30], of the English Premier Lord Salisbury [31], and even in kinetic theory, where the molecules are no longer considered as mere aggregates of point masses, but as unspecified systems, defined by generalized coordinates. He continues [27]:

Therefore we agree on the necessity for enlarging the field of every viewpoint. But none of the arguments with which Ostwald tries to show that the representations of traditional theoretical physics cannot be defended and that we must henceforth prefer those of energetics seems to me to be well founded.

It is true that the usual methods of theoretical physics have holes and are still quite far from supplying a perfectly clear and coherent description of all phenomena. But this can be said *a fortiori* of energetics. "We should deny ourselves any picture of reality." But are man's thoughts something different from these images? It is only God of whom we must not and cannot make any picture.

Boltzmann continues with some philosophical considerations about energetics, to arrive at discussing the reversibility objection taken up again by Ostwald against the mechanical interpretation of the Second Law [27]:

Ostwald concludes that, because of the complete time reversibility of the differential equations of mechanics, the mechanical conception of the world should not be able to explain why natural processes occur always in a privileged direction. It seems to me that in this way one neglects the fact that mechanical processes are defined not only by differential equations but also by initial conditions. In a manner directly opposed to Ostwald's thesis, I have shown that one of the most staggering confirmations of mechanicism is exactly that of giving an excellent picture of the dissipation of energy, if we concede that the world started from an initial state corresponding to certain initial conditions, which I designated as an improbable state [[32]]. Here I shall restrict myself to giving an idea of this matter by a rather simple example...

and he goes on talking of the mixing process of white and black marbles initially separated [27]:

The same will happen to the universe, if it started from a state in which the order of atoms and their velocities showed certain regularities: the mechanical forces will preferably provoke modifications which tend to destroy these regularities. The way in which these regularities could arise matters just as little as the way in which atoms and their laws arose.

The mechanical description of nature, granted that it is possible, poses very difficult problems; this is the reason why it often fails. Certain assumptions which have been abandoned are still useful to illustrate certain laws. The oldest assumptions leave their footprint in science and thus they survive in some way; no theory completely supplants previous ones. As examples one may quote the theory of electric and magnetic fluids and the wave theory of light before the advent of Maxwell's equations. Boltzmann concludes that although it is convenient to move, following the example of Kirchhoff, Clausius, Helmholtz, and Gibbs, towards a description of nature as unconstrained as possible by particular assumptions, the way of expression of energetics has proved so far to be insufficiently appropriate and can turn out to be pedagogically harmful. Besides the general formulations, the models of mechanistic physics are still very useful and should still be cultivated as a means not only of research but also of classifying, representing clearly, and memorizing our ideas. The supporters of well-established physics are far from believing in the permanence and perfection of the traditional forms of thought and are far from prejudging every new attempt. But the innovators too must beware, before they achieve true success, of judging as absurd, ways of thinking which have proved their value. It would be completely gratuitous to ask today what forms of thought will reveal themselves more adequate in the future. On this point, it is more than possible that the developments of energetics will turn out to be very useful to science. But its champions, who wrongly present themselves as continuers of Gibbs, do not follow his path.

Ostwald and Helm [33] replied to this rather long paper by Boltzmann, and Mach hurried his book on thermodynamics into publication (1896), but they could produce only disarming sentences against the penetrating and devastating attack on energetics. Boltzmann in turn replied with a second paper entitled *On energetics* [34], in which he starts by saying:

A discussion as this present one on energetics is not undertaken in the expectation that one side will be right and the other wrong, but with the intention that the views will be clarified. Therefore I can be satisfied with the result as regards the relations between energetics and mechanics. Helm's last essay (*Wied. Ann.* **57**, 1896, p. 646) seems to put everything perfectly straight.

Planck and Helm have shown (simultaneously, as now turns out) that the ordinary equations of motion for a system of material points can be obtained from the principle of energy if we assume that it holds separately for each of the particles in the direction of every co-ordinate or, according to Helm, in any arbitrary direction whatsoever.

On the other hand Helm goes so far as obtaining the Lagrange equations and thus the whole of mechanics by transformation of rectangular coordinates of material points and of the forces acting on them, which therefore involves the presupposition that bodies are systems of material points. This presupposition, however, evidently once more takes us completely into the area of atomism. From it follows in known ways that for long-lasting motion under the influence of forces that do not act uniformly on all material points, there must arise irregular mutual motions of the particles, which always swallow up a part of the visible kinetic energy; that if the motion is sufficiently violent, the particles creep past each other, which liquefies the body; and that particles must separate from the surface, which vaporizes the body.

These atomistic hypotheses incidentally recognize the concept of energy, too, as one of the most important; indeed, if you will, they might even be obtained from that concept by means of suitable subsidiary assumptions. If however energetics will not recognize such hypotheses on the ground that they are insufficiently attested, it would have to take quite a different path.

Just how one might construct a mechanics on the assumption that the kinetic energy of the motion is the primarily given and the moving object itself a concept derived from it, I cannot quite imagine at present. If then energetics takes the comfortable path of starting from the concept of mass, then in order to avoid the atomistic hypothesis it would have to assume that matter continuously occupies its space. From the principle of energy together with suitable auxiliary hypotheses one would then first have to obtain the equations of motion for rigid bodies, perhaps by deriving Lagrange's equations without the detour via the co-ordinates of the individual points of which the body consists and via the forces that act on them. By means of fresh auxiliary hypotheses one would have to derive from the formulae for elastic and hydrodynamic energy the corresponding equations of motion. All these derivations should be possible, indeed variously so, according as this or that auxiliary hypothesis is enlisted, and I should regard it as useful for science to attempt such derivations.

What would seem to be more difficult is to give a survey purely from the point of view of energetics of all cases where mechanical energy is transformed into heat, phenomena of melting and vaporisation, properties of gases and vapours and so on, whereas it is precisely these phenomena that become so intelligible by means of molecular theory and the special mechanical theory of heat.

Energetics seems as yet a long way from having solved all the problems sketched here. It is clear that until this has happened no judgement of how intuitive the auxiliary hypotheses needed by energetics can be formed, nor can they be compared with molecular theory over the entire range of mechanics.

Going on, Boltzmann also remarks that "…only if a clear and unobjectionable account of thermodynamics, chemistry and electricity from an energetic point of view had been achieved at least in its first basic outlines, would it be possible to ascertain what essentially new additions energetics has made to Gibbs's theory." [34]

During proof-correction, Boltzmann added that he had come across H. Ostwald's reply, from which it appears that the latter does not want to regard energy as the primary concept in mechanics, as Boltzmann himself had thought, and that [34]

he proposes to deduce mass from certain properties of it, but that he retains the concepts of the old mechanics, starting from mass and defining energy as $mv^2/2$. [...]

That H. Ostwald is personally convinced of his approach and will not let himself be shifted from it I have never doubted. However, as regards the alleged barrenness of atomism, many chemist[s] too will disagree, since they are wont to deduce the possible number of isomeric compounds and the property of rotating the plane of polarisation directly from the picture that they have formed of the position of the atoms. For my part I permit myself to point out that in justifying his theorems Gibbs must surely have used molecular ideas, even if he nowhere introduced molecules into the calculation; [...] finally that the most recent electro-chemical theory has its starting point in the purely molecular view that Nernst had of the pressure of solutions. It was only later that these propositions were severed from their molecular justification and presented as pure facts. The mathematical part of gas theory on the other hand pursues mainly the purpose of further development of mathematical method, for the valuation of which immediate practical utility was never decisive. Let the purely practical man skip this part but also forebear to criticize.

Before concluding the part of this chapter devoted to energetics, we should mention the reactions of Mach and Ostwald when clear evidence of the existence of atoms was given by Perrin (see next chapter). In the preface to the fourth edition of his *Grundriss der allgemeinen Chemie*, published in 1909, Ostwald completely reversed his views, as we can gather from this sentence: "I am now convinced that we have recently come into possession of experimental proof of the discrete or grainy nature of matter, for which the atomic hypothesis had vainly sought for centuries, even millennia."

As for Mach, we have two conflicting statements. The first is in a recollection by Stefan Meyer, written in 1950. According to this, the apparatus invented by Erstel and Geitel and by Crookes which made it possible to display the flashes made by individual alpha particles on a screen was shown to Mach in 1903 and he said: "Now I believe in the existence of atoms". But he reiterated his disbelief in atoms in the preface to his book *Physical optics* published in 1913. And when Einstein talked to him about the same time, Mach said that he would agree that the atomic theory was the best and most useful assumption for physics, without necessarily accepting the "real existence" of atoms. That is why S. Brush [35], when reporting these facts, entitles the relevant section "The unrepentant sinner".

11.5 Planck

Among the contemporaries of Boltzmann, although almost fifteen years younger, one can also mention Max Planck (1858–1945), whom we are often used to considering as a much more modern scientist, since he reached world-wide fame just at the dawn of the twentieth century. As a matter of fact we have had occasion to meet his name in a previous chapter and in the quotations from Boltzmann's writings in the present one. There is another biographical connection between the two scientists, since it was Planck who occupied Kirchhoff's chair that Boltzmann had refused in Berlin.

Until, like St Paul on the road to Damascus, he was dazzled by the revelation which appeared to him precisely during his study of black-body radiation, in the form of Boltzmann's statistical methods, Planck's position was that of a convinced follower of Mach's philosophy, as he earnestly admits [36]. He was to become one of his severest critics. On the technical side he just stated that he did not consider Mach competent with respect to the Second Law, and later that Mach's book on thermodynamics could only lead to a superficial grasp of thermodynamics. Planck opposed Ostwald and shortly after the Lübeck debate wrote a savage attack entitled "Against the new energetics" on Ostwald's ideas. Let us look at two sentences that he wrote in 1904 and 1909. The earlier [37] reads:

Though I am persuaded of the fact that Mach's system, pushed to all its logical consequences, does not contain any internal contradiction, I am no less convinced of the fact that this system has, after all, but a purely formal meaning: it is incapable of penetrating the very essence of science, and this because it is extraneous to the essential feature of any scientific research, i.e. the construction of a picture of the world which is rigorously stable, independent of the differences which mark the generations and peoples.

Even if the last attribute of rigorous stability may seem too much when picturing an immutable science, it can be accepted only if one understands it as referring to an unattainable ideal which one can approach by trial and error, without ever reaching it. However, the later passage [38] seems to be more motivated:

If Mach's economy principle had to be truly placed at the centre of every theory of knowledge, the geniuses who are yet to be born would find in it but an obstacle to their thought. The wings of their imagination would be paralysed and the progress of science would undoubtedly be spoiled in a disastrous manner. In these conditions, would not it be more "economical" to let the economy principle have a more modest place?

Planck goes so far as to call Mach a "false prophet". "By their fruits shall ye know them!" [38]. Boltzmann could count W. Nernst and S. Arrhenius, two Nobel prize winners, among his students, as well as L. Meitner, F. Hasenöhrl, P. Ehrenfest, and many other outstanding physicists, but whom could Mach name?

The next quotation [39] helps to explain at the same time Planck's development and the difficult situation in which Boltzmann was working in the last part of his life:

In the eighties and nineties of the last century, personal experience taught me how much it cost a researcher who had had an idea on which he had reflected at length to try to propagate it. He had to realize how little weight the best arguments he exhibited to that end carried, since his voice had not sufficient authority to impose it on the world of science. In those days it was a vain enterprise to try to oppose such men as Wilhelm Ostwald, Georg Helm, Ernst Mach.

11.6 Students and younger colleagues

Let us now look at the next generation, students of Boltzmann or younger colleagues, whom he held in great esteem.

We start by mentioning Arnold Sommerfeld (1868–1951), who was appointed successor to Boltzmann in Munich in 1905, after the position had remained vacant for several years. It is known that he had been recommended by Boltzmann himself and by Lorentz, who had been the first to be contacted but had refused the appointment. Sommerfeld was a great teacher and played an important role in the development of early quantum theory.

Then we should recall Walther Hermann Nernst (1864–1941), a former student of Boltzmann, who became famous because he stated the so-called Third Law of Thermodynamics, according to which entropy must attain a universal value when the absolute temperature tends to zero. He was also the instigator of the famous Solvay congress and continued Boltzmann's ideas in philosophy of science, in particular the idea usually attributed to Kuhn that physics proceeds by jumps. In the introduction to his textbook on theoretical chemistry which came out in 1893 [40], he stresses the importance of theoretical research, which leads to "new knowledge the correctness of which has to be tested by experiment only subsequently", though the theorist "is continuously in danger of being led astray by the delusive light of unfortunately chosen principles." In the preface to the English edition of 1911 [40], Nernst reminds us that

many a long-recognized law has had to undergo revision to meet the requirement of the progress of knowledge. [...] If we consider the matter more closely, it is obvious that the law in question has retained its validity over a wide range, but that the limits of applicability have been more sharply defined. It can even be said that since the development of the exact natural sciences, there is scarcely one law established by an investigator of the highest rank which has not preserved for all time a wide range of applicability, i.e. which has not remained a serviceable law of nature within certain limits. We cannot say, for example, that the electromagnetic theory of light has completely overthrown the older optical theory put forward by Fresnel and others. On the contrary, now as formerly, an enormous range of phenomena can be adequately dealt with by the older theory. It is only in special cases that the latter fails; and further, there are many relations between optical and electrical phenomena which certainly exist, but of which the older theory takes no account. Hence the electromagnetic theory implies a great advance, but by no means nullifies the successes of the older theory, [...] So scientific theories, far from dropping off like withered leaves in the course of time, appear to be endowed under certain restrictions with eternal life; every famous theoretical discovery of the day will doubtless undergo certain restrictions on future development, and yet remain for all time the essence of a certain sum of truths.

Other aspects of Nernst's discussion of the revolutionary component of the development of our theoretical understanding of nature can be found in a paper of his published in 1922 [41] and clearly influenced by both Boltzmann's approach and the new revolution brought in by General Relativity.

Another scientist to be mentioned as a student of Boltzmann and visitor to Graz is

Svante Arrhenius (1859–1927), who applied ideas from the kinetic theory of gases to chemical kinetics.

Finally, we should recall Lise Meitner (1878–1968), also a student of Boltzmann. She was very fond of Boltzmann's lectures, as we have already mentioned in Chapter 1. She fully shared his philosophical views. She followed his lectures from 1902 to 1906 and later recalled [42]:

Boltzmann had no inhibitions whatever about showing his enthusiasm while he spoke, and this naturally carried his listeners along. He was also very fond of introducing remarks of an entirely personal character into his lectures—I particularly remember how in describing the kinetic theory of gases, he told us how much difficulty and opposition he had encountered because he had been convinced of the real existence of atoms, and how he had been attacked from the philosophical side, without always understanding what the philosophers had against him. [...] Lampa was an excellent experimentalist, but as an enthusiastic follower of Mach, was rather sceptical of the modern development of physics.

This comment offers us a hint to look into the opposition between Boltzmann and Mach as seen by their students. Of course, those among them who were less gifted for scientific enquiry were upset or even offended by the way Boltzmann spoke in a rather contemptuous way about the ideas of his colleague. Thus Ludwig Flamm, who married Boltzmann's daughter Elsa, says [43]:

H. Mache confesses that he suffered at that time and for many years afterwards because he had had two teachers with such different scientific views as Boltzmann and Mach. He thinks that if Mach himself had not taught that experimental physicists need not trouble themselves too much about epistemological matters, he would have suffered even more from this discord.

Gabriele Rabel was so upset that she even sent a letter to Mach to find out whether Boltzmann was telling the truth about him: "On the part of a prominent university teacher I have heard you repeatedly characterized—and today once again—as a sensualist or even a psychomonist, as one for whom the world exists only as perceptions, and who views the psyche as the only reality." [44] To calm her down, Mach replied by saying that words can be used and understood in different ways, but he apparently did not succeed in making her feel less disturbed. In fact, almost twenty years later she published both her and Mach's letters to show how Mach's ideas were "misunderstood" by important people.

A less radical view can be found in the following declaration of a Machian student about Boltzmann's views [45]:

It was well known that the atomic theory was only a picture, but he held on to it as a convenient hypothesis. He therefore found himself in opposition to Ostwald, Ernst Mach, and most other physicists, who for the most part are energeticists, which means that they make no special assumptions with respect to type of matter, but try to understand all physical processes merely with the help of concepts of pure energy. Boltzmann is the last [!] great representative of the atomic theory in the physical world.

But it was also possible to compromise between the views of the two most famous Austrian physicists of those days. Here are some reminiscences of Philipp Frank (1884–1966) [46]:

Also, strange as it was, in Vienna the physicists were all followers of Mach *and* followers of Boltzmann. It was not the case that people would hold any antipathy against Boltzmann's theory of atoms because of Mach. And I do not even think that Mach had any antipathy. At least he did not play as important a role as is often thought. I was always interested in the problems, but it never occurred to me that because of the theories of Mach one should not pursue the theories of Boltzmann. [...] It is said that Boltzmann was so desperate about the rejection of atomic theory by physicists, resulting from Mach's attacks on it, that he took his life. As a matter of fact this could hardly be true, since Boltzmann was himself, philosophically speaking, rather a follower of Mach. Boltzmann once said to me, "You see, it doesn't make any difference to me if I say that the atomic model is only a picture. I don't mind this. I don't require that they have absolute real existence. I don't say this. 'An economic description' Mach said. Maybe the atoms are an economic description. This doesn't hurt me very much. From the viewpoint of the physicists this doesn't make a difference." Boltzmann had a philosophical viewpoint which did not require that you believe in the real existence of atoms. And there was not, I would say, any opposition to Boltzmann's physics from the viewpoint of Mach. This opposition existed only, so to speak, in the philosophical realm.

One may suspect however that Frank was rather the exception than the rule. And we do not know how much of his own he put in Boltzmann's words, which sound more like those of scientists of the twentieth century, those scientists who have been labelled as "Boltzmannized Machists" (as opposed to "pure Machists") by Mach's biographer J.T. Blackmore [47].

Before closing this chapter, let us go back to Lise Meitner. Her teacher's suicide persuaded her to move to Berlin, where she had a distinguished career in radioactivity. Later, after the annexation of Austria by Germany, she felt insecure and accepted a position in Stockholm, where she stayed for the next 22 years. Just before she left Berlin, she published, with O. Hahn and F. Strassmann, the first paper that provided preliminary evidence for transuranic elements. A few months later L. Meitner and O.R. Frisch published a letter in *Nature* announcing the discovery of nuclear fission!

12

The influence of Boltzmann's ideas on the science and technology of the twentieth century

12.1 Brownian motion

The influence of Boltzmann's ideas on the physics of the twentieth century is obvious, even if it often appears as interpreted in the writings of different authors, mainly through those of Gibbs, Planck, Einstein, and the Ehrenfests. The reasons for this seemingly singular circumstance have already been partly examined in Chapter 7, with reference to equilibrium statistical mechanics. To the arguments developed there, we can add the fact that in the last years of his life Boltzmann devoted himself more to defending his theories and his viewpoint on theoretical physics than to expounding systematically his discoveries and methods or applying these methods to new areas that were opening up (theory of electrons, black-body radiation, Brownian motion). Even his published lectures [1], and in particular the second volume, are devoted more to demonstrating the usefulness of the basic concepts of the kinetic theory of gases than to underlining the role played by their author in the development of that theory. As we have hinted at above, some applications of which Boltzmann was well aware he never developed. Perhaps the most evident case is that of Brownian motion; Boltzmann mentions this motion in a couple of places, one of them a passage already quoted in Chapter 10, but he does not develop the consequences of some of his remarks such as the following [2]: "...likewise, it is observed that very small particles in a gas execute motions which result from the fact that the pressure on the surface of the particles may fluctuate."

Let us summarize the strange history of Brownian movement. Robert Brown (1773–1858) was a great botanist, but we shall not enter into this aspect here. In 1828 he wrote a booklet entitled "A brief account of microscopical observations made in the months of June, July and August, 1827, on the particles contained in the pollen of plants; and on the general existence of active molecules in organic and inorganic bodies". This booklet was never officially "published" even though it was later reprinted. He describes his examination of strange, irregular motions of pollen particles immersed in water. He became convinced that he was seeing the "molecules" of life and observed various particles of animal and vegetable tissues. All of them showed the same kind of motion.

214

Later he began to suspect that this motion had nothing to do with life, and observed the same phenomenon with particles of any substance, from a piece of window glass to a fragment of the Sphinx.

His suggestion that small particles had a sort of perpetual motion gave rise to suspicion. Many mechanisms were suggested, from illumination to evaporation, from capillarity to heat transfer. Faraday gave a lecture in which he defended Brown; the only inaccuracy was in the use of the word "molecule", which had other meanings. Brown made further experiments to show that the proposed explanations had nothing to do with the phenomenon.

After the initial excitement, interest in Brownian motion disappeared for about thirty years. When the kinetic theory of gases reached a certain stage of development, a connection with molecular motion was suggested. However, Clausius, Maxwell, and Boltzmann are conspicuous by their absence from the debate. Giovanni Cantoni, an Italian physicist, wrote a paper in 1868 in which he claimed that Brownian motion is a "beautiful and direct experimental demonstration of the fundamental principles of the mechanical theory of heat" [3]. In 1870 W. Stanley Jevons, a British expert on political economy and scientific method, claimed that Brownian motion was related to osmosis. The idea of a connection between Brownian motion and molecular motions began to appear in several papers, but nobody attempted a calculation. In 1879 the German botanist Karl Nägeli attempted to disprove this connection, essentially by noting the enormous difference in size between a molecule and a Brownian particle, which would result in movements much slower than those actually observed. The same kind of argument was independently used by the British chemist William Ramsay in 1882. Then people started to invoke coordinated movements, among them the French physicist Léon Gouy, who pointed out that in any case, Brownian motion would violate the Second Law. This led to an important comment by Poincaré in an address at the Congress of St Louis (1904): "If this be so, to see the world return backward, we no longer have need of the infinitely subtle eye of Maxwell's demon; our microscope suffices us." [4]

12.2 Enter Einstein

It was left to Einstein to elaborate the content of this remark and to arrive at a theory which was to constitute the starting point for ascertaining the atomic structure of matter beyond any doubt. Einstein himself mentioned his own astonishment at the fact that this result had not been obtained by Boltzmann, saying in a conversation that "it is puzzling that Boltzmann did not himself draw this most perspicuous consequence, since Boltzmann had laid the foundations for the whole subject" [5].

As a motivation for the fact that Boltzmann did not pursue his remark on Brownian motion, cited above [2], we may quote from Pais's biography of Einstein, who reports the latter's puzzlement about this and comments: "However, it is hard to imagine the embattled Boltzmann evincing the serious yet playful spirit with which Einstein handled the problem of molecular reality." [6]

Let us remark here that Einstein had certainly read Boltzmann's lectures before he started his own work on statistical mechanics. It seems doubtful whether he was aware of Boltzmann's papers; he certainly covered this gap later and as a result gained a high opinion of the Austrian scientist. Concerning the first point we can quote from Einstein's autobiographical sketch [7]:

Not acquainted with the earlier investigations of Boltzmann and Gibbs which had appeared earlier and which actually exhausted the subject, I developed the statistical mechanics and the molecular-kinetic theory of thermodynamics which was based on the former. My major aim in this was to find facts which would guarantee as much as possible the existence of atoms of definite finite size. In the midst of this I discovered that, according to atomistic theory, there would have to be a movement of suspended microscopic particles open to observation, without knowing that observations concerning Brownian motion were already long familiar.

A very early judgement by Einstein on Boltzmann, based presumably on Boltzmann's lectures [1], can be found in a letter of his to his girlfriend, Mileva Marić, whom he later married. The letter was written on a Thursday, presumably 13 September 1900, from Milan and says [8, p.154]:

The Boltzmann is magnificent. I have almost finished it. He is a masterly expounder. I am firmly convinced that the principles of the theory are right, which means that I am convinced that in the case of gases we are really dealing with discrete mass points of definite finite size, which are moving according to certain conditions. Boltzmann very correctly emphasizes that the hypothetical forces between the molecules are not an essential component of the theory, as the whole energy is of the kinetic kind. This is a step forward in the dynamical explanation of physical phenomena.

Some authors [9, 10] have indicated that there was correspondence between Boltzmann and Einstein, but, as stated in Pais's biography of Einstein [6], there is no evidence for this. Boltzmann is mentioned twice more in his letters. In a letter to Marcel Grossmann, dated Friday, presumably 6 September 1901 (evidently in those days mail was delivered so quickly that omitting the date did not matter), one can read: "Lately I have been engrossed in Boltzmann's work on the kinetic theory of gases and these last few days I wrote a short paper myself that provides the keystone in the chain of proofs that he started. [...] I'll probably publish it in the Annalen." [8, p.181]

The same paper, a minor one, is presumably the one referred to in another letter to Mileva Marić [8, p.192], in which Einstein says that he will send Boltzmann the part of the paper that refers to his work. Maybe this sentence suggested to the aforementioned authors [9, 10] that there was correspondence between the two scientists.

We shall say more about Einstein's theory of Brownian motion later.

Concerning the second point (i.e. the fact that Einstein acquired more detailed knowledge of Boltzmann's work and a great admiration for him), it seems sufficient to recall that in an exposition to a wide public of the theory of relativity, a subject not at all related to Boltzmann's work, Einstein writes in the introduction: "I adhered scrupulously to the precept of that brilliant theoretical physicist L. Boltzmann, according to whom matters of elegance ought to be left to the tailor and the cobbler." [11]

12.3 Black-body radiation

Since we spoke of Gibbs, a contemporary of Boltzmann in a strict sense, and his work in detail in Chapter 7, we shall begin a more detailed treatment of influence on later developments by speaking of the influence exerted by Boltzmann on Planck, whose contributions to black-body theory—which were made between the end of the nineteenth and the beginning of the twentieth century—open in a doubly symbolic way the new era in physics that followed the period of Boltzmann's most important activity, i.e. the thirty years which close the nineteenth century. And Planck's work is indeed the connecting link between Boltzmann and quantum mechanics.

After publishing a work in five parts [12–16] on black-body radiation between 1896 and 1899, Planck presented his conclusions in a further paper with the same title, which appeared in 1900 [17]. Initially, impressed by Zermelo's paradox, he writes [12]:

Undoubtedly, kinetic theory has dealt with the task of explaining the trend to thermo-mechanic equilibrium [...] in terms of conservative effects, i.e. as a final result of all collisions between the countless molecules which, conceived of as points, interact through conservative forces. But a more accurate study shows that the molecular motions assumed by the kinetic theory of gases do not possess a time direction in any meaning of this expression, and that, in a completely general way, any state which once existed will occur again in the course of time with such a high frequency as to satisfy any desirable level of approximation. Starting from the standpoint of the kinetic theory of gases, we shall not obtain any rigorous theory of viscosity unless we make use of some additional assumption.

After this declaration of failure of kinetic theory, Planck indicates what are his own hopes of reducing the changes endowed with a time direction to conservative effects: "I believe I must recognize as an oriented process made up of completely conservative effects the influence of a resonator vibrating without friction or resistance on the wave that excites it..." [12] In fact, such a resonator would alter the field, e.g. by absorbing energy from a plane wave and re-emitting it in the form of a spherical wave, or by gradually eliminating the fluctuations, or even by altering the spectrum of frequencies. Further, since the system possesses infinitely many degrees of freedom (since it includes the field), Zermelo's paradox would be avoided.

In the first three parts of his work [12–14], Planck follows this line (which we might call "anti-Boltzmann"). It is not our intention here to examine in detail the way Planck proceeds, which has been carefully analysed in relatively recent times (see e.g. Kuhn's monograph [18]). Hence we shall dwell on those aspects of the development of Planck's thought which appear to be relevant to the theme of the present book.

Four months after the first part of Planck's work [12] was communicated, a short criticism by Boltzmann [19] was presented at the Berlin Academy itself. This paper, after recognizing the great value of the formulae obtained by Planck, remarked how the programme for which they had been developed was doomed to failure. In fact, the reversibility paradox (with which Boltzmann was clearly familiar) retained its full validity. Without invoking suitable (even if highly probable) initial conditions, it was not possible to derive an irreversible equation from reversible ones.

After noting this objection, Planck in his second paper [13] passes over the details and considers it to be the result of a "misunderstanding". In the third paper [14] however, though following the route he had undertaken, Planck gives a clear indication that he has understood the relevance of Boltzmann's criticism, since he ends the paper with these remarks:

Probably it is possible that cases in which [...] the process of radiation is disordered at the beginning but appears ordered at later moments occur as well. In such circumstances the intensity of radiation would be constant to start with, but would undergo notable changes at subsequent times. The possibility that a process of this kind really takes place in nature, or not, depends on the kind of initial state.

Starting in the fourth part of his communication to the Academy [15], Planck's programme changes completely; in fact, it opens with a statement according to which "we must mainly exclude, thanks to a precise preliminary assumption, all the radiation processes which do not exhibit the feature of irreversibility." Without any doubt this is one of those inversions in way of proceeding, or interpreting one's own way of proceeding, about which historians of science can only conjecture, in the absence of explicit admissions. What is certain is that, well before introducing the hypothesis of the quantum of radiation in 1901 [20], Planck had become converted to the ideas of kinetic theory, or, to employ again the terminology we have used previously, he had adopted a "Boltzmann-like" strategy.

In fact, the result at which he arrives (the Wien distribution, which was already known to be in reasonable agreement with experimental data) depends exclusively on an *ad hoc* definition of an entropy having a form patently analogous to Boltzmann's H (with a change of sign) and on the search for its maximum under constraints, as is acknowledged in the fifth paper [16].

At that moment, Planck perceived that he had arrived at an important result, or at least something that looked like an important result. He then prepared a paper for the *Annalen* [17] which started with a long introduction, where the parallelisms between his (new) and Boltzmann's programmes are stressed to the point of saying: "No obstacle blocks the road to a general development of the molecular chaos hypothesis. The possibility of developing the second law in all directions on the basis of the kinetic theory of gases is accordingly ensured." However, Planck was well aware of the weakness of his arguments; further, the experimental data began to show that Wien's energy distribution law was not so accurate as had been thought so far.

A complete and earnest admission of the influence of Boltzmann's criticism on his own work can be found in the talk Planck gave at the ceremony during which he was awarded the Nobel prize [21]:

I had suggested that a resonator could exert a time-oriented, hence irreversible, action on the surrounding field energy, but I thus caused energetic opposition from Boltzmann. With the more mature experience that he had in this kind of question, it did not take him long to show that, according to the laws of classical mechanics, all the phenomena I had considered could take place in both ways, i.e. that the spherical wave emitted by a resonator could inversely return to the same resonator, in the form of spherical surfaces which contract more and more until they are completely adsorbed. The resonator would thus be able to send back into space the energy which it had previously received, along the same direction from which it had arrived. True, I could neglect, in my hypothesis of radiation, notions so singular as that of a time-oriented wave, by introducing a restrictive condition, but it was no less true that all these analyses clearly showed the absence of an essential link in order to arrive at the solution of the problem.

Given these conditions, I was thus left with just one possibility, i.e. to take up the problem again in an opposite direction, I mean to attack it from the standpoint of thermodynamics, a ground where I felt myself more at ease and, so to speak, at home.

He therefore wrote a paper [22] in which, motivated by simplicity and agreement with experimental data, he postulated a simple formula for the second derivative of entropy with respect to internal energy and, using well-established relations, arrived by integration at what is known today as the Planck distribution.

At this point it was possible to derive the expressions for the energy and the entropy of the resonators which Planck had used at length in his previous researches. In particular, the expression of entropy in terms of energy lends itself to combinatorial interpretations of the kind introduced by Boltzmann [23] and discussed in detail in Chapter 6. As Planck himself says [21]: "The question viewed from this standpoint led me to consider the relation between energy and entropy, by taking up again Boltzmann's viewpoint. After a few weeks, which were certainly occupied by the most dogged work of my life, a flash of lightning lit up the dark in which I was struggling."

One must in fact transform the formula just mentioned into something containing only integers. As we saw in Chapter 6, Boltzmann had already subdivided energy into a certain number, say P, of "quanta", denoted by ϵ, thus paving the way to Planck. Let us remark that if the resonators of a given frequency are N and the energy of each of them U, then $NU = P\epsilon$. But whereas Boltzmann thought of ϵ as something then to be allowed to go to zero, Planck discovered that if he wanted to interpret the aforementioned expression of entropy with combinatorial analysis, he had to assume ϵ to be proportional to the frequency ν of the oscillator, thus obtaining the famous formula $\epsilon = h\nu$. Using Stirling's formula in the opposite direction to that in which we used it in Chapter 6, one can derive a probability (which Planck denoted by R), as follows:

$$R = \frac{(N + P - 1)!}{(N - 1)!P!}.$$

The numerical value of this expression equals the number of ways in which P indistinguishable objects can be distributed in P distinguishable boxes.

This is only the most probable reconstruction of his argument, since in ref. [22], in which the quantum of action h makes its first appearance, Planck introduces the hypotheses that he had discovered by working backward, then proceeds to the combinatorial computation, thus finding the expression R. He then hints at the search for the maximum of P with "undoubtedly very prolix" calculations and simply gives the result of his calculation, which had already been introduced on a semi-empirical basis in a previous communication [13].

In his famous paper which appeared in 1901 [20], Planck proceeds more quickly. There we can find a full adherence to Boltzmann's concept:

Entropy accounts for the lack of organization and, according to the electromagnetic theory of radiation, this lack of organization lies, for the monochromatic oscillations of a resonator—even when the latter is located in a steady radiation field—in the irregularity of its continuous changes in amplitude and phase, when one considers a time interval very large with respect to that of an oscillation but very short with respect to that of a measurement. Were the amplitude and phase absolutely constant, i.e. were the oscillations absolutely homogeneous, there could not be any entropy and the oscillation energy would be entirely convertible into work. The energy U of a single oscillator which oscillates in a steady manner should then be considered as but a time average or, what turns out to be the same, as the average—at a given time instant—of the energies of a large number N of identical resonators which are located in the steady radiation field, sufficiently far from each other to avoid mutual influence. It is in this sense that we shall talk of the average energy U of a resonator.

Planck then adopts the relation between entropy and the logarithm of probability, remarking that its simplicity "and its strict connection with an assertion of the kinetic theory of gases pleads its cause a priori." In order to find the expression of probability, "it is necessary to consider U_N [$= NU$] not as a continuous quantity which may be indefinitely subdivided, but as a quantity made up of equal parts, discrete and finite. Let us call an *energy element* such a part, in such a way that

$$U_N = P\epsilon,$$

P being a large integer."

He then introduces the combinatorial computation of the previous paper for N resonators with energy $P\epsilon$, restricting himself to thermodynamic equilibrium conditions and using Wien's general law, which had been proved on thermodynamic grounds (see Chapter 9); this implies as a consequence the relation $\epsilon = h\nu$. If one then applies the well-known thermodynamic relation according to which the partial derivative of entropy with respect to energy equals the inverse of temperature, one immediately obtains again the strange distribution law which Planck had already indicated in his previous papers [22, 24]. Here one finds for the first time in a more or less implicit way the assumption that the resonators are indistinguishable, an assumption that only much later (1924) was to be explicitly taken as an assumption by Satyendra Nath Bose (1894–1974) and was to give birth to Bose–Einstein statistics.

This theory of Planck's was explicitly approved by Boltzmann, as is apparent once again in the address at the award of the Nobel prize [21]: "After so many disillusionments, I had the particularly notable satisfaction to see that Ludwig Boltzmann declared in a letter which he wrote to me after I had sent him my paper that he fully agreed with me, on both the principles and the chain of my deductions."

12.4 Einstein again

Surprising as it may appear to us today, it is a fact that until 1905 Planck's theory received few comments, and when it did, nobody realized that it contained an assumption decidedly in contrast to classical theories. It was Einstein, in a paper of 1906 which came after one of his famous papers of 1905 [25], who first wrote [26]:

Hence we must recognize as a fundamental assumption for Planck's theory of radiation the following: the energy of an elementary resonator can only assume values which are integer multiples of $(R/N)\beta\nu$ [where, in Einstein's notation, $(R/N)\beta = h$]. During the absorption and emission the energy of a resonator varies in a discontinuous way by multiples of $(R/N)\beta\nu$.

When he wrote these sentences, Einstein was presumably one of the most expert scientists in statistical mechanics. As we already remarked, he had certainly read Boltzmann's lectures on gas theory, had independently rediscovered Gibbs's theory of the statistical ensemble [27], was an expert on thermodynamic fluctuations, and was applying the corresponding concepts to Brownian motion [28]. In his first paper of 1905 [25], as is well known, he had introduced the concept of light quanta, called photons only after 1926, and explained the mysterious photoelectric effect. The concept of light quantum was to play a fundamental role in the development of quantum mechanics, but it was practically ignored before 1922 and even explicitly rejected by the vast majority of physicists (including Niels Bohr). When news about the Compton effect (published in 1923) began to disseminate, the concept of a particle that we now call the photon began to be accepted. As is fairly widely known, Einstein got his Nobel prize not for Special or General Relativity, but, as the official commendation says, "for his services to theoretical physics and especially for his discovery of the law of the photoelectric effect".

In a subsequent paper [29], Einstein showed in an even more explicit way that Planck's theory, the only one then available which agreed with experimental data, required a break from classical concepts.

In 1907, Einstein showed that the hypothesis of the energy quantum could also be applied to the theory of specific heats [30], the first indication that quantum concepts could appear in problems not involving electromagnetic radiation. Recall that we have already mentioned in Chapter 4 that Boltzmann was the first to compute the specific heat of a solid body [31] and found a value in complete agreement with the empirical law discovered by Pierre Louis Dulong (1785–1838) and Alexis Thérèse Petit (1791–1820) in 1819. Their rule played an important role in the determination of atomic weights, but it became clear soon that the rule was not always accurate. In particular, in 1833 A. Avogadro [32] remarked that it did not hold for carbon. More accurate subsequent measurements showed that the deviation from Dulong and Petit's law was even more serious than indicated by Avogadro's data. In addition, the data did not agree among themselves. Heinrich Friedrich Weber (1843–1912), who was then in Berlin but later moved to Zurich, where he was also Einstein's teacher, remarked that the experiments were performed at different temperatures and could be reconciled if the specific heat of carbon were to vary with temperature [33]. He confirmed this conjecture experimentally [34, 35], though he had to interrupt his experiments in the good weather because there was no longer any snow for his ice calorimeter. James Dewar (1842–1923) made the same conjecture in the same year as Weber [36], and much later (1905) he measured an extremely low average value for the specific heat of carbon in the range from 20 to 85 K [37].

Let us go back to Boltzmann. He proved that Dulong and Petit's law was a consequence of his equipartition theorem [38]. He said that his theoretical result was in good agreement with experiment "for all simple solids with the exception of carbon, bromine and silicon" [31] (presumably *brom* was a misprint for *bor*). He also speculated that these deviations might be a consequence of a loss of degrees of freedom due to "sticking together" at low temperatures of atoms at neighbouring lattice points. In Chapter 8 we saw that it was clear that there are problems for the specific heats of polyatomic gases.

Einstein's work [30] is remarkable because it was the first in which quantum statistical concepts were applied to something different from thermal radiation. Writing about Planck's formula for the average energy of an oscillator, Einstein asked whether we must modify the theory of periodically vibrating structures different from radiation, and answered [30]:

In my opinion the answer cannot be in doubt. If Planck's theory of radiation goes to the heart of the matter, then we must expect to find contradictions between the present [1907!] kinetic theory and experiment in other areas of thermodynamics as well—contradictions that can be resolved by following this new way. In my opinion, this expectation is actually fulfilled.

Then he introduces a schematic model of a solid in which all the lattice atoms oscillate with a single frequency v about their equilibrium positions. The resulting formula for the specific heats contains a critical temperature $T_E = hv/k$. By comparison with experimental data, T_E is found to be about 1300 K for carbon, contrasted for example with the value of 70 K for lead. This explains why at room temperature carbon behaves in a strange way: it is well below its critical temperature!

More accurate measurements, including those of Dewar, which were unknown to Einstein, and those performed by Nernst [39] in 1911, showed that Einstein's formula was not so good. Better theories along Einstein's line of thought were obtained by Peter Debye (1884–1966) in 1912 [40] and by Max Born (1882–1970) and Theodore von Kármán (1881–1963) in 1912–13 [41, 42].

The rest of the story, i.e. how the second theory of Planck and Sommerfeld's papers paved the way for Bohr–Sommerfeld's theory's is beyond the scope of this book and is more or less known (see e.g. [18] or [43]). Boltzmann's ideas, suitably modified, led also to the new statistics of Bose–Einstein and Fermi–Dirac, which appeared almost simultaneously with wave or quantum mechanics.

Let us now return briefly to Einstein's theory of Brownian motion [28]. He found that the mean velocity of a particle in an interval τ will be inversely proportional to the square root of τ. In other words, the instantaneous velocity is meaningless or random (because of the idealizations of Einstein's model, as he remarked in a subsequent paper). Thus previous experimentalists were trying to measure something that was not well defined. This is the mathematics of the diffusion equation which had been known for at least 50 years. There also follows from a simple argument the formula for the distribution of the vertical distance of particles of density ρ in a liquid of density ρ_0. This was the formula used by Perrin to obtain the Avogadro constant and thus provide spectacular confirmation of Einstein's theory and of the existence of molecular motion.

Why did people before Einstein miss the opportunity to develop his theory? Because they underestimated the role of fluctuations. They thought that the effects of successive random impulses, each of which gives a small velocity to the particle, would cancel out, whereas this is not the case. These impulses produce a random walk which tends, through little random steps, to take the Brownian particle far from its starting point. The randomness is reflected in the fact that the distance travelled by the particle is proportional not to time but to its square root, for sufficiently long times. There is also of course a fairly high maximum speed, dictated by equipartition.

The explanation of Brownian movement was a great success for kinetic theory and convinced people like Ostwald, as we saw in the previous chapter.

12.5 The role of Boltzmann's ideas during the twentieth century

It is of course impossible to review the successes of Boltzmann's ideas during the twentieth century. We shall restrict ourselves to a very short survey of the development of the concepts and problems related to his equation.

It is remarkable that even after the theories of modern physics had been introduced, the Boltzmann equation continued to play a basic role in several areas and not only in gas theory, where it had become a practical instrument for studying the properties of dilute gases.

In 1912 the great mathematician David Hilbert (1862–1943) [44] indicated how to obtain approximate solutions of the Boltzmann equation in the form of a series expansion in powers of a parameter inversely proportional to the gas density. His paper is also reprinted as Chapter XXII of his treatise entitled *Grundzüge einer allgemeinen Theorie der linearen Integralgleichungen*. The motive for this circumstance is clearly stated in the preface to that book ("Recently I have added, to conclude, a new chapter on the kinetic theory of gases. [...]. I recognize in the theory of gases the most splendid application of the theorems concerning integral equations.")

Later, about the same year (1916–17), Sidney Chapman (1888–1970) [45] and David Enskog (1884–1947) [46] independently obtained approximate solutions of the Boltzmann equation, holding for a dilute gas. Their results were identical for practical purposes, but their methods were widely different both in spirit and in detail. Enskog introduced a systematic technique which generalized Hilbert's idea, whereas Chapman simply extended a method used previously by Maxwell to obtain the values of transport coefficients. Enskog's method was then adopted by Chapman and T.G. Cowling in their book *The mathematical theory of non-uniform gases* and thus became known as the Chapman–Enskog method.

Then for many years no essential progress took place in solving the Boltzmann equation. However, the ideas of kinetic theory penetrated into other areas, such as radiation transfer, the theory of ionized gases, and subsequently the theory of neutron transport. As we mentioned in the previous chapter, Lorentz [47] had already considered a particular case of an ionized gas, the electrons in a metal; his theory was later extended to take into account the holes as well, i.e. the carriers of positive charges, and was also applied to semiconductors.

As we have hinted in Chapter 5, the Boltzmann equation is today the object of detailed study within mathematically rigorous theory. As already mentioned in Chapter 5, this development started in 1933 with a paper [48] by Tage Gillis Torsten Carleman (1892–1949), who proved a theorem of global existence and uniqueness for a gas of hard spheres in the so-called space-homogeneous case (i.e. he proved that when the distribution function does not depend on position x but only on velocity ξ and time t, if we assign the distribution at $t = 0$, there is one and only one distribution at subsequent times that coincides with the given distribution at $t = 0$). This theorem was initially proved under the restrictive assumption that the initial data depend on the molecular velocity only through its magnitude. This restriction was removed in a book by the same author, published posthumously and edited by L. Carleson (who completed it in vital points) and O. Frostman [49].

In 1949 Harold Grad (1923–86) published a paper [50], which became famous because it contained a systematic method for solving the Boltzmann equation by expanding the solution into a series of orthogonal polynomials. As we indicated in Chapter 5 however Grad made a more fundamental contribution by introducing in a precise manner the concept of that limit, which is today called Boltzmann-Grad limit.

In the 1950s some significant results were published, such as the exact solutions found independently by C. Truesdell [51] in the United States and by V.S. Galkin [52, 53] in the then Soviet Union, whereas the existence theory was extended by D. Morgenstern [54], who proved a global existence theorem for a gas of Maxwellian molecules in the space homogeneous case. His work was extended by L. Arkeryd [55, 56] in 1972. The techniques used to obtain the most recent results have become rather complicated, even for a synthetic description. On this subject, in addition to the brief hints given in Chapter 5, we refer the interested reader to specialized treatments [57–59]. We cannot refrain from mentioning however that Pierre-Louis Lions obtained the Fields medal (the so-called "Nobel prize for mathematicians") in 1994; the commendation quotes explicitly his work with the late Ronald DiPerna on the existence of solutions of the Boltzmann equation.

The Boltzmann equation is not only a conceptual but also a practical tool. When an aerospace engineer studies the re-entry of a space shuttle, he must take into account the fact that the description of air as a continuum, usually adopted in the design of airplanes, no longer holds in the upper part of atmosphere and he must use the Boltzmann equation. If we want to study the motion of the minutest particles that pollute our atmosphere, we must again, because of the reduced size of these particles, abandon the model of air as a continuum and use the Boltzmann equation.

Engineers use suitable modifications of the same equation to study important phenomena in other fields of modern technology, from the motion of neutrons in nuclear fission reactors to that of charged particles in research on fusion reactors, from the radiation in a combustion chamber to the movement of charge carriers in the very small semiconductor chips for computers.

Boltzmann would have been pleased by these technological applications. He was deeply interested in technology (in particular, he predicted the superiority of the airplane over the airship) and more than once praised the role of technology in the development of science, as one can see from the following two quotations [60, 61]:

However much science prides itself on the ideal character of its goal, looking down somewhat contemptuously on technology and practice, it cannot be denied that it took its rise from a striving for satisfaction of purely practical needs. Besides, the victorious campaign of contemporary natural science would never have been so incomparably brilliant, had not science found in technologists such capable pioneers. [October 1902]

That is why I do not regard technological achievements as unimportant by-products of natural science but as logical proofs. Had we not attained these practical achievements, we should not know how to infer. Only those inferences are correct that lead to practical success. [January 1905]

In spite of the fact that Boltzmann's ideas have flourished and are continually being used for a better understanding of many phenomena, one must recognize that the mainstream of fundamental physics has somehow departed from his views. This is partly due to the very nature of Boltzmann's contributions: in fact his basic results are of permanent value because he was the first to give us the tools to reconcile a microscopic physics, very different from everyday experience, with the macroscopic physics that underlies this experience. But we must also admit that his models are only a first, rough approximation to those used by modern physicists. Today's atoms have shown an embarrassing richness of internal constituents, the elementary particles, which seem to constitute an unbelievably varied zoo. That the atom was not a simple object was a fact clear to Boltzmann himself, as we have seen; the findings of spectroscopy and chemistry were there to show that atoms cannot be simple, indivisible entities. But today we have learned that it is not enough to look for smaller entities. The great idea of Faraday, that the true aspect of reality lies in the field concept, has grown beyond all expectation; after the great successes of the Maxwell equations, the theory of General Relativity has added a large piece to a picture that is still hazy. The field has become space itself, which is not given a priori as a container, but is a dynamic entity that varies in a surprising way. Quantum theory has also evolved into a quantum theory of fields, and although physicists talk and write about elementary particles, they do their calculations with fields obeying strange (non-classical) rules. The Copenhagen interpretation of quantum mechanics, which was the tenet of all theoretical physicists up to thirty years ago, makes them uneasy nowadays and new interpretations are sought for [62]. Perhaps here the study of Boltzmann's philosophical ideas can be fruitful in finding a better understanding of the complex picture that modern physics puts before our eyes. Perhaps we are close to one of those great changes in our view that Boltzmann, long before Kuhn (see Chapter 10), had indicated as a necessary ingredient of scientific development.

EPILOGUE

Having come to the end of this account of the life and work of one of the most tragic figures in the history of science, the author would like to be able to say something more on the reasons that led Ludwig Boltzmann to commit suicide. We have dealt at length in Chapter 1 with the manic-depressive syndrome with which he was affected from 1888. Is this sufficient to explain his desperate act? And, if this is the case, why in that year, in that place? We shall probably never know, although we have given many details in Chapter 1. Here we simply quote the comment of his grandson D. Flamm [1]:

It is the tragedy of Boltzmann's life that he did not experience the glorious victory of his ideas. He left this world while the decisive battle was still going on. During a vacation in Duino, near Trieste, my mother, Elsa Boltzmann, of whom her father used to say that she was the sunshine of his life, found her father hanged. She was only fifteen years old.

If this were a novel or a tragedy, the author could imagine that reading the paper by Einstein [2] on the equivalence between mass and energy convinced Boltzmann of the failure of the efforts of his life to show that atoms were real. Was not one of the tenets of energetics that everything was energy? Had not Boltzmann maintained, as we saw in Chapter 11, that energetics would not be able to explain mass in terms of energy? Perhaps the author of a tragedy could also imagine that, just after Boltzmann had strangled himself, a character would enter the stage carrying a copy of the previous paper by Einstein on Brownian movement, showing that the existence of atoms could be shown experimentally [3].

But this book is not an artist's work, it is a sober exposition that tries to describe sound facts. We can conjecture, but we cannot present conjectures as proved circumstances. Thus we must leave a veil of mystery over the final act of the life of Ludwig Boltzmann.

CHRONOLOGY

1790: Birth of J. Herapath

1791: Birth of M. Faraday

1796: Birth of N. L. S. Carnot

1803: Birth of C. Doppler

1804: Birth of W. Weber

1805: Death of F. Schiller

1811: Birth of R. W. von Bunsen

1814: Birth of J. R. Mayer

1818: Birth of J. P. Joule

1821: Births of J. Loschmidt and H. L. F. von Helmholtz

1822: Births of R. Clausius, K. Krönig, and J. Thomson

1824: Births of G. Kirchhoff and W. Thomson (Lord Kelvin of Largs)

1831: Birth of J. C. Maxwell

1832: Death of N. L. S. Carnot

1835: Births of J. Stefan and A. Bäyer

1837: Births of L. Königsberger and J. D. van der Waals

1838: Births of E. Mach and F. Brentano

1839: Birth of J. W. Gibbs

1842: Births of J. W. Strutt (Lord Rayleigh), L. Sohnke, and C. von Linde

1844: Birth of L. E. Boltzmann

1849: Births of H. Seeliger and F. Klein

1850: Births of Albert von Ettingshausen and A. Pringsheim

1852: W. Thomson (Lord Kelvin) asserts the tendency to the dissipation of mechanical energy

1854: H. von Helmholtz describes the "heat death" of the universe

1853: Births of H. A. Lorentz and W. Ostwald

1854: Death of C. Doppler

1856: Births of W. F. A. von Dyck and S. Freud

1857: Birth of H. Hertz

1858: Clausius introduces the notion of mean free path. Birth of M. Planck

1859: Birth of S.A. Arrhenius. G. Kirchhoff introduces the concepts of emissive and absorptive powers of a body with respect to radiation

1860: J. C. Maxwell introduces the equilibrium distribution that bears his name

1861: Birth of P.-M. Duhem

1862: Birth of D. Hilbert

1864: Birth of W. Nernst

1865: Clausius introduces the notion of entropy

1867: Boltzmann becomes Assistant Professor. Maxwell introduces the transfer equations. Death of M. Faraday

1868: Boltzmann extends Maxwell's distribution to molecules in equilibrium in a conservative force field and is awarded the *venia legendi*. Birth of A. Sommerfeld

1869: Boltzmann obtains the chair of mathematical physics in Graz. Death of J. Herapath

1871: Boltzmann's first attempt to understand the Second Law on a mechanical basis. J. W. Gibbs is appointed Professor of Mathematical Physics at Yale

1872: Boltzmann publishes the equation named after him

1873: Boltzmann publishes a study on the relation between dielectric permittivity and refractive index and accepts a chair in Vienna as a professor of mathematics

1874: W. Thomson (Lord Kelvin) first discusses the reversibility paradox. Births of F. Hasenöhrl, G. Marconi, and A. Schoenberg

1876: Boltzmann marries Henriette von Aigentler and moves back to Graz. J. Loschmidt publishes his paper on the reversibility paradox

1877: Boltzmann publishes a paper discussing the objections related to the reversibility paradox and another paper containing the relation between entropy and probability, now named after him

1878: Boltzmann becomes Dean of the Faculty in Graz. Death of J. R. Mayer. Birth of L. Meitner

1879: Deaths of J. C. Maxwell and K. Krönig. Birth of A. Einstein

1880: Birth of R. Musil

1881: Boltzmann becomes Government Councillor. Birth of G. F. Helm

1882: Birth of M. Born

1884: Boltzmann publishes his theoretical derivation of the Stefan–Boltzmann law as well as a basic paper where he introduces (under different names) the microcanonical and canonical ensembles. Birth of D. Enskog

1885: Death of Boltzmann's mother. Boltzmann becomes a member of the Imperial Academy of Sciences

1887: H. A. Lorentz finds a flaw in Boltzmann's derivation of the H-theorem for polyatomic molecules. Death of G. Kirchhoff. Boltzmann becomes Rector of Graz University

1888: Boltzmann first accepts, then refuses, a chair in Berlin, and shows the first signs of manic-depressive syndrome. Birth of P. Ehrenfest. Death of R. Clausius

1889: Boltzmann becomes Court Councillor. Death of his son Ludwig. Birth of L. Wittgenstein. Death of J. P. Joule

1890: Boltzmann is appointed to a chair of theoretical physics at the University of Munich

1891: Death of W. Weber

1892: Death of J. Thomson. H. A. Lorentz starts to publish his papers on electrodynamics

1893: H. Poincaré first mentions the recurrence paradox as an argument against kinetic theory. Death of J. Stefan

1894: Deaths of H. L. F. von Helmholtz and H. Hertz. Boltzmann becomes PhD *honoris causa* at the University of Oxford

1895: Death of J. Loschmidt. Public debate on energetics between Boltzmann and Ostwald in Lübeck

1896: E. Zermelo publishes his paper discussing the recurrence paradox. Planck publishes his first paper on the thermodynamics of electromagnetic radiation

1897: Boltzmann replies to Zermelo's paper and criticizes Planck's approach to the thermodynamics of radiation

1898: Planck acknowledges Boltzmann's criticism and changes his approach to the thermodynamics of radiant heat

1899: Death of R. W. von Bunsen. Boltzmann makes his first trip to the United States and receives a PhD *honoris causa* at Clark University

1900: Boltzmann accepts an appointment as professor of theoretical physics in Leipzig. Planck introduces the hypothesis of the quantum of radiation

1901: Mach retires. Boltzmann undertakes a cruise in the Mediterranean

1902: Boltzmann goes back to Vienna. J. W. Gibbs publishes his book on statistical mechanics. Birth of Karl Popper

1903: Death of J. W. Gibbs. Boltzmann starts to teach the course on philosophy of science that had been Mach's

1904: Boltzmann participates in a meeting in St Louis

1905: Boltzmann's last trip to the United States (Berkeley). A. Einstein publishes his three famous papers on relativity, quantum theory, and Brownian motion

1906: Boltzmann commits suicide. Einstein first recognizes that Planck's theory on black-body radiation contains an assumption at variance with classical physics

1907: Death of W. Thomson (Lord Kelvin of Largs). Einstein shows that the quantum hypothesis can also be applied to the theory of specific heats

1915: Death of F. Hasenöhrl

1916: Deaths of E. Mach and P.-M. Duhem

1917: Deaths of Adolf Bäyer and F. Brentano

1919: Death of J. W. Strutt (Lord Rayleigh)

1921: Enskog introduces an approximate extension of the Boltzmann equation to the case of dense gases. Death of L. Königsberger

1922: Birth of Thomas Kuhn

1923: Deaths of J. D. van der Waals and G. F. Helm

1924: Death of H. Seeliger

1925: Death of F. Klein

1927: Death of S. A. Arrhenius

1928: Death of H. A. Lorentz

1932: Deaths of Albert von Ettingshausen and W. Ostwald

1933: Death of P. Ehrenfest

1934: Deaths of W. F. A. von Dyck and C. von Linde

1936: Death of Karl Kraus

1937: Deaths of L. Sohnke and G. Marconi

1938: Death of Henriette von Aigentler

1939: Death of S. Freud

1941: Deaths of A. Pringsheim and W. Nernst

1942: Death of R. Musil

1943: Death of D. Hilbert

1945: Death of M. Planck

1947: Death of D. Enskog

1951: Deaths of A. Sommerfeld, A. Schoenberg, and L. Wittgenstein

1955: Death of A. Einstein

1968: Death of L. Meitner

1970: Death of M. Born

1994: Death of Karl Popper

1996: Death of Thomas Kuhn

"A German professor's journey into Eldorado"*

by Ludwig Boltzmann

As I have already visited America several times, as well as Constantinople, Athens, Smyrna and Algiers, many requests have been made that I should publish some of my travel experiences. Everything else seemed too unimportant to write about, but my most recent journey to California was definitely something exquisite, and for this reason I shall venture a light-hearted description of it.

I do not on any account wish to say that people must go to California in order to see interesting and beautiful things and to enjoy themselves. On a tour of the breathtaking mountains of our Fatherland it is possible to feel as much pleasure and joy as any human being can bear. One can be as happy as a king with a quite simple meal, but California is Veuve Clicquot and oysters.

The first part of my journey took place in a hurry and will also be told in a hurry. On June 8th I attended the Thursday meeting of the Vienna Academy of Sciences as usual. As we were leaving, a colleague noticed that I was not turning towards the Bäckerstraße as expected, but towards the Stubenring, and he asked where I was going. To San Francisco, I answered laconically.

In the restaurant of the Nordwestbahnhof I consumed a leisurely meal of tender roast pork, cabbage and potatoes and drank a few glasses of beer. My memory for figures, otherwise tolerably accurate, always lets me down when I am counting beer glasses.

No-one at all well-travelled will be surprised that I talk of food and drink. It is not just an important factor, it is the pivot. The essential thing when travelling is to keep the body healthy when confronted with a multitude of unfamiliar influences, to preserve above all the stomach, especially the fastidious Viennese stomach. No Viennese can eat his last 'Gollasch mit Nockerl' unmoved, and where the Swiss concentrate their homesickness in the memory of 'Kuhreihen' [a dance of the Alpine cowherds], and cowbells, the Viennese think of 'Geselchte mit Knödel' [chimney-smoked pork with bread dumplings]. 'Sagt nicht, das Alter machte mich kindisch, es fand mich eben noch als ein wahres Kind.'

As I finished my meal, my wife and children appeared with my luggage. After farewells I was on my way, first to Leipzig for the joint conference of the Academies,

*Translated by Margaret Malt from L. Boltzmann, *Populäre Schriften*, Barth, Leipzig, 1905, and reprinted with permission from *Annals of Nuclear Energy*, pp. 147–159, 1977, Elsevier Science Ltd, Oxford, England.

which was to begin next day at 10 a.m. I made myself as respectable as possible in the train, (for which I could have done with the 'washing rooms' on American trains!), climbed into a cab as soon as we arrived in Leipzig and reached the first meeting on time.

At the door of the Aula I met my colleague Credner, who was also on his way to the conference and kindly helped me to carry my luggage up to the anteroom, for I had been unable to deposit it anywhere.

I was going to this joint conference in a state of some anxiety: a matter was to be discussed which might prove painful for me.

Will it bore the reader if I guide him for a while around a workshop of scientific activity, to show him the external structure and to explain the mechanism? I hope not. Nowadays almost every educated person, directed by his *Baedeker*, has visited some factory making household goods, and I find the satisfaction of curiosity about how glassware or ironmongery reach their final familiar forms amusing as well as instructive. Why should I not assume that there is the same curiosity about the mechanics of a factory which is, I might say, more important for culture than the largest glass-works, though not, I hope, involving more hot air?

Several German scientific academies and learned societies have banded together to hold annual conferences and discuss matters of general importance. This is the 'Akademie-Kartell'. The conference decided years ago to lend material support to the publication of a massive work, the *Encyclopedia of the Mathematical Sciences*. Mathematics has increased its scope considerably in the last century; besides which, every author uses his own special terms and often writes so obscurely that only his closest colleagues can understand him, and then only with great difficulty. Yet in this complicated, often impenetrable mathematical literature there lies buried a wealth of useful, even indispensable, material awaiting practical application.

The task of the Encyclopedia under discussion is to collect and present all this material in a form both well-ordered and easy to understand. It should make everything achieved in mathematics readily accessible to mathematicians and at the same time bridge the gap between theorist and practitioner. The need (*Bedürfnis*) for such an encyclopedia of mathematical literature is so glaringly obvious, that Professor Klein in Göttingen has described it as a mathematical public convenience (*Bedürfnisanstalt*).

Such an undertaking would not be so very difficult, if it were just a matter of citing the most conspicuous achievements without much criticism and registering the most important and best known developments. To extract everything really useful from obscure writings, weed out the insignificant, achieve the fullest possible coverage of the literature and, at the same time, arrange all the information in a clear and simple manner for the benefit of readers seems an almost terrifying task to anyone who has studied mathematical writings. The aforesaid Professor Klein, however, was attracted by these problems. The Academies are providing money for printing costs, contributors' fees and travelling expenses, Klein and his team are doing the work.

The first step is to seek out the world expert in each specialist field. Indeed, Germans and Frenchmen, Russians and Japanese are working harmoniously together. The chosen expert is often a great man who has enough money and little time, perhaps little desire for the work, but more than his share of obstinacy. He must first be induced to promise a contribution, then instructed and prevailed upon with every means of persuasion to write his article in a manner compatible with the framework of the whole, and 'last but not least' to keep his promise within a time limit.

Many hours are taken up with deliberations over whether to include an article right away because it has actually arrived, even though it would be better placed later in the work, while those which should come first are still unwritten. Unlimited sums are spent in sending Klein and his apostles to all corners of the earth, so that contributors are not spared the pressure of a personal interview. One gap remained for a long time because the man chosen to fill it, a mathematical Russian officer, was locked up in Port Arthur. I have often taken part in these meetings to discuss the Encyclopedia; German dramatists could benefit from the tension and excitement of them.

To return to myself. When Klein approached me to write an article for the Encyclopedia, I refused for a long time. In the end he wrote to me: 'If you do not do it, I shall give it to Zermelo.' This man holds an opinion diametrically opposed to my own. So that Zermelo's theories should not be presented as the fashionable view in the Encyclopedia I wrote by return of post: 'Ehe der Pestalutz es macht, mache ichs'. (All the quotations are in inverted commas; they are taken mostly from Schiller, as a supplementary celebration of the Schiller centenary. Readers can look them up for themselves!)

Now is the time when my contribution is supposed to be ready. I should have preferred to rest in the country in September, to recover from the exertions of travelling, but I have given my word and must therefore rummage about in mathematical literature and prepare the article with a small army of Viennese physicists. 'Ewigkeit geschworenen Eiden.'

Professor Wirtinger seems to have suffered similar treatment, for he sketched as the emblem of the Encyclopedia a mousetrap: attracted by the bait, the Professor is caught.

Where does the irresistible attraction of the whole work lie? There is no particular honour to be gained, except the knowledge of having done something useful; there is no money to be made. How does Klein manage to find the weak point in everyone upon whom he fixes his eye? How does he find the key to overcoming each individual's reluctance with a psychological insight which philosophers might envy him? It is his idealism which gives him this skill, and if we open our eyes, we can see idealism everywhere, as far away as the Pacific Ocean. There two sturdy white towers, the Lick Observatory, greet us, the work of an idealist and a multi-millionaire; but more of this later. I have pondered for a long time on which is more remarkable, that in America millionaires are idealists, or that idealists are millionaires. How fortunate is the country where millionaires think ideally, and idealists can become millionaires! Yet smoked meat and dumplings should be held in high esteem; idealists everywhere need good digestion.

The idealism of Klein and his co-workers bore good fruit. Right after the first part appeared the print-run had to be increased; a French translation has begun, an English one will soon follow. The Academies have scored a success and the bookseller is prospering.

Unfortunately the Berlin Academy of Sciences does not belong to the Kartell and is taking no part in the operation. They were not represented at the Meteorologists' Congress in Southport or the Congress on Sun Research in St. Louis. I am afraid that this principle of participating in nothing which they have not started will damage themselves and Germany more than science. It annoyed me that in Southport and St. Louis the

French were given priority among the 'foreigners' (non-English). We Germans really had no need to take second place to them! But what could I do as an Austrian? If only Hann had been present at the meteorology conference, where he was sorely missed. But he cannot be persuaded to travel!

When I reach my travelogue, I shall give my tongue a completely free rein. At this point I shall not conceal that an American colleague has talked of the decline of Berlin. It is a fact that in the time of Weierstraß, Kronecker, Kummer, Helmholtz and Kirchhoff, American mathematicians and physicists mostly went to Berlin to study, but now they prefer Cambridge or Paris. Because fewer people are gaining a closer knowledge of the Germans, America and, consequently, the world is taking a step backwards. The same colleague also maintained that much would have improved if I had not refused the chair in Berlin. Certainly my lectures would have been the least improvement, but one man can carry tremendous weight in discussions on appointments and the creation of new posts, if he works with Klein's idealism and Klein's assurance. Many who were not available for posts could have been, if they had been approached in the right way. A little wheel which always functions correctly in the correct place can achieve a great deal.

If I stop this long in every town the size of Leipzig, I shall not get very far; 'aber man muß die Einwohner nicht zählen, sondern wägen.' That refers, of course, to their intellectual significance.

After several very enjoyable private dinners and one official one, where I met in person for the first time the Saxon Minister of Education, Seydewitz, under whom I had been a Professor for two years, I travelled on to Bremen and then with a Hohenzollern Prince to New York. That is not to say that I had the honour of accompanying this prince on a voyage to America, but simply that he carried me on his back. It was *Kronprinz Wilhelm* on the way out, and *Kaiser Wilhelm II* on the way back.

Dear Reader, I am in great haste, but I cannot dismiss the sea-passage from Bremen to New York with such a weak joke. The great ocean-going steam ships are among the most admirable of man's inventions and each voyage on them seems more enjoyable than the last. The wonderful rolling seas, each day different and each day more amazing! Today foaming white, raging wildly. Look at that ship there! Now it has been swallowed up by the waves! But no! The hull is rising again, victorious. Tomorrow the weather is calm again, the sea smooth but misty grey; the sky is misty grey, too, the way Melancholy is painted. Then the sun breaks through the haze and yellow and red sparks dance on the waves between the dark expanses in the shadow of clouds; the golden light is wedded to the darkness. And then the whole sky is blue again—the sea, azure and white, sparkles so brightly that I have to shut my eyes. Only on special days does the sea wear its most beautiful ultramarine dress, a colour so deep and yet so brilliant, edged with milk-white foam like lace. I once laughed when I read that an artist had searched many days and nights for a particular colour; now I no longer laugh about it. I wept at the sight of this colour in the ocean; how can a mere colour bring tears to the eyes? Then again, consider the effect of moonlight or phosphorescence on a pitch-dark night! It is necessary to be an artist to convey all these wonders, and even then it would be impossible.

If there is one thing more worthy of admiration than this natural beauty, it is the skill of the men who won the battle with the endless sea begun long before the time of the Phoenicians. How mercilessly the keel slices through the water, how wildly Neptune foams around the churning screw! In truth, the greatest wonder of nature is the skilful brain of man!

If I were asked, like Solon, who was the happiest of mortals, I would name Columbus without hesitation. Not that other discoveries were not equal to his: that of the German, Gutenberg, to name but one. Happiness, however, depends on the effect on the senses, which must have been greatest in the case of Columbus. I can never set foot in America without a certain feeling of envy towards him, or perhaps of rapture because I share a little part of his joy. Obviously Columbus did not make the crossing with the *Kronprinz Wilhelm*, nor did he see New York with his own eyes, but in his mind's eye he may have visualized more than we do, New York one hundred or two hundred years later.

Columbus has become the prototype of discoverers. His 'onwards, onwards towards the West' stands for their perseverance, his 'land, land' for the joy of success, and his whole exploit for the conviction that life itself is not the greatest possession. 'Setzest du nicht das Leben ein, nie kann dir das Höchste gewonnen sein.'

All the senses, not just one's sense of beauty, find complete satisfaction on a sea-voyage. Abundant good food takes care of the palate, a nice little orchestra the hearing. Our Viennese composers often crop up there, obviously not the really great ones, but Strauß, Ivanovici, Waldteufel among others. The waves of the Danube are loudly applauded on the waves of the Atlantic, and if we think of Haydn, Mozart, Schubert or Beethoven, we can say of the Danube as Schiller once said of the Ilm, that 'ihre leiseren Wellen im Vorüberziehen manch unsterbliches Lied erlauscht haben.'

Thus there is nothing more comfortable than life on board ship, particularly for those to whom God has granted freedom from sea-sickness and the ability to observe with an untroubled mind all those people stretched out on deck. The enjoyment reaches its climax when some hydrodynamic accident unexpectedly sprays a wave over the deck and the dozing sunbathers spring up with shrill cries.

Whenever I enter the harbour at New York I am filled with a sort of ecstasy. These towering buildings and the Statue of Liberty with her torch dominating everything. And all the time the ships whistling and singing to each other: one gives a blunt warning, another takes fright, shrieking, one whistles happily, another laments sadly in fourths. The inimitable tones of the sirens echo around as well. If I were a musician, I would compose a symphony: New York Harbour.

On that occasion, however, I had no time for sentimentality. I hired a cab immediately in Hoboken, which was to take me first to the Southern Pacific Railroad office and from there straight to the station, all for 3 dollars. But in the office I learned that the special express, for which I had a reduced-price ticket, only runs twice a week and I would have to wait in New York for two days. I re-directed my cab to the Westminster Hotel and had time to explore New York.

It is not boring in New York either. Just a simple trip on a tram offers plenty of material for amusement and observation. There are no tickets, there are no rules about overcrowding or differing fares. The conductor spies out each new passenger with the eye of a hawk; the passenger presses 5 cents into his hand; one tug at a cord and the payment is receipted by a cash-register in the roof of the tram. Simultaneously a bell rings, heard by all present. If one has a place near the driver, one can marvel at driving skills which must rival the leadership of Napoleon I or Moltke. The crazy speed with which they cover a straight stretch, how they pull up short for automobiles (for these simply overtake the coaches and tramcars in a sharp curve), all this and many other sights can be seen in real life in New York.

On the third day my journey speeded up: I travelled from New York to San Francisco in four days and four nights. The traveller is just catapulted, shot forward, as it were. The bumps and bangs received when walking through the endless train to the restaurant or the observation-car are not very pleasant. The observation-cars are open at the back, and it is possible to sit on the railing or lean out over it, but then one must be careful not to fall off when there is a sharp jolt.

The countryside was admittedly monotonous on the whole, yet the observation of travelling at speed is interesting in itself. If one looks backwards from the observation-car, the railway lines seem like an endless ribbon which is being pulled out from beneath the carriage at a furious rate. It was also interesting to cross the Salt Lake on a giant wooden dam and see the fields before and after the lake covered with salt crystals like snow. Crossing the Sierra Nevada towards the end of the journey was also very beautiful; it reminded me of the Semmering, clearly not quite so picturesque, but much more impressive in length and the height of the mountains.

As a result of the delay in New York I arrived in Berkeley late. The Summer School began on the 26th and I did not arrive until the evening. Since that day had been frittered away with introductory speeches, registration etc., I would not have wasted a single hour if I had started next day at 9 a.m. Unfortunately I was quite unable to manage it. The effect of four days' rattling and bumping came to light at this point I was incapable of walking steadily on firm ground and at night in bed I woke up constantly in terror, because I was not being shaken about and yet dreamt that I was.

Now I must confess that I always suffer a little from stage-fright before the first lecture, particularly here, where I had to speak English. I had fewer opportunities on the journey for English conversation than I had hoped. The Germans who could have spoken English changed back to German after a few words and the genuine Englishmen did not speak at all.

This is the form my English conversation took:

I: *When will lunch be served?*

He: *ieeöö.*

I: *I beg you, could you say me, at what hour lunch will be served?*

His splutterings sink a good fifth deeper: *aoouu.*

I realize the error in my plan of attack and cry despairingly: *lönch, lanch, lonch, launch* and so on. I produce vowels which would never be found in Gutenberg's type-cast. Now his face shows a glimmer of understanding: *ah, loanch?* Now the bridge of comprehension has been built:

I: *When? At what hour? When o'clock?*

He: *Half-past one!*

We have understood each other. And now I was supposed to give thirty lectures in this language! I therefore declared myself unfit for work on Tuesday June 27th and began on Wednesday. During the first lecture I was somewhat timid, but by the second I was more relaxed and when I finally heard that the students could understand me well, indeed found my presentation lucid and distinct, I soon felt at home.

At this point I cannot but express my thanks for this success to my English teacher in Vienna, Miss May O'Callaghan. Without her tireless efforts to assist my protesting tongue, I should never have managed it. How proudly I used the words 'blackboard' and 'chalk' as if they came naturally, when I needed to obtain some usable chalk and a large enough blackboard! How well I succeeded in pronouncing 'algebra', 'differential calculus', 'chemistry', 'natural philosophy' etc.!

Thanks to my industry I also obtained an excellent lobster salad. On the menu it said 'lobster salad'. At once I remembered the lesson when I could scarcely believe that 'Hummer' was translated 'lobster'; so I ate the lobster and it tasted delicious.

The University of [California at] Berkeley, where I was working, is the most beautiful place imaginable. A park a kilometre square, with trees which must be centuries old, or is it millennia? Who can tell at a moment's notice! In the park there are splendid modern buildings, obviously far too small already; new ones are under construction, however, since both space and money are available.

A certain philosophical aura surrounds the university. Berkeley is the name of a distinguished English philosopher, who is even credited with the discovery of the greatest folly ever produced by the human brain, namely philosophical idealism, which denies the existence of the material world; idealism in a quite different sense from the way I use it. At Berkeley philosophy has its own teaching structure; not a system built of phrases and fantasies—sorry, I mean to say logical conclusions and rational concepts—but a real structure of stone and wood where the psyche is explored with tuning-forks, spectra, kymographs and recording cylinders.

Another building was more important to me. A speculative landlord had read in an encyclopedia that Berkeley was an English bishop, whose residence was called Cloyne Court, so he built a hostel for professors which he called Cloyne Court; this is where I stayed. This man attached little importance to any outward resemblance to an English Bishop's palace of an earlier age; Cloyne Court stood in Euclid Avenue and was a precise parallelepiped without a trace of anything non-Euclidean. The interior, however, was comfortable. I had a small bedroom, a slightly larger study and a bathroom, all with electric light. Hot water could be made to circulate around the rooms through thick pipes, to give reasonable warmth; this was often welcome in July, on a latitude south of Palermo, because the wind blowing from the Pacific was sometimes icy. On the other hand, winter in Berkeley is only slightly colder than summer; there is heavy rainfall in winter, none in summer.

The food was good. At least, it was usually possible to force down one of the dishes offered. There was no printed menu-card. Before each meal the menu was drawled out by the waitresses, most of whom wore glasses; it sounded more like a monotonous song delivered in an undertone.

'Doch des Lebens ungemischte Freude ward keinem Irdischen zuteil', not even on a trip to Berkeley. First my stomach shall have its say. Until I came there I had drunk neither from open bottles of water, nor from the sealed bottles filled with water and carbonic acid, and thus kept my stomach healthy despite the unfamiliar food. But Berkeley is teetotal: to drink or retail beer and wine is strictly forbidden. As I did not want to die of thirst I tried the water, without ice—perhaps it might be healthier in Berkeley than in New York and St. Louis. Unfortunately not! My stomach rebelled and after once having to keep my clothes on all night, to avoid dressing and undressing too often, I ventured to ask a colleague about the location of a wine-merchant. The effect my question produced reminded me of a scene in the smoking-car of a train between Sacramento and Oakland. An Indian had joined us, who asked quite naïvely for the address of a... well, as he was an Indian, let's say the address of a house with bayadères in San Francisco. Most of the people in the smoker were from San Francisco and there are certainly girls there with the motto: 'Give me money, I give you honey', but everyone was startled and embarrassed. My colleague reacted in exactly the same way when I asked about a wine-merchant. He looked about anxiously in case someone was listening, sized me up to see if he could really trust me and eventually came out with the name of an excellent shop selling Californian wine in Oakland. I managed to smuggle in a whole battery of wine bottles and from then on the road to Oakland became very familiar. My stomach also said amen to this and recovered amazingly quickly, although my diet remained otherwise unchanged. I always had to drink my glass of wine surreptitiously after meals, so that I myself almost felt that I was addicted to a vice. Temperance is well on the way to creating a new kind of hypocrisy, of which there are already quite enough in the world.

My stomach was scarcely pacified before other ailments set in. Previously my troublesome asthma had always disappeared as soon as I set foot on the steamer and stayed away until I returned to European soil. The same thing happened this time, until I reached California: the much-praised moist coolness of the climate brought out the unwelcome guest, asthma, to plague me.

Next I developed a boil under my arm (as a result of putting on a new shirt unwashed, I think) I had to have it lanced in the Roosevelt Hospital. It was extremely interesting to get to know so thoroughly an American hospital which rivalled the *Kaiser Wilhelm II* in elegance (I mean, of course, the steam' ship) but cost 35 dollars. This was the most expensive luxury I allowed myself during the whole trip, and thus robbed myself of a less doubtful pleasure. Tuesday July 4th was Independence Day, the greatest American festival, and as I had no lectures on Saturday or Sunday I needed only to leave out Monday's lectures, or make them up later, in order to have four days free for a quick visit to the Yosemite Valley. This had to be abandoned, but instead on Sunday July 2nd I did at least hear the 'half hour of music' which was performed every Sunday free in the Greek Theatre. This is a replica of the Sophoclean theatre in Athens, except, it seemed to me, that it had been somewhat enlarged. As it never rains in the summer in Berkeley and there is little sunshine because of the constant mist, the open-air theatre does a good trade. The music, however, was consistently weak in this architecturally beautiful place, surrounded by eucalyptus and 'lifeoaks'. It would have been a fine setting for Mahler and the Philharmonic Orchestra playing the Third Symphony in such a way that the trees would have trembled in ecstasy, and the Pacific Ocean, pricking up its ears, would have become even calmer; the people would not have understood it, though.

On Tuesday from the roof of Cloyne Court I watched the superb fireworks which were let off, as every year, to celebrate Independence Day. Cloyne Court is situated on high ground and affords a view of the Bay of San Francisco, the Golden Gate, Mount

Tamalpais etc.; the English bishop can hardly have had a more beautiful view from his Cloyne Court.

God Himself seemed to take pleasure in the celebrations, for at sundown He took the lead with a firework display worthy of His own greatness and His creation. As so often before on my American tour, I wished that I could paint!

As the last streaks of sunset disappeared and the lights of San Francisco greeted each other brightly across the bay, the earthly fireworks began. Sometimes a brilliant light flared at our feet, sometimes a glowing star shone forth on the horizon. Which way should I look? Berkeley and San Francisco are ablaze with fireworks; over in Oakland there is another attractive display. As soon as I look that way, I have missed an even better one in Alameda. I made up my mind to hold a small firework display in my garden every year on July 4th. After all, historically, the struggle of Washington and his followers has not only local but world significance.

Schiller once said, 'Noch ein paar tausend solche Kerle wie ich, und aus Deutschland soll eine Republik werden, gegen die Rom und Sparta Nonnenklöster waren'. Clearly that did not come to pass. A few thousand fellows like you? The world has not produced even one. But ideas do not die. The republic, beside which Rome and Sparta were nunneries, does exist beyond the ocean, and how colossal it is and how it grows! 'Die Freiheit brütet Kolosse aus.'

Subsequently I was invited somewhere or other every Saturday and Sunday. The first invitation was to Mrs Hearst's splendid estate near Livermore. Who is Mrs Hearst? It is not easy to explain to a European. The nearest to the truth would be that she is the University of [California at] Berkeley. In Europe the Alma Mater is a classical idealized figure, in America she is a real person, and what is most important, has real millions of dollars, some of which she donates every year for the expansion of the university; my American visit was paid for with her money, naturally. The President of the University (what we should call the Rector, but he has the post for life) is only the executive agent of the Trustees, at whose head stands Mrs Hearst. The present incumbent had to stipulate all kinds of rights and privileges on taking up the appointment, so that he could at least do something for the university on his own authority.

Even worse—yet how can I say 'worse'?—how could anyone who, like me, owes many many happy hours to Mrs Hearst's hospitality, maintain that such an Alma Mater might be something bad?—So—even harsher are the conditions at the Leland Stanford Junior University, to which I made a one-day visit.

Mr Leland Stanford senior was in charge of the construction of the first Pacific railroad, that is, the first railway which established a continuous rail-link between the Atlantic and the Pacific. He was also influential in Congress: he knew how to present the importance of the undertaking in the right light and induced Congress to contribute half the building costs, reserving certain sovereign rights for himself, but leaving the income mainly to the contractors. Then Mr Stanford founded a company with a completely different name, which was to supply all the materials and labour to the railroad company. Since he was the head of both companies he easily managed it so that one bought from the other at doubled prices, and the State had to pay nominally half, in practice the whole cost, while he received all the income.

Just when he had become enormously rich a sudden misfortune snatched away his only son, for whom all his wealth was intended. He and, more particularly, Mrs Stanford fell into a kind of religious mania. In Europe when an old lady goes slightly crazy, she buys herself a dozen cats or a parrot; here she summoned a first-class master-builder (what can you not obtain for money?) and built a university which will surely remain a blessing for future generations.

While Berkeley is constructed as a series of pavilions, Stanford has been laid out according to a unified, architecturally attractive plan which nevertheless seems unsuitable for teaching purposes. Architects are the same everywhere. The University Church is especially magnificent, richly decorated with wall and ceiling murals, stained glass and sculptures. The organ, which was played in my presence, sounded so wonderful that I willingly became a devout listener to such music. Then she died, too, but not before providing generously for the university in her will.

A Professor of Political Economy at Stanford is once supposed to have inveighed against the founder's fraud; the President thought he would do himself some good in Mrs Stanford's eyes by dismissing this professor on the spot. Mrs Stanford, however, was so magnanimous that she recalled him and reprimanded the over-zealous President.

It goes without saying that in such universities male and female students and staff have equal rights, and I shall quote one drastic example to prove the far-reaching power of the female element. One of my teaching colleagues—I remember her name: Miss Lilian Seraphine Hyde, rather a nice lady—gave a two-hour lecture on the preparation of salads and desserts which was announced just like mine. I can still show you the lecture-schedule today.

All the rooms in the university are crowded with women; there must be almost as many female students as male. It is particularly noticeable that there is a lady's hat deposited in every room. In the professors' room—a lady's hat; in the room which serves as a washroom, telephone booth and more besides—a lady's hat; in the dark room—a lady's hat; indeed, as I left hospital after the operation already mentioned, feeling somewhat weak and confused, in my absent-mindedness I nearly put on a lady's hat instead of my own.

Now let us return to Mrs Hearst, the Alma Mater Berkleyensis. She had, as already stated, invited me and a number of other professors lecturing at the Summer School to her estate near Livermore, this is a jewel which only luxury, wealth and good taste know how to create in such a lavishly equipped manner. The carriages met us at the railway station and soon we passed through a distinctly fanciful but not unattractive entrance-gate into a park filled with magnificent trees and beautiful flowers. Wealth here is converted into water and where this is not spared, in California a show of flowers springs up which bloom in both summer and winter. Slowly, though all too quickly for me, we crossed the park, which also afforded the most beautiful views of Mount Diable and Mount Hamilton. Eventually we reached the house. It is built in the Portuguese-Mexican style, a circle of buildings around a courtyard sealed off with heavy iron gates; obviously a kind of fortress. The centre-piece of the courtyard is an antique marble fountain which our hostess bought in Verona and had transferred to the Pacific. The whole estate is named after it; 'Hacienda del pozzo di Verona'.

One of my companions in the carriage explained that the proprietress had summoned a German architect called Schweinfurt, who created this place after studying all the old

Spanish and Portuguese buildings in Mexico. I remarked, 'He must have had very good taste', to which my companion replied, 'Yes, and he died as a result of his good taste,' 'How did that happen?' I asked. He answered, 'Californian wines tasted too good to him and he drank on until he died.' these Californians have a terrible idea of their admittedly very strong wines. It was not so very sad after all. One day I, too, shall die and then stop drinking, thus I shall also go on tippling until I die.

The interior of the hacienda is a treasure-chest full of the most superb works of art and curios, which the owner has collected from every corner of the Old and New World: a most original mixture of Greek, Roman, medieval, Mexican, Chinese, Japanese and Indian curiosities.

At table I sat on Mrs Hearst's right, as I was the only European present. The first course was blackberries. I declined them. There followed a melon which my hostess had most appetizingly salted for me with her own hands. I declined again. Then came oatmeal, an indescribable paste on which people might fatten geese in Vienna—then again, perhaps not, since I doubt whether Viennese geese would be willing to eat it. I had already noticed the displeased look of the Alma Mater when I refused the melon. Even an Alma Mater is proud of her cooking. Therefore I retched with my head turned away and, thank God, I was not actually sick. That is the unpleasant thing about accepting invitations in America. In hotels one may leave what one cannot cat, but what can one do when faced with a housewife who is proud of the high quality of American cooking in general and her own in particular? Fortunately poultry, compôte and various other things followed with which I could cover the taste of oatmeal.

After dinner we went into the music room which is, I would estimate, about as large as the Bösendorfer Saal, but what fantastic-baroque decoration! In terms of beauty I could not compare any of the smaller Viennese concert halls with this. Tidings of my less than adequate piano-playing had reached the hacienda. I was called upon to open the concert. After some resistance I seated myself at the grand piano, a Steinway from the most expensive price-range. Without misgivings I struck the keys; I had perhaps heard a piano with such beautiful tone in a concert, but my fingers had never touched one. If the hardships which beset my Californian visit had ever made me regret it, from then on they ceased to do so. I played a Schubert sonata; obviously at first I found the mechanics strange, but how quickly one becomes accustomed to good things. The second part of the first movement went well and in the second movement, an Andante, I forgot myself completely: I was not playing the melody, it was guiding my fingers. I had to hold myself back forcibly from playing the Allegro as well, which was fortunate because there my technique would have faltered. A pupil of Barth in Berlin played next, with as much technique as understanding of the music. Among those present was a Professor of Music from Milwaukee, a martially masculine figure, certainly a splendid bear-hunter but also with a good musical education. He, too, had practised music with Barth, one cannot say that he had studied. He knew that Beethoven wrote nine symphonies and that the ninth is the last. He paid me an undeserved compliment: in the course of a debate over whether music can also be humorous, he requested me to play the Scherzo from the Ninth Symphony. Should I admit in front of a Professor from Milwaukee that I cannot? I became humorous and said, 'Willingly, but I entreat you to play the kettle-drum as it works better if someone else joins in.' After that he was silent.

At night in the hacienda I slept in a wonderful bedroom with adjoining bathroom and my own personal Negro to look after me, who also polished the shoes. A guardian angel of perfect beauty hung right above my bed. I am especially fond of art when it provides food for thought. What use to me in the home is a picture of the Battle of Abukir, no matter how splendid? But a guardian angel at the head of the bed expresses to a certain extent the host's wish that I should sleep well in his house. Yes, I am superstitious. Just at that time I was suffering from asthma and had already thought twice about going to the hacienda. This guardian angel comforted me and the asthma's hold was definitely broken after that night.

On the following day there was no end to the inspection of interesting objects in the house and courtyard, woods and fields. Among other things we drove to see huge 'life oaks' [California live oak, *Quercus agrifolia*, an evergreen oak] with enormous outspread boughs. A whole cabin had been built on the branches of one of them—a kind of Hunding hut on the first floor. I did not return until the afternoon and reached Berkeley late in the evening, ready to appear punctually in the lecture-halls next day.

The following Saturday and Sunday were devoted to the Lick Observatory. In fact I travelled on Friday afternoon to the friendly little town of San José, where many of the streets are lined with palms. People do not just stroll under palms there, they travel by tram, bicycle and car under them. Next day at 7 a.m. I stepped into the somewhat dilapidated mail-coach for the drive up Mount Hamilton, which rises about as high above sea-level as the Semmering, but seems higher because the journey begins at a point only just above sea-level. The road is very good and zigzags slowly and steadily upwards among vineyards and orchards, woods and meadows. The latter are covered with hay at this time of year. The cows eat hay there in the summer and fresh grass in the winter.

My coachman, a gnarled, grumpy old man, is also the postman. Immediately after leaving the hotel he sorts out, with much muttering, the post which lies in sacks at our feet. Soon we have left the town. Outside the gate of a large fenced property a lovely dog greets us with eager barks. My coachman puts some letters into a bundle of newspapers and skilfully throws them into the mongrel's mouth; the dog slips away under the fence with them at once. This kind of postal service is repeated at several entrance-gates. At others, a wooden pole with a large nail in it has been erected. Without stopping the carriage my coachman deftly removes the items to be posted and puts the newly arrived post in their place. Only twice, when there were baskets of groceries or other large packages to be handed over, were maids waiting, whose nationality I am not anthropologically qualified to decide. We changed horses twice and ate lunch once (but please do not ask me what the lunch was like!).

We arrived high up at the Observatory at about half-past one. Only the younger astronomers were present, with Dr Tucker in charge, because the Director, Campbell, and the senior ones were already in Spain preparing for the total eclipse of the sun. As at this time I still had a mind to see this for myself, I asked Dr Tucker where the eclipse was to be observed. 'In Daroca-Ateca-Almazan', he answered. I was somewhat startled and said involuntarily that it was 'ein spanisches Dorf' [i.e. all Greek to me]. He replied calmly that it was indeed a Spanish village north-east of Madrid. The name did not make a very favourable impression on me; Spain began to seem somewhat foreign to me.

They showed me all the resources of this magnificently equipped observatory, which can be used most fruitfully on account of its favourable position. The most remarkable thing is the giant 28-inch lens polished by Alvan Clark—they just call it 'the big glass'—with which one of the most interesting astronomical discoveries of recent times was made: that of Mars' twin moons. Inside the giant pier which supports the telescope is buried Citizen Lick, who had the whole observatory built with his private means. Is that not idealistic? I can read his mind. He certainly knew that it would be a matter of indifference to him where his bones rested, but he wanted to give the world a striking indication of what a millionaire's final goal should be. I declare, he has bought himself immortality with his wealth.

If I were a poet, I should like to describe the meeting of Schiller and Lick in Heaven under the title 'Two Idealists'. Schiller makes Wisdom say to Wealth: 'I do not need you'. Lick proves the opposite. Obviously the thrill which one gets from money is only a second-rate joy; love obtained through money is not even a third-rate joy; nevertheless, a Steinway piano, an Amati violin, a Böcklin and now immortality as well can be obtained with money.

At this point I wish to tell another story which relates to idealism and money-making. The great American physicist, Rowland, had made a speech saying that scholars should not strive to make money. A year later he felt ill, was examined by a doctor and learned that he had at most three years to live. He had a wife and four children as yet unprovided for. In the conflict of loyalties, love of his family won. He invented a tele-printer, patented it and amassed a 200,000 dollar fortune for his widow; he really did die a short while afterwards. He gives the lie to his own principle, however. Gentle Reader, do you know what I admire most about Rowland? That he had such a lucrative invention in readiness. May he also shake hands with Schiller in Heaven!

American merchants are decidedly realistic when it comes to making money. So it was that a very intelligent businessman, to whom I explained the purpose of my visit, simply could not grasp why I had travelled as far as San Francisco when my income only covered my travelling expenses.

I had stood and day-dreamed in front of the pier which supports the telescope and guards Mr Lick's bones in its base. Then we continued through all the sections of the observatory. The very best use is made of a wealth of material here. Each region of the sky has its own cabinet, and in these each star has its own drawer so that all observations can be traced quickly and used. The contents of the store grow very quickly. It is hardly surprising that time passes quickly for the astronomers in this mountain retreat. It goes without saying that there are pretty lady-astronomers there as well.

In the evening, after I had observed Mars through the great telescope, large and bright, almost like the face of the moon, I made my way back into the valley. The noteworthy thing about the journey was the demarcation line of the fog. We had the stars above us and the mist below like a smooth sea. All at once the coach entered the fog, the stars disappeared and the light of the coach-lamps penetrated only a short distance ahead.

The next day (Sunday) was taken up with a tour of San José, but I returned in the afternoon to Berkeley in order to have plenty of time for sleep and the preparation of lectures.

My next Sunday excursion was no less interesting: a visit to the seaside resorts of Monterey, Pacific-Grove and Santa Cruz. I had already driven out this far from San Francisco to enjoy the view of the great ocean. This time I had the opportunity to admire at length the cliff-edged shores and dancing waves of the Pacific. But more than all this, the thing which interested me most was a little house in Pacific Grove where Professor Löb has his laboratory.

How great a difference there is between the giant factories of industry and the workshops of science! What impressive colossi the ocean-going steamers are! Once one has made several voyages, it becomes clear that the officers, engineers and sailors always do the same jobs. In the passenger quarters the same people talk about the same things, stretch out on the same deck-chairs, all aim with the same quoits at the same targets on the upper deck. Huge masses, but not a new thought to be found! I agree that in science many things have been achieved through mass development (we have seen it in the Lick Observatory); but truly great achievements (clearly our Minister of Education ought not to hear this) are always produced with the smallest means.

It must be a great achievement to control the expenditure of millions for the good of a great nation, to win battles at the head of hundreds of thousands of people. To me, however, it seems even more important to discover truths in a modest room with modest funds, truths which will remain the foundation of our knowledge when the memory of those battles will only be preserved laboriously in history books. What is it from the whole of Greek and Roman culture which is still alive today and has even more use and influence than before? The warriors of Marathon have been surpassed at Vionville and Liaoyang. People who read Homer or Sophocles for pleasure are dying out, but Pythagoras' theorem and Archimedes' principle are truly immortal.

These are my opinions in general; how far they can be applied to Pacific Grove in particular will only be shown by future developments. The discovery made there caused me great embarrassment some years ago when it was still new. Full of ardour I explained it in company, never dreaming that something so factual, not intended to conjure up lascivious feelings, indeed incapable of doing so, could be considered improper. The first indication was the sudden, somewhat conspicuous exit of the lady sitting next to me at table. Later this same lady sang a very dubious song by Aletter. I could not refrain from saying that I was surprised that this song was considered proper, but my subject was not. 'Yes', the lady said, we do not understand your subject' and I answered involuntarily, 'But you understand Aletter'. That is one of our old hypocrisies, to which the teetotallers want to add another. I shall have to choose my words carefully in order to make clear the object of Löb's research without giving offence.

For a long time it was believed that all chemical compounds found in living organisms, the so-called organic compounds, could only be generated by means of a special force: vital force. Today it is known that numerous organic compounds can be synthesized from their chemical elements through ordinary chemical reactions, without a sign of vital force. Many people believe, however, that life itself is something special, totally separate from its accompanying chemical processes, and that life's special workings can never be brought about by the inanimate. This opinion is certainly far from being disproved by Löb's research, but a new weight has been added to the scales to its disadvantage.

It is well known that there are animal genera of which the female egg can develop without fertilization in certain circumstances (parthenogenesis). Löb was now working with animal genera where this never happens sea-urchins and starfish—and showed that the reactions usually produced in their eggs by male sperm can also be brought about by completely nonliving acids, so that eggs subjected to the effects of carbonic acid, butyric acid or valeric acid can, in appropriate conditions, develop in exactly the same way as normally fertilized eggs.

It is easy to see the importance of the discovery that a process which was previously seen as the consequence of a specific vital action can also take place through purely chemical reagents. If this is true not only of sea-urchins but also of higher forms of life, even human, what social revolutions will result! Women's emancipation such as the suffragettes of today have never even dreamed of. Men will simply become superfluous; a little bottle filled with carefully mixed chemicals will replace them completely. It will be possible to organize heredity on a far more rational basis than now, when it is subject to such chance. It will not be long before someone discovers which mixture produces boys, which girls, and since the former will be totally unnecessary, only a few examples will be reared for zoos. Clearly by that time wine will be superfluous too.

From Pacific Grove I drove to the resort of Santa Cruz, which is especially remarkable for the fact that a large number of its visitors do not stay in houses, but in canvas tents with tiny linen windows which are leased like holiday cottages. Others stay in small wooden cabins erected on boats, which can be rowed about in the shallow bays and river estuaries.

Everywhere I was struck by the small size of the houses, almost all of which are made of wood. In Berkeley there are many houses which remind me of those around Graz belonging to the Schnitzelbauer [smallholders skilled in carpentry] which they have constructed out of wooden boards on their small plots of land and decorated with carvings.

On another occasion I had a look at the so-called 'big-trees', which are thousands of years old. A dead tree-stump is displayed from whose rings one can tell exactly how many years old it is. I have forgotten the answer.

The other days—excluding Saturday and Sunday—were devoted to work, but not entirely devoid of entertainment. There were several social gatherings, including a couple of very formal ones. In one case a colleague who was fetching me had warned me in advance to wear evening dress, showing a truly English concern for propriety. As he entered I called out to him, 'Don't I look splendid?' But no, alas! I had forgotten to polish my shoes. My colleague knew exactly what to do. He led me down to the basement, took off his jacket, waistcoat and cuffs, found the necessary equipment in a cupboard and polished my shoes most skilfully. Afterwards he drank out of the glass from which he had previously dripped water on to the wax. American!

There was no lack of women either in the social circle. The wives of Berkeley professors joined the visiting lecturers, as well as the wife of the proprietor [at Cloyne Court] with her delightful daughters, one of whom sang charmingly, and other female friends. I was often in such company and was struck down by another disease, as yet unmentioned: versititis. As I have already described my other ailments, I shall also try to give an idea of this one; there follows one of its products:

An meine Frau
[To my wife]

Sol ich mit fremden Fraun in der Fern mich nicht unterhalten?
Sind sie von Allem, was hier, dir denn am ähnlichsten nicht?
Oft schon küßt' ich dein Bild auf Pappe: Oh so verzeihe,
Wenn ich dein Bild auch geküßt, fand ichs in Fleisch und in Blut!
Zudem bin Theoretiker ich von der Zeh' bis zum Scheitel
Und so vertrau daß ich auch nur theoretisch geküßt.

[Shall I not amuse myself with strange women in distant parts? Are they not, among all things here, the most similar to you? I have kissed your picture on cardboard often. So forgive me, although I kissed your picture, I found it in flesh and blood as well! All the same, I am a theoretician from head to toe and so trust me to have kissed only in theory.]

So, apologies on all sides, also because of the kissing. It was no more than a poetic necessity. I should like to see the man who could produce a poem with nothing but companionable walks, conversation, tennis and music-making!

The women in California are strikingly large and strongly built, and as the growth of their beards often leaves little to be desired, I had to agree with a colleague when he said, 'Don't you find that American women are somewhat masculine?' On the other hand he refused to concur when I replied, 'And the men somewhat feminine.' The latter is true only with regard to their beardlessness; they show their masculinity in powerful wills, courage, the spirit of enterprise and strength of character.

Among the events which brought variety into the weekdays was the visit of the American Secretary of War, who was going to the Philippines via San Francisco. Miss Roosevelt is supposed to have been in his entourage, but I did not catch sight of her. The Secretary of War visited a public meeting in the great live oak plantation belonging to the University at Berkeley. You should have heard the naïve bluntness, eagerness and enthusiasm of the speeches! Just one example: the President of Berkeley introduced the Secretary of War after a short address with the following words, 'This is Mr Taft! A good Secretary of War, a good citizen and above all in every way "ein guter, alter Kerl".' In English it sounds even more familiar: a good old fellow.

Yes, America will achieve great things still, I believe in this nation, even though I saw it engaged in an occupation at which it does not excel: integrating and differentiating in a theoretical physics seminar. There they behaved more or less as I do when jumping over ditches and toiling over hills, of which there are so many on the Berkeley campus.

Finally the evening came when I listened to the sing-song voices of the bespectacled waitresses for the last time. As I finished cutting up the last omelette, a colleague near me examined the pieces with a hawklike eye and pointed out that there was half a minute left for each piece. Then the railroad seized me and dragged me away, first to Portland (two nights on the train). Although an exhibition there tempted me, I travelled straight on to Livingstone (another two nights on the train). The journey was beautiful, if only it had been day all the time! The most magnificent sight was Mount Shasta, with its high snow-capped peak rising out of sub-tropical vegetation. We passed many lakes surrounded by mountains, ringed by forests, which make the Gmundnersee and Attersee seem insignificant. There are no houses on their shores; I do not even know whether all of them have names. I shall say nothing about Yellowstone Park. It is a marvel and I doubt that its equal exists anywhere. Read about it in *Baedeker*, or look at good pictures of it or best of all, see it in reality, if you have sufficient time, money and good humour. Do not do what I did. Go at the beginning of June when the heat is not so intense, and devote a fortnight or, even better, a month to it, in order to see everything at leisure and progress from amazement to enjoyment.

I had overburdened myself with too many good things to do. Now I was to spend another four nights on the train and my ability to enjoy was exhausted like my supply of clean clothes. In addition, there was the appalling heat to contend with. I continually clutched a towel to wipe away the perspiration; fortunately plenty of hand-towels are always available on American trains. Now I understand what a sweat-rag is. On top of all this, the Americans like to seal their railway carriages hermetically, not from fear of draughts, which they do not know, but to avoid soot. On this line there are no pleasant observation-cars right at the back where there is less soot. I once left the window open for some time in my compartment, which was further forward, but became so black that I should not be surprised if in the next century some scholar postulated the hypothesis that Negroes turned so black because they were always employed as railway personnel.

My stomach troubles returned at this point. It is possible to get wine in the restaurant-car, but grudgingly, only after the meal when most of the diners—especially the ladies—have left. The first items brought are a glass of iced water and a slip of paper, on which everything required must be written down at once. It takes an eternity before the note is removed and then one has to sit in front of the iced water with a parched throat. I succumbed to temptation (angels would have fallen) and drank some of the poison.

Then suddenly I could not get wine at all. This was explained as follows: the whole State of North Dakota is dry and no wine may be sold while the train is passing through it. 'What do I care about North Dakota', I protested, 'I only want to reach Vienna. As far as I am concerned you can take me via Pepperland.'* 'Oh, a lot of pepper is grown round here', they answered. Good Lord! In this country our best curses are ruined. Obviously by means of tips and more besides I did obtain wine, but it had to be paid for surreptitiously and could not be entered on the 'bill of fare'.

The American railroad authorities usually add another carriage only when the others are full; nevertheless, the trains are always enormously long. The individual carriages have names like ships, for otherwise passengers would get quite lost I travelled consecutively in Sant Jesabel, Pembina and Vernedal. They were crowded with people of all nationalities, often thinly clothed because of the terrible heat. An infant lay completely naked on the velvet seat the whole time and reminded me of the Christ child, who obviously never lay in a Pullman car. I wanted to say this to the mother as a compliment, but how horrible it all sounds in English: ä de tscheild tschises kreist!

It is lucky that I was not born an Englishman. I would never have managed to find a bride. As one might imagine, I was rather embarrassed when it came to a declaration of love. 'Der schaut drein, als müßt er in den Hörsaal hinein', says Mephisto. Indeed, if

---

*Das Land, wo der Pfeffer wächst, i.e. Jericho, somewhere far away.

only I had nothing more to do than to enter a lecture room, if only it was just the ladies Physics and Metaphysics who stood before me! But a chaming young girl stood before me. The going was hard, but when I reached the *punctum saliens* the wisdom and good taste of our ancestors helped me, for they found the best-sounding word for the noblest of feelings: Liebe. Just as when I played the Steinway, I did not rule my tongue, my words hastened me forward and I scored a success. If I had had to say the same thing in English: 'Ei lowff ju', my chosen bride would have run away like hens before the goitrous Styrian as he pauses for breath while trying to catch them.

This colourful assortment of travellers has to be packed into bed at night. The arrangement of American sleeping-cars is as follows: a fairly narrow corridor runs down the middle of the carriage; on both sides there are upholstered seats with room for two people on each. Every traveller is directed to one of these seats. In the evening each pair of seats facing one another on the same side of the gangway is fumed into two-tier bunks, shut off from the corridor by curtains. The beds lie lengthwise in the same direction as the train is going. It is possible to have two bunks to oneself—a whole compartment—but that costs double. Since there are no drawers, nightwear, slippers and so on have to be packed in a brief-case which the attendant then puts on the bed. Everyone has to undress behind their curtain, find a place for clothes and briefcase somewhere in the bed itself and then sleep without suffocating.

Each sleeping compartment has only a small ventilation hole covered with fine mesh, leading into the open air, and in hot weather it became so steamy that I slept in my birthday suit, thus sparing myself the effort of unpacking. Once I left a whole window open overnight, but after I got up next morning the black attendant addressed me as 'Brother'.

To avoid losing my watch, wallet, spectacles etc. I put my hat in the wide-meshed luggage net (the only storage space in the whole sleeping compartment), and placed all the small items in it. The Negro always hung my hat up at an unreachable height when making the beds, and it was comical to see how perplexed he was that I needed my hat in bed.

The most critical moment is when the seats are converted into beds. It is impossible to sit on the seats and the beds are not yet ready. I flee into the washroom, but there one passenger is busy brushing clouds of dust out of this clothes, another is spraying water in all directions as he washes. I try to reach the saloon-car which, incidentally, not all trains have; the only problem is that I would have to pass through seven or eight carriages already prepared for the night with curtains drawn. These curtains are alive: an arm, a foot or something softer deals the passer-by a blow from the concealment of each bunk. On top of all this, one is continually falling over luggage jutting out on the floor. Finally I find room in a compartment which still has its seats intact. On the other side of the corridor there is already a bed, with curtains constantly on the move. There is definitely some truth in the concept of the Eternal Woman. When I ask for a show of hands in a lecture hall in Vienna they are the only part of the students that I can see, but I always recognize every female hand immediately. Therefore I was convinced that this curtain concealed female limbs; eventually it billowed out when the occupant made a careless movement while undressing and I saw that I was right. In the morning I avoided much discomfort because of my habit of getting up early. Thus I was able to dress and wash alone, then I called out to those pushing their way into the washroom from Mozart's *Entführung* already quoted by Bismarck: 'Mich zu hintergeh'n, müßt ihr früh autsteh'n'.

In the end I was so weary as a result of the heat, soot, upset stomach and thirst that I not only decided to forgo the eclipse of the sun, but tried to make the connection with the *Kaiser Wilhelm II*, which would carry me home most quickly. A telegraphists' strike had just started, with the result that we were six hours late. I was furious about this, but the phlegmatic behaviour of the Americans was incredible. They look at the raging man almost with pity, as if they are saying to themselves that the poor man imagines his rage will help. In my case the conductor said shortly, 'We don't want to risk a collision'.

In Chicago I had only twelve minutes to spare and had to get from Canal Street Union Station to Nickelplate Station. Laden with luggage I hurried aimlessly to and fro. Two people from whom I asked information did not answer. A young lady noticed this and asked me charmingly what I wanted. She could not give me directions, perhaps she did not even understand my question, but she pointed out a policeman whom I had not noticed in my agitation, despite his size. As I called out to her 'You are an angel' from the depths of my heart, not as empty flattery, I noticed that she was exactly like the guardian angel in the 'pozzo di Verona'. Why should a belief in guardian angels be pure fantasy? How does a fairy-tale atmosphere fit into the halls of Canal Street Union Station in Chicago? So I moved from the guardian angel to the guardian of the peace, who quickly gave me the necessary instructions so that I reached the other station on time.

In New York I had another surprise. The beautiful pier which had led from the railway to the ferry had burned down and I had to stumble over charred pieces of wood, still weighed down with my luggage.

In spite of all these hindrances I reached the *Kaiser Wilhelm II* in good time. How my spirits lifted when I and all my luggage were on board!

> 'Stimmet an die frohen Lieder,
>
> Denn dem väterlichen Herd
>
> Ist das Schiff nun zugekehrt
>
> Und zur Heimat geht es wieder.'

We made the return voyage in superb weather. The excellent shipboard food restored my stomach to complete health. I drank not one drop of water and not much beer, but plenty of noble Rüdesheimer. That is the great advantage of a ship: if one wobbles a little, people put it all down to the swell.

All that remains is the trifling railway journey from Bremen to Vienna, a stylish ride in a Viennese cab and I am at home again. Such a journey offers both interesting and glorious experiences: California is beautiful, Mount Shasta magnificent, Yellowstone Park wonderful, but by far the loveliest part of the whole tour is the moment of homecoming.

Notes

1. Boltzmann makes a number of factual errors, for example, Berkeley was Irish, not English. The Hearst Greek Theatre is modelled on the amphitheatre at Epidaurus, not Athens. The 'Universität Berkeley' should be the University of California.

2. Additional material to aid the translation is given in square brackets.

3. Round brackets are used by Boltzmann in the original.

4. Inverted commas around English words indicate that Boltzmann used the English word(s) himself.

5. The original German word/phrase has been included when the pun/idiom is untranslatable.

6. Quotations have not been translated.

APPENDICES

These appendices are identified by a double number, n.m. The first indicates the chapter to which each appendix refers, the second the number of the appendix to that particular chapter. The references are listed together with those of the relevant chapter.

Appendix 3.1
Calculation of pressure in a rarefied gas

Let P be a point on the wall of a vessel containing a gas in an equilibrium state. Without loss of generality, we can assume the wall to be flat. Let us take the x-axis in the direction of the normal to the wall, pointing towards the wall itself. Then a molecule with mass m hitting the wall with velocity $\boldsymbol{\xi}$, having components ξ_1, ξ_2, ξ_3 ($\xi_1 > 0$) will transfer a momentum $m\xi_1$ to the wall; and a molecule recoiling from the wall with velocity $\boldsymbol{\xi}$, having components ξ_1, ξ_2, ξ_3 ($\xi_1 < 0$), will transfer, a momentum $m|\xi_1|$ to the wall. If one constructs a cylinder on a piece of the wall of area ΔS with side $\boldsymbol{\xi}\Delta t$ (Fig. A3.1), all molecules with velocity $\boldsymbol{\xi}$ in this cylinder will strike or have struck (according to the sign of ξ_1) that piece of wall in a time interval Δt.

Since the volume of the cylinder is $|\xi_1|\Delta t\Delta S$, we conclude that if all the molecules had velocity $\boldsymbol{\xi}$, the total amount of momentum transferred to the wall in time Δt would be $n(|\xi_1|\Delta t\Delta S)(m|\xi_1|) = nm\xi_1^2 \Delta S\Delta t$, where n denotes the number of molecules per unit volume. If the molecules have different velocities, we must take an average over their distribution of velocities and obtain $nm\overline{\xi_1^2}\Delta S\Delta t$, where the overbar denotes the average value of a quantity. The transfer of momentum equals the impulse exerted on the area ΔS in time Δt; hence a force per unit area $nm\overline{\xi_1^2}$ is exerted by the wall on the gas and hence by the gas on the wall. In order to proceed, at this point we need an assumption of symmetry: if the gas is in a statistical equilibrium in a container macroscopically at rest, all the velocity components have the same probability distribution and hence

$$\overline{\xi_1^2} = \overline{\xi_2^2} = \overline{\xi_3^2}. \tag{A3.1}$$

Thus

$$3\overline{\xi_1^2} = \overline{\xi_1^2} + \overline{\xi_2^2} + \overline{\xi_3^2} = \overline{\xi_1^2 + \xi_2^2 + \xi_3^2} = \overline{|\boldsymbol{\xi}|^2} \tag{A3.2}$$

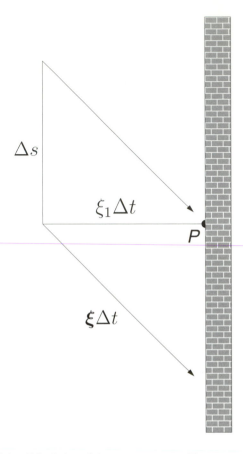

FIG. A3.1.   Calculation of the pressure exerted by a gas on a wall.

and the force per unit area, which is none other than the gas pressure p, will be given by:

$$p = \tfrac{1}{3}nm\overline{|\boldsymbol{\xi}|^2}. \tag{A3.3}$$

We cease our mathematical treatment for a little while in order to make a comment on two aspects of the calculation so far which may puzzle some readers. The first point is for those who have not seen this calculation before or have forgotten it. Why does the square of ξ_1 appear in the calculation of the momentum exchange between the wall and the gas, if momentum is proportional to the velocity and not to its square? The answer is very simple: we are computing not momentum but momentum flow. In fact the square arises from the product of two factors, each of which proportional to the velocity component ξ_1; one of these gives the momentum of each particle, the other the particle flow, i.e. the number of particles hitting the surface, clearly proportional to ξ_1. This is how the square arises.

The second comment is for those have seen the argument before and remember it, at least vaguely. In almost every book it is said that the momentum exchange per particle is $2m|\xi_1|$, and here we seem to be missing the factor 2. This is a rather subtle point. The usual argument assumes that the molecules are specularly reflected by the wall and hence each particle hitting the wall will carry a momentum $m|\xi_1|$ but will, because of the recoil effect, exchange a momentum which is twice as much. The average is then taken just over the arriving particles. Here we avoid using any (unphysical) assumption of specular reflection and so we count the momentum exchange in a slightly different way because we average over both arriving and departing molecules. Each of them carries a momentum $m\xi_1$ in the direction normal to the wall, but this momentum will have opposite signs for arriving and departing molecules. The flow of particles is also in opposite directions, so that we obtain a square (independent of the sign) at the end. The factor 2 is not lost, because the same molecule will be counted twice, once as a departing and once as an arriving particle. This point is rather subtle and we advise the reader to reflect on this, because a good understanding of this point will help him in the next chapter, where more complicated, but conceptually analogous, calculations will be performed.

Let us now proceed with our considerations. If V is the volume of the gas and N the total number of molecules, it follows from the last equation upon multiplication by V, since $nV = N$, that

$$pV = \tfrac{1}{3}Nm\overline{|\boldsymbol{\xi}|^2} = \tfrac{2}{3}Me,$$ (A3.4)

where e is the (kinetic) energy per unit mass and M the total mass. Thus the product of pressure and volume depends only on the number of molecules and the average kinetic energy of a molecule. But according to an empirical law, the law of Boyle (1660) and Mariotte (1676), at constant temperature the product of the pressure and volume of a given amount of ideal gas is constant; we see that this law is reproduced if we make the reasonable assumption (which may be rigged to make it look tantamount to a definition of the temperature of a rarefied gas according to kinetic theory) that the average kinetic energy e depends only on the (absolute) temperature T. In fact if we take into account the relation which combines the Boyle–Mariotte law with that of Gay-Lussac and Charles,

$$pV = MRT,$$ (A3.5)

where R is the gas constant, we conclude that

$$e = \tfrac{3}{2}RT.$$ (A3.6)

We remark that R is constant for a given gas but is related to the molecular mass by $R = k/m$, where k is the Boltzmann constant ($k = 1.38 \times 10^{-23}\,\mathrm{J/K}$); this follows by considering a mixture of two different gases in the same vessel [3].

Here we meet for the first time the Boltzmann constant k. We remark that Boltzmann never used it. The first scientist who used k was Planck (see Chapter 12). The reason why Maxwell and Boltzmann could dispense with this constant is that it does not have any deep meaning and its use can be avoided in purely theoretical considerations. Essentially it is a constant useful to convert units. We could measure temperatures in energy units (Joules) and then k could be taken to be unity or, say, 2/3. If we want to retain kelvins for temperature, then we need k. Apparently in the nineteenth century this was considered to be the last step in a calculation, not worth publication in a scientific paper. At the beginning of the twentieth century however, it became clear that this innocent constant was a key to finding an accurate value of Avogadro's constant, the number of molecules in one mole of gas. It is thus not surprising to find in a not so well-known paper by Einstein [30], written in 1904, a section entitled "On the meaning of the constant κ in the kinetic atomic energy", where κ is half the Boltzmann constant. He was turning from a study of the statistical foundations of thermodynamics to the determination of atomic constants. He was still struggling with problems that were to lead him to his basic paper on Brownian motion in 1905 (see

Chapter 12). At that time he was enquiring about the meaning of k and the possibility of measuring it.

We also remark that what has been said above applies only to monatomic gases, which are well modelled by perfectly rigid, smooth, and elastic spheres, and sufficiently rarefied so that in a neighbourhood of the wall we may neglect intermolecular collisions (with respect to those with the wall). A more careful analysis would show that the degree of rarefaction required for the argument to be valid is such that the product of the number density n by the cube of the molecular diameter σ is negligible compared with unity.

Appendix 4.1
The Liouville equation

In order to obtain eqn (4.6) from eqn (4.5) and hence the Liouville equation, we can now exploit the fact that the set of values of $z \in D$ coincides with the set of points z_t obtained by the time evolution of z_0 with $z_0 \in D_0$ and change the integration variables on the left-hand side from z to z_0. We obtain

$$\int_D P(z, t)\, \mathrm{d}z = \int_{D_0} P(z_t, t)\, J(z_t/z_0)\, \mathrm{d}z_0, \tag{A4.1}$$

where $J(z_t/z_0)$ is the Jacobian determinant of the old variables with respect to the new ones (which turns out to be positive; by continuity, if no collisions occur, but also in the presence of collisions, see below). Comparison of eqns (4.4) and (4.5) of the main text gives, due to the arbitrariness of D_0,

$$P(z_t, t)\, J(z_t/z_0) = P_0(z_0). \tag{A4.2}$$

If we assume that no forces act on the molecules, the Jacobian turns out to be unity (it is an easy exercise, because the velocities do not change and each position vector undergoes a translation which actually depends upon the corresponding velocity but makes non-zero only the matrix elements lying on one side of the diagonal and equal to 1 those on the diagonal). If assigned forces act on the molecules, one can see, in a less simple way (see e.g. [10]), that the result holds true if the forces do not depend on the velocity, or, if they do, the dependence is such that the divergence with respect to velocity vanishes (Liouville's theorem). We have now to examine what happens at a collision.

When two spheres collide, conservation of momentum and energy must hold. Thus the velocities of the two molecules after the impact ($\boldsymbol{\xi}_1, \boldsymbol{\xi}_2$) and before the impact ($\boldsymbol{\xi}'_1, \boldsymbol{\xi}'_2$) satisfy:

$$\begin{aligned} \boldsymbol{\xi}_1 + \boldsymbol{\xi}_2 &= \boldsymbol{\xi}'_1 + \boldsymbol{\xi}'_2, \\ |\boldsymbol{\xi}_1|^2 + |\boldsymbol{\xi}_2|^2 &= |\boldsymbol{\xi}'_1|^2 + |\boldsymbol{\xi}'_2|^2. \end{aligned} \tag{A4.3}$$

Let us introduce a unit vector $\boldsymbol{n}$ directed along $\boldsymbol{\xi}_1 - \boldsymbol{\xi}'_1$; this direction, see Fig. A4.1, bisects the directions of $\boldsymbol{V} = \boldsymbol{\xi}_1 - \boldsymbol{\xi}_2$ and $-\boldsymbol{V}' = -(\boldsymbol{\xi}'_1 - \boldsymbol{\xi}'_2)$ (therefore $\boldsymbol{n} = (\boldsymbol{x}_1 - \boldsymbol{x}_2)/|\boldsymbol{x}_1 - \boldsymbol{x}_2|$ is the unit vector directed along the line joining the centres of the spheres, since the change of momentum at the moment of the impact between two smooth spheres must be directed along such line). It is a simple exercise to see that eqns (A4.3) imply

$$\begin{aligned} \boldsymbol{\xi}'_1 &= \boldsymbol{\xi}_1 - \boldsymbol{n}[\boldsymbol{n} \cdot (\boldsymbol{\xi}_1 - \boldsymbol{\xi}_2)], \\ \boldsymbol{\xi}'_2 &= \boldsymbol{\xi}_2 + \boldsymbol{n}[\boldsymbol{n} \cdot (\boldsymbol{\xi}_1 - \boldsymbol{\xi}_2)]. \end{aligned} \tag{A4.4}$$

We remark that the relative velocity

$$\boldsymbol{V} = \boldsymbol{\xi}_1 - \boldsymbol{\xi}_2 \tag{A4.5}$$

satisfies

$$\boldsymbol{V}' = \boldsymbol{V} - 2\boldsymbol{n}(\boldsymbol{n} \cdot \boldsymbol{V}), \tag{A4.6}$$

i.e. undergoes a specular reflection at the impact. This means that if we split $\boldsymbol{V}$ at the point of impact into a normal component $\boldsymbol{V}_n$ directed along $\boldsymbol{n}$ and a tangential component $\boldsymbol{V}_t$ (in the plane normal to $\boldsymbol{n}$), then $\boldsymbol{V}_n$ changes sign and $\boldsymbol{V}_t$ remains unchanged in a collision.

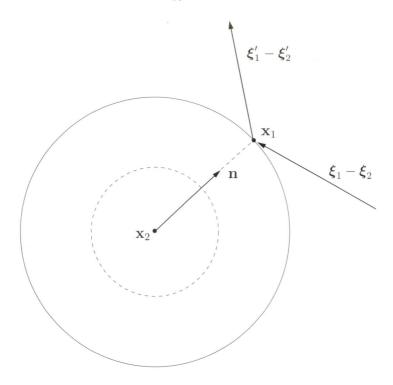

FIG. A4.1.   The angle between the relative velocities before and after the impact is bisected by
the line joining the centres of the colliding molecules.

It is now easy to compute the Jacobian of the velocities after the impact with respect to those
before. The easiest way is to transform first each set of variables into the corresponding variables
V (the relative velocity) and $\bar{\xi} = \frac{1}{2}(\xi_1 + \xi_2)$ (the velocity of the centre of mass). These linear
transformations are easily seen to have unit Jacobian. The Jacobian matrix of the transformation
from $(\bar{\xi}, V)$ to $(\bar{\xi}', V')$ is now diagonal if we adopt normal and tangential components; it differs
from the unit matrix, because one entry (corresponding to the normal component) is -1 rather than
1. Hence the Jacobian is -1.

We now need the Jacobian of the position variables after the impact with respect to those before
the impact. It is now clear that this Jacobian is -1, because the volume elements change their
relative orientation upon impact (see Fig. A4.2), as does the image of a glove in a mirror.

The Jacobian J of the transformation occurring in phase space when a collision occurs is clearly
the product of the Jacobians corresponding to the transformations undergone by space and velocity
variables, respectively. Hence $J = (-1) \cdot (-1) = 1$.

Thus we conclude that, in the absence of bulk forces acting on the molecules during their
movement between two subsequent collisions (but the result remains true even in other cases, e.g.
if the forces depend upon position only), eqn (4.5) in the main text simply becomes

$$P(z_t, t) = P_0(z_0). \tag{A4.7}$$

In other words, P is constant along the trajectory of z in phase space.

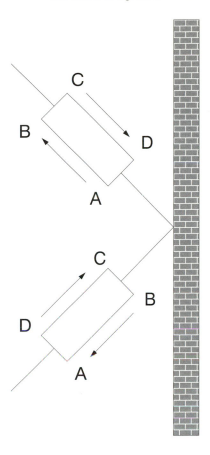

FIG. A4.2. A volume changes its orientation upon specular reflection.

P is defined on the set $\Omega^N \times \mathfrak{R}^{3N}$, where Ω is a subset of $\mathfrak{R}^3$ where the N molecules move; however, it vanishes at the points of this set which satisfy:

$$\exists i, j \in \{1, 2, \ldots, N\} \quad (i \neq j): \quad |\mathbf{x}_i - \mathbf{x}_j| < \sigma, \tag{A4.8}$$

where σ is the sphere diameter. In fact, if z is a point in the set defined by eqn (A4.8), the ith and jth molecules would partly overlap, which is impossible, since they are assumed to be hard spheres. Accordingly, it is convenient, as mentioned in the main text, to consider the set Λ obtained by deleting from Ω^N the points satisfying eqn (A4.8).

If P is a differentiable function of the variables z, t in $\Lambda \times \mathfrak{R}$, eqn (A4.7), in the absence of bulk forces, implies that

$$\frac{\partial P}{\partial t} + \sum_{i=1}^{N} \boldsymbol{\xi}_i \cdot \frac{\partial P}{\partial \mathbf{x}_i} = 0 \qquad (z \in \Lambda). \tag{A4.9}$$

In fact, since $z = T^t z_0$ describes a rectilinear motion of all the molecules inside Λ, it is obviously true that

$$\mathrm{d}z/\mathrm{d}t = (\boldsymbol{\xi}_1, \boldsymbol{\xi}_2, \ldots, \boldsymbol{\xi}_N, 0, 0, \ldots, 0) \tag{A4.10}$$

and eqn (A4.9) follows from eqn (A4.7) by differentiation. Equation (A4.9) is called the *Liouville equation* for the system under consideration.

Equation (A4.9) is a partial differential equation, and as such, must be accompanied by suitable initial and boundary conditions. The initial condition simply assigns the value of P at $t = 0$. As for the boundary conditions, they are present even if the gas is free to move without bounds in space (i.e., if $\Omega = \Re^3$). In fact, we had to introduce boundaries in order to define Λ (where (A4.9) holds); these are the boundaries with the regions where the spheres would partly overlap. At these boundary points we must impose the condition dictated by eqn (A4.7): since P is always constant along the trajectories in Λ (boundary included), but the velocity variables undergo a discontinuous transformation there, we must impose that P is the same at z and z', where z and z' indicate points of the boundary of $\Lambda \times \Re^3$ which are transformed one into the other by the transformation associated with an impact:

$$P(z, t) = P(z', t) \qquad (z \in \partial\Lambda \times \Re^3), \tag{A4.11}$$

or, in more detail:

$$
\begin{aligned}
&P(x_1, \xi_1, \ldots, x_i, \xi_i, \ldots, x_j, \xi_j, \ldots, x_N, \xi_N, t) \\
&\quad = P(x_1, \xi_1, \ldots x_i, \xi_i - n_{ij}(n_{ij} \cdot V_{ij}), \ldots, x_j, \xi_j + n_{ij}(n_{ij} \cdot V_{ij}), \ldots, x_N, \xi_N, t), \\
&\quad \text{if} \quad |x_i - x_j| = \sigma \quad (i \neq j),
\end{aligned}
\tag{A4.12}
$$

where $V_{ij} = \xi_i - \xi_j$ and n_{ij} is the unit vector directed as $x_i - x_j$.

If Ω does not coincide with the entire space $\Re^3$, then there are additional boundary points corresponding to those z for which at least one x_i is on $\partial\Omega$. A suitable boundary condition must be assigned at these points as well. Frequently one assumes specular reflection ($\xi_i' = \xi_i - n_i(n_i \cdot \xi_i)$, where n_i is the normal at x_i), but other boundary conditions are used in practice [10]. If Ω is a box, periodicity conditions are very popular; in that case one can avoid mentioning the boundaries and talk about a flat torus (after identification of opposite faces).

An important point to be mentioned is the circumstance that the initial values that we shall allow are symmetric upon interchange of any two molecules (since the molecules are identical):

$$
\begin{aligned}
&P_0(x_1, \xi_1, \ldots, x_i, \xi_i, \ldots, x_j, \xi_j, \ldots, x_N, \xi_N) \\
&\quad = P_0(x_1, \xi_1, \ldots x_j, \xi_j, \ldots, x_i, \xi_i, \ldots, x_N, \xi_N) \qquad \forall(i, j).
\end{aligned}
\tag{A4.13}
$$

Since the time evolution is consistent with this symmetry (as one can easily check), eqn (A4.8) shows that the same symmetry is preserved for $t > 0$.

Appendix 4.2
Calculation of the effect of collisions of one particle with another

We shall now compute the effect of the collisions of particle 2 with particle 1.

Let x_2 be a point of a sphere such that the vector joining the centre of the sphere with x_2 is σn, where n is a unit vector. A cylinder with height $|V_2 \cdot n| dt$ and base area $dS = \sigma^2 dn$ (where dn is the area of a surface element of the unit sphere about n) will contain the molecules with velocity ξ_2 hitting the base dS in the time interval $(t, t+dt)$; its volume is $\sigma^2 dn |V_2 \cdot n| dt$. Thus the probability of collision of molecule 2 with molecule 1 in the ranges $(x_1, x_1 + dx_1)$, $(\xi_1, \xi_1 + d\xi_1)$, $(x_2, x_2 + dx_2)$, $(\xi_2, \xi_2 + d\xi_2)$, $(t, t + dt)$ occurring at points of dS is $P^{(2)}(x_1, x_2, \xi_1, \xi_2, t) dx_1 d\xi_1 d\xi_2 \times \sigma^2 dn |V_2 \cdot n| dt$. If we want the probability of collisions of molecule 1 with 2, when the range of the former is fixed but the latter may have any velocity ξ_2 and any position x_2 on the sphere (i.e. any n), we integrate over the sphere and all the possible velocities of molecule 2 to obtain (we use the notation l, L, g, L introduced in the main text to denote the loss and gain contributions to time evolution due to collisions; the small letters refer to the contribution of collisions with a given molecule, the capitals to the effect of all molecules):

$$l\, dx_1 d\xi_1 dt = dx_1 d\xi_1 dt \int_{\Re^3} \int_{S_-} P^{(2)}(x_1, \xi_1, x_1 + \sigma n, \xi_2, t) |V_2 \cdot n| \sigma^2\, dn\, d\xi_2, \qquad (A4.14)$$

where S_- is the hemisphere corresponding to $V_2 \cdot n < 0$ (the molecules are moving towards one another before the collision). Thus we have the following result:

$$L = (N-1)\sigma^2 \int_{\Re^3} \int_{S_-} P^{(2)}(x_1, \xi_1, x_1 + \sigma n, \xi_2, t) |(\xi_2 - \xi_1) \cdot n|\, d\xi_2\, dn. \qquad (A4.15)$$

The calculation of the gain term G is exactly the same as the one just performed for L, except for the fact that we are looking at molecules which have velocities ξ_1 and ξ_2 after collision and hence we have to integrate over the hemisphere S_+, defined by $V_2 \cdot n > 0$ (the molecules are moving away from one another after the collision). Thus we have:

$$G = (N-1)\sigma^2 \int_{\Re^3} \int_{S_+} P^{(2)}(x_1, \xi_1, x_1 + \sigma n, \xi_2, t) |(\xi_2 - \xi_1) \cdot n|\, d\xi_2\, dn. \qquad (A4.16)$$

We thus could write the right-hand side of eqn (A4.15) as a single expression:

$$G - L = (N-1)\sigma^2 \int_{\Re^3} \int_B P^{(2)}(x_1, \xi_1, x_1 + \sigma n, \xi_2, t)(\xi_2 - \xi_1) \cdot n\, d\xi_2\, dn, \qquad (A4.17)$$

where now S is the entire unit sphere and we have eliminated the bars of absolute value on the right-hand side.

Equation (A4.17), although absolutely correct, is not so useful. It turns out that it is much more convenient to keep the gain and loss terms separated. Only in this way, in fact, can we insert in eqn (4.8) of the main text the information that the probability density $P^{(2)}$ is continuous at a collision (a consequence of (A4.12)). This in turn as we shall see, will permit us to use the essential circumstance that molecules that are about to collide are statistically independent (the famous *Stosszahlansatz* of Boltzmann, or assumption of molecular chaos), while those that have just collided are not. In order to use eqn (A4.12), we remark that if we write $i = 1$, $j = 2$ and integrate with respect to the positions and velocities of the remaining $N - 2$ molecules, we have:

$$P^{(2)}(x_1, \xi_1, x_2, \xi_2, t) = P^{(2)}(x_1, \xi_1 - n(n \cdot V), x_2, \xi_2 + n(n \cdot V), t)$$

$$\text{if} \quad |x_1 - x_2| = \sigma, \qquad (A4.18)$$

where we have written V for $V_{12} = \xi_1 - \xi_2$ and n for $-n_{12}$ (in agreement with the notation used above). For brevity, we write (in agreement with eqn (A4.4)):

$$\xi_1' = \xi_1 - n(n \cdot V), \qquad \xi_2' = \xi_2 + n(n \cdot V). \tag{A4.19}$$

Inserting eqn (A4.18) into eqn (A4.16), we thus obtain:

$$G = (N-1)\sigma^2 \int_{\Re^3} \int_{S_+} P^{(2)}(x_1, \xi_1', x_1 + \sigma n, \xi_2', t)|(\xi_2 - \xi_1) \cdot n|\,d\xi_2\,dn, \tag{A4.20}$$

which is a frequently used form. Sometimes n is changed to $-n$ in order to have the same integration range as in L; the only change (in addition to the change in the range) is in the third argument of $P^{(2)}$, which becomes $x_1 - \sigma n$. If we make the simplifying assumptions, which arise from the idea of molecular chaos or are natural in the Boltzmann–Grad limit, as mentioned in the main text, we obtain for the gain and loss terms:

$$G = N\sigma^2 \int_{\Re^3} \int_{S_+} P^{(1)}(x_1, \xi_1', t) P^{(1)}(x_1, \xi_2', t)|(\xi_2 - \xi_1) \cdot n|\,d\xi_2\,dn, \tag{A4.21}$$

$$L = N\sigma^2 \int_{\Re^3} \int_{S_-} P^{(1)}(x_1, \xi_1, t) P^{(1)}(x_1, \xi_2, t)|(\xi_2 - \xi_1) \cdot n|\,d\xi_2\,dn. \tag{A4.22}$$

By inserting these expressions in eqn (4.8) of the main text and expanding the time derivative along the trajectory, as we did in the previous appendix for the Liouville equation (but now with $N = 1$), we can write the *Boltzmann equation* in the following form:

$$\frac{\partial P^{(1)}}{\partial t} + \xi_1 \cdot \frac{\partial P^{(1)}}{\partial x_1} = N\sigma^2 \int_{\Re^3} \int_{B_-} [P^{(1)}(x_1, \xi_1', t) P^{(1)}(x_1, \xi_2', t)$$
$$- P^{(1)}(x_1, \xi_1, t) P^{(1)}(x_1, \xi_2, t)]|(\xi_2 - \xi_1) \cdot n|\,d\xi_2\,dn. \tag{A4.23}$$

Appendix 4.3
The BBGKY hierarchy

In this appendix we shall deal with the equations satisfied by the s-molecule distribution functions $P^{(s)}$ as a consequence of the Liouville equation (A4.11), which we rewrite here for convenience of the reader:

$$\frac{\partial P}{\partial t} + \sum_{i=1}^{N} \boldsymbol{\xi}_i \cdot \frac{\partial P}{\partial \boldsymbol{x}_i} = 0 \qquad (z \in \Lambda). \tag{A4.24}$$

A rigorous derivation of these equations involves some subtleties, which we shall not enter into here; we refer the interested reader to Chapter IV of ref. [11]. Here we shall assume that P is a smooth function, so that all the steps to be performed are justified.

We first state the relation between $P^{(s)}$ and P, which follows from their definition and is similar to eqn (A4.14):

$$P^{(s)}(\boldsymbol{x}_1, \boldsymbol{\xi}_1, \boldsymbol{x}_2, \boldsymbol{\xi}_2, \ldots, \boldsymbol{x}_s, \boldsymbol{\xi}_s, t) =$$
$$\int_{\Omega^s \times \Re^{3s}} P(\boldsymbol{x}_1, \boldsymbol{\xi}_1, \boldsymbol{x}_2, \boldsymbol{\xi}_2, \ldots, \boldsymbol{x}_N, \boldsymbol{\xi}_N, t) \prod_{j=s+1}^{N} \mathrm{d}\boldsymbol{x}_j \, \mathrm{d}\boldsymbol{\xi}_j. \tag{A4.25}$$

The first step to be performed in order to derive an evolution equation for $P^{(s)}$ is now rather obvious: we integrate eqn (A4.24) with respect the variables $\boldsymbol{x}_j$ ($s+1 \le j \le N$) over $\Omega^s \times \Re^{3s}$. It is convenient to keep the terms in the sum appearing in eqn (A4.24) with $i \le s$ separated from those with $i > s$.

$$\frac{\partial P^{(s)}}{\partial t} + \int \sum_{i=1}^{s} \boldsymbol{\xi}_i \cdot \frac{\partial P}{\partial \boldsymbol{x}_i} \prod_{j=s+1}^{N} \mathrm{d}\boldsymbol{x}_j \, \mathrm{d}\boldsymbol{\xi}_j + \sum_{k=s+1}^{N} \int \boldsymbol{\xi}_k \cdot \frac{\partial P}{\partial \boldsymbol{x}_k} \prod_{j=s+1}^{N} \mathrm{d}\boldsymbol{x}_j \, \mathrm{d}\boldsymbol{\xi}_j = 0, \tag{A4.26}$$

where integration with respect to the velocity variables extends to the entire space $\Re^{3N-3s}$, while it extends to Ω^{N-s} deprived of the spheres $|\boldsymbol{x}_i - \boldsymbol{x}_j| < \sigma$ ($i = s+1, \ldots, N, i \ne j$) with respect to the position variables. It is also expedient to call k rather than i the dummy suffix in the second sum.

A typical term in the first sum in eqn (A4.26) contains the integral of a derivative with respect to a variable, $\boldsymbol{x}_i$, over which one does not integrate; however, it is not possible simply to change the order of integration and differentiation to obtain a derivative of $P^{(s)}$, even if the function P is assumed to be smooth, because the domain has boundaries ($|\boldsymbol{x}_i - \boldsymbol{x}_j| = \sigma$) depending on $\boldsymbol{x}_i$. To obtain the correct result, a boundary term has to be added:

$$\int \boldsymbol{\xi}_i \cdot \frac{\partial P}{\partial \boldsymbol{x}_i} \prod_{j=s+1}^{N} \mathrm{d}\boldsymbol{x}_j \, \mathrm{d}\boldsymbol{\xi}_j = \boldsymbol{\xi}_i \cdot \frac{\partial P^{(s)}}{\partial \boldsymbol{x}_i} - \sum_{k=s+1}^{N} \int P^{(s+1)} \boldsymbol{\xi}_i \cdot \boldsymbol{n}_{ik} \mathrm{d}\sigma_{ik} \, \mathrm{d}\boldsymbol{\xi}_k, \tag{A4.27}$$

where $\boldsymbol{n}_{ik}$ is the outer normal to the sphere $|\boldsymbol{x}_i - \boldsymbol{x}_k| = \sigma$ (with centre at $\boldsymbol{x}_k$), $\mathrm{d}\sigma_{ik}$ the surface element on the same sphere, and $P^{(s+1)}$ the $(s+1)$-molecule distribution function with arguments $(\boldsymbol{x}_j, \boldsymbol{\xi}_j)$ ($j = 1, 2, \ldots, s, k$).

A typical term in the second sum in eqn (A4.26) can be immediately integrated by means of the Gauss theorem, since it involves the integration of a derivative taken with respect to one of the integration variables. We find:

$$
\int \boldsymbol{\xi}_k \cdot \frac{\partial P}{\partial x_k} \prod_{j=s+1}^{N} dx_j \, d\boldsymbol{\xi}_j = \sum_{i=1}^{s} \int P^{(s+1)} \boldsymbol{\xi}_k \cdot \boldsymbol{n}_{ik} d\sigma_{ik} \, d\boldsymbol{\xi}_k
$$

$$
+ \sum_{\substack{i=s+1 \\ i \neq k}}^{N} \int P^{(s+2)} \boldsymbol{\xi}_k \cdot \boldsymbol{n}_{ik} \, d\sigma_{ik} \, d\boldsymbol{\xi}_k \, dx_i \, d\boldsymbol{\xi}_i + \int P^{(s+1)} \boldsymbol{\xi}_k \cdot \boldsymbol{n}_k \, dS_k \, d\boldsymbol{\xi}_k, \tag{A4.28}
$$

where dS_k is the surface element of the boundary of Ω in the three-dimensional subspace described by x_k, and $\boldsymbol{n}_k$ is the unit vector normal to such a surface element and pointing into the gas. The last term in eqn (A4.28) is the contribution from the solid boundary of Ω; if the boundary conditions are of the form described at the end of Appendix 4.1, after eqn (A4.12) (in particular if there are specular reflection or periodicity boundary conditions), the term under consideration is zero; henceforth it will be omitted.

Inserting eqns (A4.27) and (A4.28) into eqn (A4.26), we find:

$$
\frac{\partial P^{(s)}}{\partial t} + \sum_{i=1}^{s} \boldsymbol{\xi}_i \cdot \frac{\partial P^{(s)}}{\partial x_i} = \sum_{i=1}^{s} \sum_{k=s+1}^{N} \int P^{(s+1)} \boldsymbol{V}_{ik} \cdot \boldsymbol{n}_{ik} \, d\sigma_{ik} \, d\boldsymbol{\xi}_k
$$

$$
+ \frac{1}{2} \sum_{\substack{i,k=s+1 \\ i \neq k}}^{N} \int P^{(s+2)} \boldsymbol{V}_{ki} \cdot \boldsymbol{n}_{ik} \, d\sigma_{ik} \, d\boldsymbol{\xi}_k \, dx_i \, d\boldsymbol{\xi}_i, \tag{A4.29}
$$

where $\boldsymbol{V}_{ik} = \boldsymbol{\xi}_i - \boldsymbol{\xi}_k$ is the relative velocity of the ith molecule with respect to the kth molecule and we have taken into account that $\boldsymbol{\xi}_k \cdot \boldsymbol{n}_{ik}$ can be replaced by $\frac{1}{2} \boldsymbol{V}_{ki} \cdot \boldsymbol{n}_{ik}$ in the second sum, because of the fact that $\boldsymbol{n}_{ik} = -\boldsymbol{n}_{ki}$. Now the last integral in eqn (A4.29) is easily shown [10, 11, 13] to be zero. Further, the first integral in eqn (A4.29) is the same whatever is the value of the dummy index k; thus we can eliminate this index and write $x_*, \boldsymbol{\xi}_*$ in place of $x_k, \boldsymbol{\xi}_k$. Summarizing, we have:

$$
\frac{\partial P^{(s)}}{\partial t} + \sum_{i=1}^{s} \boldsymbol{\xi}_i \cdot \frac{\partial P^{(s)}}{\partial x_i} = (N - s) \sum_{i=1}^{s} \int P^{(s+1)} \boldsymbol{V}_i \cdot \boldsymbol{n}_i \, d\sigma_i \, d\boldsymbol{\xi}_*, \tag{A4.30}
$$

where $\boldsymbol{V}_i = \boldsymbol{\xi}_i - \boldsymbol{\xi}_*, \boldsymbol{n}_i = (x_i - x_*)/\sigma$, and the arguments of the $(s + 1)$-particle distribution function, $P^{(s+1)}$, are $(x_1, \boldsymbol{\xi}_1, \ldots, x_s, \boldsymbol{\xi}_s, x_*, \boldsymbol{\xi}_*, t)$.

It should be clear that the partial differential operator on the left-hand side of eqn (A4.30) should be complemented with the boundary conditions on the boundary of Λ^s. This operator is the generator of the free motion of s molecules. The physical meaning of eqn (A4.30) should be transparent: the s-molecule distribution function evolves in time according to the s-molecule dynamics, corrected by the effect of the interaction with the remaining $(N - s)$ molecules. The effect of this interaction is described by the right-hand side of eqn (A4.30).

We remark that for $s = 1$, eqn (A4.30) reduces to eqn (4.8) of the main text when the right-hand side $G - L$ is written in the form (A4.17). That expression is thus rigorously justified for functions P which are smooth enough. However, as we remarked in the previous appendix in the particular case $s = 1$, the form (A4.30) is not the most convenient for the right-hand side of the equation. It is better to keep the contributions from the two hemispheres $\pm \boldsymbol{V}_i \cdot \boldsymbol{n}_i > 0$ separated. As we did before, here we separate in eqn (4.8) the contributions from the two hemispheres S_+^i and S_-^i, defined by $\boldsymbol{V}_i \cdot \boldsymbol{n}_i > 0$ and $\boldsymbol{V}_i \cdot \boldsymbol{n}_i < 0$, respectively. In addition, we remark that $d\sigma_i = \sigma^2 d\boldsymbol{n}_i$

(where dn_i is the surface element on the unit sphere described by n_i) and write:

$$\frac{\partial P^{(s)}}{\partial t} + \sum_{i=1}^{s} \xi_i \cdot \frac{\partial P^{(s)}}{\partial x_i}$$

$$= (N - s)\sigma^2 \left(\sum_{i=1}^{s} \int_{\Re^3} \int_{S_+^i} P^{(s+1)} |V_i \cdot n_i| \, dn_i \, d\xi_* - \sum_{i=1}^{s} \int_{\Re^3} \int_{S_-^i} P^{(s+1)} |V_i \cdot n_i| \, dn_i \, d\xi_* \right).$$

$$(A4.31)$$

Now, exactly as we did in the previous appendix for the particular case $s = 1$, we use the laws of elastic impact and the continuity of the distribution functions (a consequence of the analogous property enjoyed by P) to obtain

$$\frac{\partial P^{(s)}}{\partial t} + \sum_{i=1}^{s} \xi_i \cdot \frac{\partial P^{(s)}}{\partial x_i}$$

$$= (N - s)\sigma^2 \sum_{i=1}^{s} \left(\int_{\Re^3} \int_{S_+} P^{(s+1)'} |V_i \cdot n_i| \, dn_i \, d\xi_* - \int_{\Re^3} \int_{S_-} P^{(s+1)} |V_i \cdot n_i| \, dn_i \, d\xi_* \right),$$

$$(A4.32)$$

where $P^{(s+1)'}$ means that in $P^{(s+1)}$ we replace the arguments ξ_i and ξ_* by ξ_i' and ξ_*', given by:

$$\xi_i' = \xi_i - n_i(n_i \cdot V_i), \qquad \xi_*' = \xi_* + n_i(n_i \cdot V_i). \qquad (A4.33)$$

We may transform the two integrals extended to B_+^i and S_-^i into a single integral by changing, e.g. n_i to $-n_i$ in the second integral; we may even abolish the index i in n_i, provided that the argument x_* in the second integral of the ith term is replaced by

$$x_* = x_i - n\sigma, \qquad (A4.34)$$

whereas x_* is replaced by $x_i + n\sigma$ in the first integral, of course. Thus we have:

$$\frac{\partial P^{(s)}}{\partial t} + \sum_{i=1}^{s} \xi_i \cdot \frac{\partial P^{(s)}}{\partial x_i}$$

$$= (N - s)\sigma^2 \sum_{i=1}^{s} \int_{\Re^3} \int_{S} (P^{(s+1)'} - P^{(s+1)}) |V_i \cdot n| \, dn \, d\xi_*. \qquad (A4.35)$$

This system of equations is usually called the BBGKY hierarchy for a gas of hard spheres (from the initials of the names of Bogolyubov, Born, Green, Kirkwood, Yvon).

Appendix 4.4
The Boltzmann hierarchy and its relation to
the Boltzmann equation

Let us consider the Boltzmann–Grad limit ($N \to \infty$ and $\sigma \to 0$ in such a way that $N\sigma^2$ remains finite). Then we obtain (for each fixed s) that if each $P^{(s)}$ tends to a limit (which we denote by the same symbol) and this limit is sufficiently smooth, the finite hierarchy of eqn (A4.35) becomes in the limit:

$$\frac{\partial P^{(s)}}{\partial t} + \sum_{i=1}^{s} \boldsymbol{\xi}_i \cdot \frac{\partial P^{(s)}}{\partial x_i} = N\sigma^2 \sum_{i=1}^{s} \int_{\Re^3} \int_{S_+} (P^{(s+1)'} - P^{(s+1)})|V_i \cdot \boldsymbol{n}| \, \mathrm{d}\boldsymbol{n} \, \mathrm{d}\boldsymbol{\xi}_*, \tag{A4.36}$$

where the arguments of $P^{(s+1)'}$ and $P^{(s+1)}$ are the same as above, except for the fact that $\boldsymbol{x}'_* = \boldsymbol{x}_* = \boldsymbol{x}_i$ in agreement with eqn (A4.34) for $\sigma \to 0$. Equation (A4.36) gives a complete description of the time evolution of a Boltzmann gas (i.e. the ideal gas which is obtained in the Boltzmann–Grad limit), provided that the initial-value problem is well posed for this infinite system of equations, which appears to have been first written in ref. [13] and is usually called *the Boltzmann hierarchy*.

As we already know, eqn (A4.36) is not equivalent to the Boltzmann equation, unless a special assumption on the initial data is made. Indeed, as discussed by Spohn [14], the solutions of the Boltzmann hierarchy describe the evolution of a Boltzmann gas when the chaos assumption given by eqn (3.1) of the main text is not satisfied by the initial data at $t = 0$. The solutions of the Boltzmann hierarchy in the case when the factorization property is not fulfilled for $t = 0$ are discussed in Chapter IV of ref. [11]. Here we shall assume that the data satisfy eqn (4.11) of the main text, which we rewrite here for $t = 0$:

$$P^{(s)}(\boldsymbol{x}_1, \boldsymbol{\xi}_1, \boldsymbol{x}_2, \boldsymbol{\xi}_2, \dots, \boldsymbol{x}_s, \boldsymbol{\xi}_s, 0) = \prod_{j=1}^{s} P^{(1)}(\boldsymbol{x}_j, \boldsymbol{\xi}_j, 0). \tag{A4.37}$$

It is now a simple remark, made in ref. [13] (see also refs. [10, 11]), that if eqn (A4.37) is satisfied and the Boltzmann equation, given by eqn (A4.23) or shortly ($V = \boldsymbol{\xi} - \boldsymbol{\xi}_*$) by

$$\frac{\partial P^{(1)}}{\partial t} + \boldsymbol{\xi} \cdot \frac{\partial P^{(1)}}{\partial x} = N\sigma^2 \int_{\Re^3} \int_{S_+} (P^{(1)'} P^{(1)'}_* - P^{(1)} P^{(1)}_*)|V \cdot \boldsymbol{n}| \, \mathrm{d}\boldsymbol{\xi}_* \, \mathrm{d}\boldsymbol{n}, \tag{A4.38}$$

admits a solution $P^{(1)}(\boldsymbol{x}_s, \boldsymbol{\xi}_s, t)$ for given initial data $P^{(1)}(\boldsymbol{x}_s, \boldsymbol{\xi}_s, 0)$, then the Boltzmann hierarchy, eqn (A4.36) has at least one solution, given by

$$P^{(s)}(\boldsymbol{x}_1, \boldsymbol{\xi}_1, \boldsymbol{x}_2, \boldsymbol{\xi}_2, \dots, \boldsymbol{x}_s, \boldsymbol{\xi}_s, t) = \prod_{j=1}^{s} P^{(1)}(\boldsymbol{x}_j, \boldsymbol{\xi}_j, t). \tag{A4.39}$$

Therefore the chaos assumption, eqn (A4.39), is not inconsistent with the dynamics of rigid spheres in the Boltzmann–Grad limit; actually, if the Boltzmann hierarchy has a unique solution for data which satisfy eqn (A4.37), then eqn (A4.39) necessarily holds at any time if it holds at $t = 0$. Then the Boltzmann equation is justified.

However, we stress the fact that we have made several assumptions (existence of limits, their smoothness, an existence theorem for the Boltzmann equation, a uniqueness theorem for the Boltzmann hierarchy) that might not be satisfied. A few cases in which these properties have been shown to hold and thus the Boltzmann equation has been shown to be valid are discussed in Chapter 5.

We end this appendix with a few remarks on the Boltzmann equation, eqn (A4.38). First, we can omit the superscript (1), which is no longer needed, and rewrite eqn (A4.38) as follows:

$$\frac{\partial P}{\partial t} + \boldsymbol{\xi} \cdot \frac{\partial P}{\partial \boldsymbol{x}} = N\sigma^2 \int_{\mathfrak{R}^3} \int_{S_+} (P'P'_* - PP_*)|\boldsymbol{V} \cdot \boldsymbol{n}| \, d\boldsymbol{\xi}_* \, d\boldsymbol{n}. \tag{A4.40}$$

Then it should be clear that the arguments of P are $\boldsymbol{x}, \boldsymbol{\xi}, t$, those of P_* are $\boldsymbol{x}, \boldsymbol{\xi}_*, t$, those of P' are $\boldsymbol{x}, \boldsymbol{\xi}', t$, and those of P'_* are $\boldsymbol{x}, \boldsymbol{\xi}'_*, t$, where

$$\boldsymbol{\xi}' = \boldsymbol{\xi} - \boldsymbol{n}(\boldsymbol{n} \cdot \boldsymbol{V}), \qquad \boldsymbol{\xi}'_* = \boldsymbol{\xi}_* + \boldsymbol{n}(\boldsymbol{n} \cdot \boldsymbol{V}), \qquad \boldsymbol{V} = \boldsymbol{\xi} - \boldsymbol{\xi}_*. \tag{A4.41}$$

Finally we observe that the integral in eqn (A4.40) is extended to the hemisphere S_+, but could be equivalently extended to the entire sphere S, provided that a factor $1/2$ is inserted in front of the integral itself. In fact changing $\boldsymbol{n}$ to $-\boldsymbol{n}$ does not change the integrand.

The considerations of this and the previous appendices could be extended to the case when an external force per unit mass X acts on the molecules; the only difference would be to add a term $X \cdot \partial P/\partial \boldsymbol{\xi}$ on the left-hand side of eqn (A4.40): since we shall usually consider cases when the external action on the gas, if any, is exerted through solid boundaries (surface forces), we shall not usually write the aforementioned term, although it should be kept in mind that such simplification implies neglecting, *inter alia*, gravity.

Extensions of the Boltzmann equation to molecular models different from hard spheres are possible. This line began with Boltzmann himself, who, following previous calculations made by Maxwell, considered molecules modelled as point masses which repel each other with a central force. It is not hard to write a Boltzmann equation for this case [10], but since the rigorous theory for these molecular models is at a very preliminary stage, we shall not consider them in these technical appendices. The situation becomes more complicated in the case of polyatomic molecules, which were first dealt with by Boltzmann in his basic 1872 paper [1]. We return to this aspect in Chapter 8.

We mention finally that it is possible to retain some of the effects of the finite size of the molecules which disappear in the Boltzmann–Grad limit, as shown by Enskog in 1921 [15]. The relation between the Enskog equation and the Liouville equation is unclear from a rigorous standpoint. Once accepted however, the Enskog equation gives results in reasonably good agreement with experiment and lends itself to interesting mathematical investigations.

Appendix 4.5
The Boltzmann equation in the homogeneous isotropic case

As indicated in the main text, when Boltzmann first wrote his equation in the first part of ref. [1], he restricted himself to the case when the solutions do not depend on x, but only on time and the magnitude of the molecular velocity $\boldsymbol{\xi}$, i.e. on the kinetic energy per unit mass $|\boldsymbol{\xi}|^2/2$, which we shall simply denote by E. It is then immediate to see that the left-hand side of the Boltzmann equation (A4.40) reduces to just the derivative with respect to t, whereas on the right-hand side the collision integral can be considerably simplified. As a matter of fact, besides E and E_* one can use as a third variable E' and express E_*' as $E + E_* - E'$. The fivefold integral can then be reduced to a double integral by first expressing it as an integral with respect to E_* and over the surfaces of the unit spheres described by n and $m = \boldsymbol{\xi}/|\boldsymbol{\xi}|$. Three of the four integrations over the spherical surfaces can be performed; the fourth can be reduced to an integration with respect to E'. It is clear that the final form of the equation takes the form:

$$\frac{\partial P}{\partial t} = N\sigma^2 \int_0^\infty \int_0^{E+E'} (P'P_*' - PP_*)\, b(E, E_*, E')\, \mathrm{d}E_*\, \mathrm{d}E'. \tag{A4.42}$$

The only thing that remains to be determined is then the expression of $b(E, E_*, E')$. To this end let us remark that this kernel is obtained by performing the fourfold integral

$$I = \int_{S_+} \int_S V \cdot n\, \mathrm{d}n\, \mathrm{d}m, \tag{A4.43}$$

where we have omitted the bars of absolute value, since the integral with respect to n is over the hemisphere where $V \cdot n$ is positive. Taking now as polar axis for m the direction of n and as polar axis for the latter the direction of $\boldsymbol{\xi}$, the integral takes a convenient form, by noting that $V \cdot n = \boldsymbol{\xi} \cdot n - \boldsymbol{\xi}_* \cdot n = \rho\mu - \rho_*\mu_*$, where μ and μ_* are the cosines of the angles between n and m and their respective polar axes, whereas ρ and ρ_* denote $(2E)^{1/2}$ and $(2E_*)^{1/2}$ respectively. The calculation can now be performed without any major difficulty, but it turns out to be tedious and is omitted here, especially because it is restricted to the case of hard spheres. The result reads as follows:

$$
\begin{aligned}
b(E, E_*, E') &= 8\pi^2 & \text{for} \quad & E' \geq E,\ E_* \geq E \\
b(E, E_*, E') &= 8\pi^2 (E_*/E)^{1/2} & \text{for} \quad & E' \geq E,\ E_* \leq E \\
b(E, E_*, E') &= 8\pi^2 (E'/E)^{1/2} & \text{for} \quad & E' \leq E,\ E_* \geq E \\
b(E, E_*, E') &= 8\pi^2 [(E' + E_*)/E - 1]^{1/2} & \text{for} \quad & E' \leq E,\ E_* \leq E.
\end{aligned}
\tag{A4.44}
$$

Note however that eqn (A4.42) is not the same as that used by Boltzmann, who utilized a distribution function f for energy, defined in such way that, apart from constant factor, $\int f\, \mathrm{d}E = \int P\, \mathrm{d}\boldsymbol{\xi}$. It follows, again apart from a constant factor, that $P = (E)^{-1/2} f$, and this produces the square roots mentioned in the main text.

Appendix 5.1
Collision-invariants

Before embarking on the study of the properties of the Boltzmann equation, which will culminate in a discussion of the H-theorem, we remark that the unknown in the equation under consideration is not always chosen to be a probability density as in the previous chapters; in fact, we can multiply it by a suitable factor and transform it into an (expected) number density or an (expected) mass density. Both densities must be understood not in ordinary space but in phase space. With this change, the only thing that must be modified is the factor in front of the term describing the effect of collisions in eqn (A4.40), which is no longer $N\sigma^2$. In order to avoid any commitment to a special choice of that factor, we replace $N\sigma^2$ by a constant α and the unknown P by another symbol f (which is also the symbol most commonly used to denote the one-particle distribution function, no matter what its normalization is). In some physically interesting situations in which the gas domain is the entire $\Re^3$ and the total mass (or total number of particles) is only locally finite (i.e. $\int_{\Omega \times \Re^3} f(x, \xi) \, dx \, d\xi < +\infty$, for any bounded $\Omega \in \Re^3$) the distribution function cannot even be normalized (i.e. $\int_{\Re^6} f(x, \xi) \, dx \, d\xi = +\infty$). Let us then rewrite eqn (A4.40) in the following form:

$$\frac{\partial f}{\partial t} + \xi \cdot \frac{\partial f}{\partial x} = \alpha \int_{\Re^3} \int_{S_+} (f'f'_* - ff_*)|V \cdot n| \, d\xi_* \, dn. \tag{A5.1}$$

The right-hand side contains a quadratic expression $Q(f, f)$, given by:

$$Q(f, f) = \int_{\Re^3} \int_{S_+} (f'f'_* - ff_*)|V \cdot n| \, d\xi_* \, dn. \tag{A5.2}$$

This expression is called the collision integral or simply the collision term, and the quadratic operator Q goes by the name of collision operator. The fact that one usually writes $Q(f, f)$ rather than simply $Q(f)$ is meant to underline the fact that we are dealing with a second-degree (and hence non-linear) operator, with which one can usually associate a bilinear operator $Q(f, g)$ that acts linearly on f and g separately and reduces to $Q(f, f)$, when $g = f$ [37, 38, 76]. Here we study some elementary properties of Q.

Our first aim is to study the eightfold integral:

$$\int_{\Re^3} Q(f, f)\phi(\xi) \, d\xi = \int_{\Re^3} \int_{\Re^3} \int_{S_+} (f'f'_* - ff_*)\phi(\xi))|V \cdot n| \, d\xi \, d\xi_* \, dn, \tag{A5.3}$$

where f and ϕ are functions such that the indicated integrals exist and the order of integration does not matter. A simple interchange of the asterisked and unasterisked variables (with a glance at eqn (A4.41)) shows that

$$\int_{\Re^3} Q(f, f)\phi(\xi) \, d\xi = \int_{\Re^3} \int_{\Re^3} \int_{S_+} (f'f'_* - ff_*)\phi(\xi_*)|V \cdot n| \, d\xi \, d\xi_* \, dn. \tag{A5.4}$$

Next we consider another transformation of variables, the exchange of primed and unprimed variables (which is possible because the transformation in eqn (A4.41) is, for a fixed n, its own inverse). This gives

$$\int_{\Re^3} Q(f, f)\phi(\xi) \, d\xi = \int_{\Re^3} \int_{\Re^3} \int_{S_+} (ff_* - f'f'_*)\phi(\xi')|V \cdot n| \, d\xi' \, d\xi'_* \, dn. \tag{A5.5}$$

(Actually, since $V' \cdot n = -V \cdot n$, we should write S_- in place of S_+; changing n into $-n$, however, gives exactly the expression written here.)

The absolute value of the Jacobian determinant for the change of variables from $\boldsymbol{\xi}, \boldsymbol{\xi}_*$ to $\boldsymbol{\xi}', \boldsymbol{\xi}'_*$ is unity; thus we can write $d\boldsymbol{\xi}\, d\boldsymbol{\xi}_*$ in place of $d\boldsymbol{\xi}'\, d\boldsymbol{\xi}'_*$ and eqns (A5.4) and (A5.5) become

$$\int_{\Re^3} Q(f, f)\phi(\boldsymbol{\xi})\, d\boldsymbol{\xi} = \int_{\Re^3} \int_{\Re^3} \int_{S_+} (ff_* - f'f'_*)\phi(\boldsymbol{\xi}')|V \cdot \boldsymbol{n}|\, d\boldsymbol{\xi}\, d\boldsymbol{\xi}_*\, d\boldsymbol{n}, \tag{A5.6}$$

$$\int_{\Re^3} Q(f, f)\phi(\boldsymbol{\xi})\, d\boldsymbol{\xi} = \int_{\Re^3} \int_{\Re^3} \int_{S_+} (ff_* - f'f'_*)\phi(\boldsymbol{\xi}'_*)|V \cdot \boldsymbol{n}|\, d\boldsymbol{\xi}\, d\boldsymbol{\xi}_*\, d\boldsymbol{n}. \tag{A5.7}$$

Equations (A5.4), (A5.6), (A5.7) differ from eqn (A5.3) because the factor $\phi(\boldsymbol{\xi})$ is replaced by $\phi(\boldsymbol{\xi}_*)$, $-\phi(\boldsymbol{\xi}')$, and $-\phi(\boldsymbol{\xi}'_*)$ respectively. We can now obtain further expressions for the integral on the left-hand side by taking linear combinations of the four different expressions available. Of these, the most interesting is the expression obtained by taking the sum of eqns (A5.3), (A5.4), (A5.6), and (A5.7) and dividing by four. The result is:

$$\int_{\Re^3} Q(f, f)\phi(\boldsymbol{\xi})\, d\boldsymbol{\xi}$$
$$= \frac{1}{4} \int_{\Re^3} \int_{\Re^3} \int_{S_+} (f'f'_* - ff_*)(\phi + \phi* - \phi' - \phi*')|V \cdot \boldsymbol{n}|\, d\boldsymbol{\xi}\, d\boldsymbol{\xi}_*\, d\boldsymbol{n}. \tag{A5.8}$$

This relation expresses a basic property of the collision term, which is frequently used. We now observe that the integral in eqn (A5.8) is zero, independently of the particular function f, if the relation

$$\phi + \phi_* = \phi' + \phi'_* \tag{A5.9}$$

holds almost everywhere in velocity space. Since the integral appearing on the left-hand side of eqn (A5.8) is the rate of change of the average value of the function ϕ due to collisions, the functions satisfying eqn (A5.9) are called "collision-invariants".

The first discussion of eqn (A5.9) is due to Boltzmann [77, 78], who assumed ϕ to be twice differentiable and arrived at the result that the most general solution of eqn (A5.9) is given by

$$\phi(\boldsymbol{\xi}) = a + \boldsymbol{b} \cdot \boldsymbol{\xi} + c|\boldsymbol{\xi}|^2. \tag{A5.10}$$

After Boltzmann, the matter of finding the solutions of eqn (A5.9) was investigated by Gronwall [79, 80] (who was the first to reduce the problem to Cauchy's functional equation for linear functions), Carleman [3], and Grad [23]. All these authors assumed ϕ to be continuous and proved that it must be of the form given in eqn (A5.10). Slightly different versions of Carleman's proof are given in refs [37] and [81]. In the latter monograph the authors assert that the solution is of the form (A5.10), even if the function ϕ is assumed to be measurable rather than continuous. In fact they use a result on the solutions of Cauchy's equation:

$$f(u + v) = f(u) + f(v) \qquad (u, v \in \Re^n \text{ or } \Re_+) \tag{A5.11}$$

holding for measurable functions. When passing from continuous to (possibly) discontinuous functions however, one should insist on the fact that eqn (A5.9) is satisfied almost everywhere and not everywhere in $\Re^3 \times \Re^3 \times S^2$, as (implicitly) assumed in ref. [22]. It should be possible, although this has never been attempted, to transform the proof under discussion into a proof that the collision-invariants are the classical ones under the assumption that eqn (A5.9) holds only almost everywhere.

The problem of solving eqn (A5.9) was tackled by the author of the present book [82] with the aim of proving that eqn (A5.10) gives the most general solution of eqn (A5.9), when the latter is satisfied almost everywhere in $\Re^3 \times \Re^3 \times S^2$, under the assumption that ϕ is in the

Hilbert space H_ω of the square integrable functions with respect to a Maxwellian weight $\omega(|\boldsymbol{\xi}|) = (\beta/\pi)^{3/2} \exp(-\beta|\boldsymbol{\xi}|^2)$, $\beta > 0$. The first step was to show that the linear manifold of the solutions possessed a polynomial basis. After that it was sufficient to look for smooth solutions. The existence of these can be made very simple if we look for C^2 solutions, as done by Boltzmann [77, 78].

A completely different proof of the same result (under the assumption that $\phi \in L^1_{\text{loc}}$) was contained in a paper by Arkeryd [83], but remained largely ignored in the literature. As shown in a paper by Arkeryd and Cercignani [84], Arkeryd's argument, when combined with the proof for C^2 functions given by Boltzmann and made more precise in ref. [82], allows a very simple proof of the fact that eqn (A5.10) is the most general solution when $\phi \in L^1_{\text{loc}}$ and eqn (A5.9) is satisfied almost everywhere.

In ref. [84], which together with the monograph [38] contains a complete discussion of the problem, a detailed proof of the result is given under the weaker assumption that f is measurable and finite almost everywhere and that eqn (A5.9) holds almost everywhere.

Here we shall restrict ourselves to Boltzmann's proof in its modern version [82] and refer the reader to the previously quoted literature for proofs holding under more general assumptions. Let us remark, however, that Wennberg [85] has recently shown that Boltzmann's proof, interpreted in the framework of distribution theory, yields a very general proof in which ϕ can be simply assumed to be a generalized function or distribution.

In order to show that eqn (A5.10) follows from eqn (A5.9), when ϕ is twice differentiable, let us first note that clearly one must have:

$$\phi(\boldsymbol{\xi}) + \phi(\boldsymbol{\xi}_*) = F(\boldsymbol{\xi} + \boldsymbol{\xi}_*, E), \tag{A5.12}$$

where

$$E = \tfrac{1}{2}(|\boldsymbol{\xi}|^2 + |\boldsymbol{\xi}_*|^2). \tag{A5.13}$$

In fact, given $\boldsymbol{\xi} + \boldsymbol{\xi}_*$ and E, $\boldsymbol{\xi}'$ and $\boldsymbol{\xi}'_*$ are completely arbitrary, provided that $\boldsymbol{\xi} + \boldsymbol{\xi}_*$ and E are conserved. For a more detailed proof of this point see the paper by Arkeryd and Cercignani [84].

If we take the gradient of eqn (A5.12) with respect to $\boldsymbol{\xi}$ and subtract the analogous gradient with respect to $\boldsymbol{\xi}_*$ from the result, we obtain:

$$\frac{\partial \phi}{\partial \boldsymbol{\xi}} - \frac{\partial \phi}{\partial \boldsymbol{\xi}_*} = \frac{\partial F}{\partial E}(\boldsymbol{\xi} - \boldsymbol{\xi}_*), \tag{A5.14}$$

where the arguments are made clear by the variables with respect to which a function is differentiated. Equation (A5.14) implies:

$$\left(\frac{\partial \phi}{\partial \xi_i} - \frac{\partial \phi}{\partial \xi_{*i}}\right)(\xi_k - \xi_{*k}) = \left(\frac{\partial \phi}{\partial \xi_k} - \frac{\partial \phi}{\partial \xi_{*k}}\right)(\xi_i - \xi_{*i}) \quad (i, k = 1, 2, 3). \tag{A5.15}$$

If we now differentiate with respect to ξ_r, we obtain

$$\left(\frac{\partial \phi}{\partial \xi_i} - \frac{\partial \phi}{\partial \xi_{*i}}\right)\delta_{kr} + \frac{\partial^2 \phi}{\partial \xi_i \partial \xi_r}(\xi_k - \xi_{*k})$$
$$= \left(\frac{\partial \phi}{\partial \xi_k} - \frac{\partial \phi}{\partial \xi_{*k}}\right)\delta_{ir} + \frac{\partial^2 \phi}{\partial \xi_k \partial \xi_r}(\xi_i - \xi_{*i}), \tag{A5.16}$$

where δ_{kr} denotes the Kronecker delta ($\delta_{kr} = 1$ if $k = r$, $\delta_{kr} = 0$ if $k \neq r$). A further differentiation with respect to ξ_{*j} yields:

$$\frac{\partial^2 \phi}{\partial \xi_i \partial \xi_r}\delta_{kj} + \frac{\partial^2 \phi}{\partial \xi_{*i} \partial \xi_{*j}}\delta_{kr} = \frac{\partial^2 \phi}{\partial \xi_k \partial \xi_r}\delta_{ij} + \frac{\partial^2 \phi}{\partial \xi_{*k} \partial \xi_{*j}}\delta_{ir}. \tag{A5.17}$$

If we let i, k, r take three different values ($i = 1, k = 2, r = 3$) and $j = k$, say, we obtain:

$$\frac{\partial^2 \phi}{\partial \xi_i \partial \xi_r} = 0 \qquad (i, r = 1, 2, 3; i \neq r). \tag{A5.18}$$

If we now take $i = r, k = j, i \neq k$ in eqn (A5.17), we obtain:

$$\frac{\partial^2 \phi}{\partial \xi_i^2} = \frac{\partial^2 \phi}{\partial \xi_{*k}^2} \qquad (i \neq k). \tag{A5.19}$$

Since the right-hand side cannot depend on $\boldsymbol{\xi}$, we conclude that both sides are equal to a constant; this constant does not depend upon the index, since we can change the values of i and k, provided that we keep $i \neq k$. From eqns (A5.18) and (A5.19) we conclude:

$$\frac{\partial^2 \phi}{\partial \xi_i \partial \xi_r} = 2c\delta_{ir} \quad (i, r = 1, 2, 3; c = \text{ constant}). \tag{A5.20}$$

Equation (A5.20) immediately yields eqn (A5.10).

An elementary application of the notion of collision-invariant can be given, as an example, in the space-homogeneous case. If we multiply the Boltzmann equation (A5.1) (where of course we omit the term with the derivatives with respect to $\boldsymbol{x}$) by 1, $\boldsymbol{\xi}$, $|\boldsymbol{\xi}|$, and integrate over the entire velocity space, we obtain (since the contribution from the right-hand side vanishes):

$$\frac{\partial \rho}{\partial t} = 0, \tag{A5.21}$$

$$\frac{\partial}{\partial t}(\rho v_j) = 0 \qquad (j = 1, 2, 3), \tag{A5.22}$$

$$\frac{\partial}{\partial t}(w) = 0, \tag{A5.23}$$

where we defined the gas density in ordinary space

$$\rho = \int_{\mathfrak{R}^3} f \, d\boldsymbol{\xi}, \tag{A5.24}$$

the bulk velocity $\boldsymbol{v}$ of the gas (e.g. the air velocity that we feel when the wind blows)

$$\boldsymbol{v} = \frac{\left(\int_{\mathfrak{R}^3} \boldsymbol{\xi} f \, d\boldsymbol{\xi}\right)}{\left(\int_{\mathfrak{R}^3} f \, d\boldsymbol{\xi}\right)}, \tag{A5.25}$$

and the energy density per unit volume:

$$w = \tfrac{1}{2} \int_{\mathfrak{R}^3} |\boldsymbol{\xi}|^2 f \, d\boldsymbol{\xi}; \tag{A5.26}$$

Equation (A5.21), (A5.22), and (A5.23) tell us that the mass and energy densities and the bulk velocity remain constant in the homogeneous case (conservation of mass, momentum and energy).

Appendix 5.2
Boltzmann's inequality and the Maxwellian distribution

In this appendix we investigate the existence of positive functions f which give a vanishing collision integral:

$$Q(f, f) = \int_{\Re^3} \int_{S_+} (f' f'_* - f f_*) |V \cdot n| \, d\xi_* \, dn = 0. \tag{A5.27}$$

As we said in the main text, this problem is one of the motivations laid down by Boltzmann in the introduction to his basic paper [1]. In order to solve eqn (A5.27), we prove a preliminary result which plays an important role in the theory of the Boltzmann equation and in the proof of the *H*-theorem, which was the second declared aim of his paper [1]: if f is a non-negative function such that $(\log f) Q(f, f)$ is integrable and the assumptions of the previous section hold when $\phi = \log f$, then the following *Boltzmann inequality* holds:

$$\int_{\Re^3} \log f \, Q(f, f) \, d\xi \leq 0; \tag{A5.28}$$

further, the equality sign applies if and only if $\log f$ is a collision-invariant, or, equivalently, f is given (almost everywhere) by

$$f = \exp(a + b \cdot \xi + c |\xi|^2), \tag{A5.29}$$

where a, b, c are constant with respect to ξ. To prove eqn (A5.28) it is enough to use eqn (A5.8) with $\phi = \log f$ (and the well-known properties of the logarithmic function):

$$\int_{\Re^3} \log f Q(f, f) \, d\xi =$$
$$-\frac{1}{4} \int_{\Re^3} \int_{\Re^3} \int_{S_+} \log(f' f'_*/f f_*)(f' f'_* - f f_*) |V \cdot n| \, d\xi \, d\xi_* \, dn \tag{A5.30}$$

and eqn (A5.28) follows thanks to the elementary inequality (please note that the two factors are positive, negative, or zero at the same time):

$$(y - z) \log(y/z) \geq 0 \quad (y, z \in \Re+). \tag{A5.31}$$

Equation (A5.31) becomes an equality if and only if $y = z$; thus the equality sign holds in eqn (A5.28) if and only if

$$f' f'_* = f f_* \tag{A5.32}$$

applies almost everywhere. But, taking the logarithms of both sides of eqn (A5.32), we find that $\phi = \log f$ satisfies eqn (A5.9) and its most general expression is thus given by eqn (A5.10). Then $f = \exp(\phi)$ is given by eqn (A5.29).

We remark that in the same equation c must be negative, since $f \in L^1(\Re^3)$. If we let $c = -\beta, b = 2\beta v$ (where v is another constant vector), eqn (A5.29) can be rewritten as follows:

$$f = A \exp(-\beta |\xi - v|^2), \tag{A5.33}$$

where A is a positive constant related to $a, c, |b|^2$ (β, v, A constitute a new set of constants). The function appearing in eqn (A5.33) is the so-called *Maxwell distribution* or *Maxwellian*, which we have frequently mentioned before. Frequently one considers Maxwellians with $v = 0$

(non-drifting Maxwellians), which can be obtained from drifting Maxwellians by a change of the origin in velocity space.

Let us return now to the problem of solving eqn (A5.27). Multiplying both sides by $\log f$ gives eqn (A5.28) with the equality sign. This implies that f is a Maxwellian almost everywhere, by the result which has just been proved. Suppose now that f is (almost everywhere) a Maxwellian; then $f = \exp(\phi)$, where ϕ is a collision-invariant and eqn (A5.32) holds; eqn (A5.27) then also holds. Thus there are functions which satisfy eqn (A5.27) and they are all Maxwellians, given by eqn (A5.33).

Appendix 5.3
The *H*-theorem

Let us consider the most famous application of the properties of the collision term $Q(f, f)$ of the Boltzmann equation:

$$\frac{\partial f}{\partial t} + \boldsymbol{\xi} \cdot \frac{\partial f}{\partial x} = \alpha Q(f, f). \tag{A5.34}$$

If we multiply both sides of this equation by $\log f$ and integrate with respect to $\boldsymbol{\xi}$, we obtain:

$$\frac{\partial \mathcal{H}}{\partial t} + \frac{\partial}{\partial x} \cdot \boldsymbol{J} = \mathcal{S}, \tag{A5.35}$$

where we have written (as permitted) $\boldsymbol{\xi}$ after the symbol of partial differentiation and also interchanged the order of differentiation and integration. We have also introduced the following abbreviations:

$$\mathcal{H} = \int_{\Re^3} f \log f \, \mathrm{d}\boldsymbol{\xi}, \tag{A5.36}$$

$$\boldsymbol{J} = \int_{\Re^3} \boldsymbol{\xi} f \log f \, \mathrm{d}\boldsymbol{\xi}, \tag{A5.37}$$

$$\mathcal{S} = \alpha \int_{\Re^3} \log f \, Q(f, f) \, \mathrm{d}\boldsymbol{\xi}. \tag{A5.38}$$

We know however that the Boltzmann inequality, eqn (A5.28), implies:

$$\mathcal{S} \le 0 \quad \text{and} \quad \mathcal{S} = 0 \quad \text{iff} \quad f \quad \text{is a Maxwellian.} \tag{A5.39}$$

Because of this inequality, eqn (A5.35) plays an important role in the theory of the Boltzmann equation. Let us illustrate the role of eqn (A5.39) in the case of space-homogeneous solutions. In this case the various quantities do not depend on x and eqn (A5.35) reduces to

$$\frac{\partial \mathcal{H}}{\partial t} = \mathcal{S} \le 0. \tag{A5.40}$$

This implies Boltzmann's *H*-theorem (for the space-homogeneous case): $\mathcal{H}$ is a decreasing quantity, unless f is a Maxwellian (in which case the time derivative of $\mathcal{H}$ is zero). Remember now (from the final example in Appendix 5.1) that in this case the densities ρ, ρv, and w are constant in time; we can thus build a constant Maxwellian M which has, at any time, the same ρ, v, and w as any solution f corresponding to given initial data. Since $\mathcal{H}$ decreases unless f is a Maxwellian (i.e. $f = M$), it is tempting to conclude that f tends to M when $t \to \infty$. The temptation is strengthened when we realize that $\mathcal{H}$ is bounded from below by $\mathcal{H}_M$, the value taken by the functional $\mathcal{H}$ when $f = M$. In fact, the inequality $z \log z - z \log y + y - z \ge 0$, holding for non-negative values of y and z, and the fact that $\log M$ is a collision-invariant and hence $\int_{\Re^3} \log M(f - M) \, \mathrm{d}\boldsymbol{\xi} = 0$, imply the bound from below which has just been mentioned, by letting $z = f$, $y = M$, and integrating with respect to $\boldsymbol{\xi}$. Since $\mathcal{H}$ is decreasing, its derivative is non-positive (where it exists; but it exists almost everywhere because of monotonicity) unless the function coincides with $\mathcal{H}_M$; one feels that $\mathcal{H}$ cannot do anything but tend to $\mathcal{H}_M$! However, this conclusion is unwarranted, without a more detailed consideration of the source term $\mathcal{S}$ in eqn (A5.40). Let us sketch this rather technical proof in the case of continuous data. In this case the solution is continuous in $\boldsymbol{\xi}$, uniformly with respect to t [2].

Since $\mathcal{H}$ is bounded from above and from below, we can find a sequence of time instants $t_n, t_n \to \infty$, such that

$$\lim_{t \to \infty} \frac{d\mathcal{H}}{dt}(t_n) = 0. \tag{A5.41}$$

For a convergence criterion, we can extract from this sequence a converging sequence (still denoted by $f(t_n) \equiv f_n$) tending to some f_∞. In order to prove the result sought for, it is enough to show that the following equation holds:

$$f'_\infty f'_{\infty*} - f_\infty f_{\infty*} = 0. \tag{A5.42}$$

In fact, were this not true, there would exist a non-zero measure set $D \in \mathfrak{R}^3 \times \mathfrak{R}^3 \times S_+$ where one would have

$$|f'_\infty f'_{\infty*} - f_\infty f_{\infty*}| \geq \rho > 0. \tag{A5.43}$$

Since f_n converges uniformly to f_∞, it would then be possible to find an n_0 such that for $n > n_0$ one would have in D:

$$|f'_n f'_{n*} - f_n f_{n*}| > \rho/2 > 0. \tag{A5.44}$$

But then, if f is bounded by a constant C (as occurs for Carleman's solutions) one would have

$$|f'_n f'_{n*} - f_n f_{n*}| \left| \log \frac{f_n f_{n*}}{f'_n f'_{n*}} \right| > \tfrac{1}{2} \rho \log[1 + \rho/(2C^2)] > 0 \tag{A5.45}$$

and hence, because of eqns (A5.41) and (A5.30),

$$\lim_{t \to \infty} \frac{d\mathcal{H}}{dt}(t_n) = 0 \leq -\tfrac{1}{8} \rho \log[1 + \rho^2/(2C^2)] \int_D |V \cdot n| \, d\boldsymbol{\xi} \, d\boldsymbol{\xi}_* \, d\boldsymbol{n}, \tag{A5.46}$$

which is contradictory if D has non-zero measure. Equation (A5.42) is thus proved and f_∞ turns to be a Maxwellian (uniquely singled out by the initial data and the conservation laws discussed at the end of Appendix 5.1).

Note that the difficult part is to prove that $\mathcal{H}$ tends to $\mathcal{H}_M$; if this is allowed, then it is easy to show that f tends to M, thanks to an improved version of a previously used inequality which takes the following form:

$$f \log f - f \log M + M - f \geq g \left(\frac{|f - M|}{M} \right) |f - M|, \tag{A5.47}$$

where c is a constant (independent of f) and

$$g(z) = \begin{cases} z & \text{if } 0 \leq z \leq 1 \\ 1 & \text{if } z \geq 1 \end{cases}. \tag{A5.48}$$

Integrating both sides of (A5.47) gives

$$\mathcal{H} - \mathcal{H}_M \geq c \left[\int_{L_t} |f - M| \, d\boldsymbol{\xi} + \int_{S_t} |f - M|^2 M^{-1} \, d\boldsymbol{\xi} \right], \tag{A5.49}$$

where L_t and S_t denote the (time-dependent) sets where $|f - M|$ is respectively larger and smaller than M. Since we concede that $\mathcal{H}$ tends to a $\mathcal{H}_M$, it follows that both integrals tend to zero when $t \to \infty$. The fact that the second integral tends to zero implies, by Schwarz's inequality, that

$$\int_{S_t} |f - M| \, d\boldsymbol{\xi} \le \left[\int_{S_t} |f - M|^2 M^{-1} \, d\boldsymbol{\xi} \right]^{1/2} \left[\int_{S_t} M \, d\boldsymbol{\xi} \right]^{1/2} \to 0. \tag{A5.50}$$

Then

$$\int_{\mathfrak{R}^3} |f - M| \, d\boldsymbol{\xi} = \int_{L_t} |f - M| \, d\boldsymbol{\xi} + \int_{S_t} |f - M| \, d\boldsymbol{\xi} \tag{A5.51}$$

also tends to zero and f tends (strongly) to M in L^1.

If the state of the gas is not space-homogeneous, the situation becomes more complicated. In this case it is convenient to introduce the quantity

$$H = \int_\Omega \mathcal{H} \, d\boldsymbol{x}, \tag{A5.52}$$

where Ω is the space domain occupied by the gas (assumed here to be time-independent). Then eqn (A5.35) implies

$$\frac{dH}{dt} \le \int_{\partial\Omega} \boldsymbol{J} \cdot \boldsymbol{n} \, d\sigma, \tag{A5.53}$$

where $\boldsymbol{n}$ is the inward normal and $d\sigma$ the measure on $\partial\Omega$. Clearly, several situations may arise. Among the most typical ones, we quote the following:

(a) Ω is a box with periodicity boundary conditions (the so-called flat torus). Then there is no boundary, $dH/dt \le 0$, and one can repeat about H what was said about $\mathcal{H}$ in the space-homogeneous case. In particular, there is a natural (space-homogeneous) Maxwellian associated with the total mass, momentum, and energy (which are clearly conserved, as can be shown by arguing as in the homogeneous case at the end of Appendix 5.1).

(b) Ω is a closed and bounded domain with boundary conditions of specular reflection. In this case the boundary term also disappears because the integrand of $\boldsymbol{J} \cdot \boldsymbol{n}$ is odd in $\boldsymbol{v} \cdot \boldsymbol{n}$ on $\partial\Omega$ and the situation is similar to that in case (a). There might seem to be a difficulty in the choice of the natural Maxwellian because momentum is not conserved, but a simple argument shows that the total momentum must vanish when $t \to \infty$. Thus M is a non-drifting Maxwellian for a box or a general domain. An exception is provided by domains which are rotationally symmetric with respect to some axis, if the angular momentum with respect to this axis, Γ, is initially non-zero. In fact in this case Γ remains constant in time and the natural Maxwellian will have a bulk velocity corresponding to a solid body rotation about the said axis, the angular velocity following by a computation of Γ for the said Maxwellian.

(c) Ω is the entire space. Then the asymptotic behaviour of the initial values at ∞ is of paramount importance. If the gas is initially more concentrated at finite distances from, say, the origin of a system of Cartesian coordinates, one physically expects and can mathematically prove [38] that the gas escapes through infinity and the asymptotic state is a vacuum.

If the initial data are different, almost anything can happen: for example, convergence to a Maxwellian (if the data are bounded, there are enough collisions and the Maxwellian is uniquely determined), or even lack of solution after a certain $\bar{t}$ (implosion with increasing speed because of a high inflow of gas from space infinity). In the latter case, the Boltzmann equation ceases to hold when the density becomes so large that the conditions prescribed by the Boltzmann–Grad limit are no longer satisfied.

(d) Ω is a compact domain but the boundary conditions on $\partial\Omega$ are different from specular reflection. Then the asymptotic state may be completely different from a Maxwellian.

For a more complete discussion of the various cases we refer to the treatise [38] and the literature quoted therein.

Appendix 5.4
The hourglass model

To illustrate the nature of the Loschmidt paradox and the mathematical significance of the growth of entropy, in this appendix we shall use, as indicated in the main text, a mathematical model of an hourglass. This model gives a simplified description of the way this device works by ignoring energy, momentum, etc. of the sand grains and paying attention only to the number of grains in one of the two halves of the device. The model we are going to use is described by B. Touscheck and G. Rossi [86] in the case of infinitely many grains and by M. Ageno [51] in the case of finitely many grains, say N. Neither author gives any reference to previous papers. We shall start from the model with N grains and shall later study what happens in the limit $N \to \infty$.

Let us describe the model. We assume that the holes between the upper and lower bulbs is so small that only one grain of sand can pass through at a time. Let us denote by λ the probability per unit time that a grain actually goes through. We shall assume that λ is a given constant. Now let $p_s(t)$ be the probability that there are s grains in the lower container. The time evolution of $p_s(t)$ is given by:

$$
\begin{aligned}
\dot{p}_0 &= -\lambda p_0, \\
\dot{p}_s &= \lambda(p_{s-1} - p_s) \qquad (s = 1, 2, \ldots, N-1), \\
\dot{p}_N &= \lambda p_{N-1},
\end{aligned}
\tag{A5.54}
$$

where the dot denotes the time derivative. This system of course needs initial conditions. The typical set will be $p_0 = 1$ and $p_s = 0$ for $s = 1, 2, \ldots, N$ (there are no grains in the lower bulb).

Let us ask now what is the probability q_s of having s grains in the upper bulb in terms of the probability p_s of having s grains in the lower bulb. Since there are s grains in the upper bulb if and only if there are $N - s$ in the lower one, we obtain $q_s = p_{N-s}$. Thus it is sufficient to replace s by $N - s$ in the last system to obtain the system governing the probabilities q_s:

$$
\begin{aligned}
\dot{q}_0 &= \lambda q_1, \\
\dot{q}_s &= -\lambda(q_s - q_{s+1}) \qquad (s = 1, 2, \ldots, N-1), \\
\dot{q}_N &= \lambda q_N,
\end{aligned}
\tag{A5.55}
$$

where we have written the last equation of the previous system as the first one and the previous first as the last one.

We now remark that our system is not time-reversible, but it is almost so (when N is sufficiently large). In fact, the transformation which changes t into $-t$, and p_s into q_s gives the system:

$$
\begin{aligned}
\dot{q}_0 &= \lambda q_0, \\
\dot{q}_s &= -\lambda(q_{s-1} - q_s) \qquad (s = 1, 2, \ldots, N-1), \\
\dot{q}_N &= -\lambda q_{N-1},
\end{aligned}
\tag{A5.56}
$$

which differs from the previous one because the indices on the right-hand side are one smaller. Thus the time reversal is not an exact symmetry of our system, but only an approximate one. Somebody who turns over the hourglass and thus exchanges the two halves so quickly that during the operation no grain passes through the little hole separating them performs an operation of time reversal, while the "little universe" under consideration can be in any state and thus Loschmidt's paradox occurs before our eyes. It is clear that the Zermelo paradox does not show up in this model, nor do fluctuations. The slight inaccuracy is to be blamed on the fact that there is also a probability that a grain is exactly in the hole, as well as on the fact that there is a preferential direction of motion, imposed by gravity, which we disregarded.

The world of an hourglass has a beginning and an end. The beginning is when we put it on a table and the end is when all the grains are in the lower half. Usually at the beginning all the sand grains are in the upper half; but we can envisage different initial conditions. The operation of time reversal which we have just discussed can produce these different, unusual, initial conditions.

Let us now examine the behaviour of entropy or its opposite, H. In agreement with the aforementioned books [51, 86], we let

$$H = \sum_{s=0}^{s=N} p_s \log p_s \tag{A5.57}$$

and compute its time derivative with the help of (A5.54). We have:

$$\dot{H} = -\lambda p_0 \log p_0 + \lambda \sum_{s=1}^{s=N-1} p_{s-1} \log p_s - \lambda \sum_{s=1}^{s=N-1} p_s \log p_s + \lambda p_{N-1} \log p_N.$$

$$= -\lambda \sum_{s=1}^{s=N} p_{s-1} \log \frac{p_{s-1}}{p_s}. \tag{A5.58}$$

We now recall the elementary inequality

$$x \log x \geq x - 1 \qquad (x > 0) \tag{A5.59}$$

and apply it with $x = p_{s-1}/p_s$ to eqn (A5.58). We have:

$$\dot{H} \leq -\lambda \sum_{s=1}^{s=N} (p_{s-1} - p_s) = \lambda(p_N - p_0). \tag{A5.60}$$

Thus we see that H will decrease as long as p_N is less than p_0. In particular, if we start from the (typical) initial conditions $p_0 = 1$ and $p_s = 0$ for $s = 1, 2, \ldots, N$, the entropy will increase as long as the probability of having N grains is entirely negligible compared with that of having no grains in the lower bulb. We remark that p_0 decreases exponentially, but, from the explicit solution of the system (A5.54) [51], we see that p_N becomes comparable with it only when $t \cong N/(2\lambda)$. Thus for a time which increases with N we obtain an increase of H.

The situation changes after times of this order and we obtain that H increases (for the standard initial condition). To see this, we remark that each p_s tends to increase unless it is larger than p_{s-1}; thus eventually (if all p_s, except p_0, are initially zero) they tend to a situation where they are ordered in increasing size ($p_s \geq p_{s-1}$). Then (A5.58) shows that $\dot{H} \geq 0$, and the entropy starts to decrease (or H to increase).

What happens in the case of infinitely many particles? The time required for the entropy to reach its maximum during the time evolution will tend to infinity. Then the ordering of p_s by increasing size will be just a trend, but this situation will be never reached; they will tend to equalize (becoming smaller and smaller, because they are infinitely many). This is what the analysis in ref. [86] shows by using a different approach.

The analysis in ref. [51], though interesting, is misleading when the author tries to apply its conclusions to Boltzmann's arguments. Ageno correctly argues that normal hourglasses reach equilibrium in a relatively short time, and hence letting $N \to \infty$ does not describe the entire process correctly. He suggests that λ should also tend to infinity, and this would be correct from the point of view of the theory of an hourglass, but not so interesting for our purposes here. In fact we want to look at this simple device as at a model of the universe analogous to the picture that

Boltzmann and Poincaré had in mind, with different views on its validity. Here we have a system in which, starting from appropriate *highly ordered* initial conditions, we observe an increase in entropy during a first stage of the time evolution of the system. After a time interval, the size of which tends to infinity with the number of particles, the entropy starts to decrease and eventually we reach a highly ordered state, though different from the initial one (the sand grains are in the lower bulb rather than in the upper one).

Ageno [51] argues that the entropy of the hourglass is not related to the function H defined above and that the "entropy" (not otherwise defined) remains zero. This is used to provide support for the main thesis of the book [51], i.e. that the H-theorem is false and the increase in entropy is due to the interaction of the system with the walls and hence the rest of the universe. We need not discuss this argument, which is based on a misunderstanding not uncommon in this area, widely discussed in the main text. There is an interaction with the universe, as stressed in Section 5.5, but this is needed only to guarantee that all the time arrows agree with each other.

Appendix 6.1
Likelihood of a distribution

In this appendix we first wish to examine in more detail the starting formula for the discrete model, in which the distribution function is replaced by a collection of integers $n_0, n_1, n_2, \ldots, n_p$. If, following Boltzmann, we give the name "complexion" to a repartition with assigned values of these integers, we wish to show that the number P of the complexions compatible with a given distribution is given by eqn (6.2) of the main text:

$$P = \frac{n!}{n_0! n_1! n_2! \ldots n_p!}.$$

Readers familiar with combinatorics can skip this first part, up to the sentence containing eqn (A6.1).

The starting point deals with the number of permutations of n objects. Suppose we have n different letters and we wish to put them in all possible orders. What is the number of possible alignments in which none of the n letters is repeated? This number is called the number of permutations of n distinguishable objects.

This number increases incredibly fast with n. Thus, if we have three letters, say e, a, t, the possible alignments are six:

eat, eta, ate, aet, tea, tae.

If there are four letters, the possible alignments are 24. Thus, if the letters are a, e, l, t, we have:

aelt, aetl, alte, alet, atle, atel, elta, elat, etla, etal, eatl, ealt,
leat, leta, late, laet, ltea, ltae, tael, tale, tlae, tlea, tela, teal.

If we start thinking how to produce the alignments, we also quickly discover the rule for making them. In fact, suppose we fix the first letter; then all the permutations of the remaining letters will give all the possible permutations starting with that letter. Thus if p_n gives the number of permutations of n distinguishable objects, it will be n times the number of permutations of $n - 1$ distinguishable objects, because we can order the alignments by putting first all those which have a given letter (which can be chosen in n ways) and then permuting the remaining $n - 1$. Thus, in the second list above there are six "words" starting with "l" and the letters following "l" are those which appear in the previous list.

Then we compute easily: if we have one letter, there will be just one "word"; if we have two letters, the "words" will be $2 \cdot 1 = 2$; if we have three letters, the "words" will be $3 \cdot 2 \cdot 1 = 6$; if we have four letters, the "words" will be $4 \cdot 3 \cdot 2 \cdot 1 = 24$; if we have five letters, the "words" will be $5 \cdot 4 \cdot 3 \cdot 2 \cdot 1 = 120$. Thus if we have n letters, the "words" will be $n \cdot \ldots 5 \cdot 4 \cdot 3 \cdot 2 \cdot 1$, the product of the first n natural numbers, which is denoted by $n!$. Now if the objects are not necessarily distinguishable, either because some of them are the same or their difference is not of interest, the situation changes. If we have five a's, three e's, four t's, we have twelve objects, but we are a long way from being able to form the rather large number of 12! (close to 500 million) "words" that we can produce with twelve different letters. How many of these 12! "words" will be different? It is easy to answer; if we think for a moment of distinguishing the equal letters (e.g. by painting them in different colours), then we shall have 12! different "words". However, they are artificially different: we can permute the five a's, the three e's, the four t's in a "word" and obtain again the same "word" (apart from colours). So we must divide 12! by 5!3!4! in order to eliminate the spurious replicas. We obtain

$$P = \frac{12!}{5!3!4!} = \frac{6 \cdot 7 \cdot 8 \cdot 9 \cdot 10 \cdot 11 \cdot 12}{2 \cdot 3 \cdot 2 \cdot 3 \cdot 4} = 27\,720,$$

quite a large number but very far off 500 million!

Thus the problem of computing the permutations without repetition of n objects when there are $n_0, n_1, n_2, \ldots, n_p$ replicas of p distinguishable objects is solved by eqn (6.2) of the main text.

Passing now to the case of continuous variables, we wish to show that if P is a probability density, then

$$V(P) = -\overline{\log P} = -\int P \log P \, d\mu \qquad (A6.1)$$

(where $d\mu$ is the volume element in the state space $\mathcal{M}$, the probability density of which is P) is a measure of the likelihood of P. In other words, if we take many P's "at random", if positive and normalized, most of them will be close to the P's for which $V(P)$ has a maximum. In order to give a meaning to the expression "at random", let us start by subdividing the state space $\mathcal{M}$ into n little cells Ω_i of volume μ_i, while replacing P by n numbers P_i, the averages of P over the cells:

$$P_i = \frac{1}{\mu_i} \int_{\Omega_i} P \, d\mu \qquad \left(\sum_i P_i \mu_i = 1 \right). \qquad (A6.2)$$

Let us take then N objects and distribute them at random in the cells. If N_i of them are in Ω_i $(0 \leq N_i \leq N)$, let us take $P_i = N_i/(N\mu_i)$ as the probability for the cell Ω_i. Analogously, given a probability density P, we can represent it in terms of a distribution of N objects in n cells, with arbitrary accuracy, provided that we take n and N sufficiently large. For a *given order of approximation*, however, we have just a finite, though huge, number of possible distributions. If we distribute the objects at random, there are $W(P) = N!/(N_1!N_2!\ldots N_n!)$ ways of obtaining the distribution $P = (P_1, P_2, \ldots, P_n)$. The sum $\sum W(P)$ that gives the number of ways of obtaining any possible distribution equates to n^N, thanks to the formula for the Nth power of a polynomial, as applied to $(1 + 1 + \ldots + 1)^N$ (for details, see e.g. [27]).

It is then reasonable to take as a measure of the likelihood of the discretized distribution $P = (P_1, P_2, \ldots, P_n)$ the quantity $W(P)/n^N$ or its logarithm divided by N (in order to obtain a finite limit when $N \to \infty$):

$$V(P)) = \frac{1}{N} \log \left(\frac{N!}{N_1!N_2!\ldots N_n!n^N} \right). \qquad (A6.3)$$

Let us now compute the limit when $N \to \infty$, which should give us the appropriate expression for $V(P)$ when the probabilities averaged over each cell, P_i, take all the admissible real values. When $N \to \infty$, the following estimate holds:

$$\log N! = N \log N - N + o(N), \qquad (A6.4)$$

where $o(N)$ indicates a quantity such that $o(N)/N$ tends to zero when $N \to \infty$. Equation (A6.4) follows from Stirling's formula [28] or the inequality

$$2 < \frac{N!e^N}{N^N} < 3N. \qquad (A6.5)$$

In turn, (A6.5) follows from the fact that, if we let $a_N = N!e^N/N^N$, then $a_{N+1} > a_N$ and $a_{N+1}/(N+1) < a_N/N$, thanks to the well-known elementary inequality:

$$\left(1 + \frac{1}{N} \right)^N < e < \left(1 + \frac{1}{N} \right)^{N+1}. \qquad (A6.6)$$

The fact that $a_{N+1} > a_N$ and $a_{N+1}/(N+1) < a_N/N$ can in fact be proved to follow from (A6.6) by induction, since $a_1 = \mathrm{e} > 2$ and $a_1/1 = \mathrm{e} < 3$. If we now use (A6.4) in (A6.3) we obtain:

$$V(P)) = \log N - \sum_{i=1}^{n} \frac{N_i}{N} \log N_i - \log n + \frac{o(N)}{N}$$

$$= -\sum_{i=1}^{n} \frac{N_i}{N} \log \left(\frac{N_i}{N} \right) - \log n + \frac{o(N)}{N}$$

$$= -\sum_{i=1}^{n} P_i \mu_i \log(n P_i \mu_i) + \frac{o(N)}{N}, \tag{A6.7}$$

where due account has been taken of the fact that $\sum_{i=1}^{n} N_i = N$ and $P_i = N_i/(N\mu_i)$. When $N \to \infty$, the last term in (A6.7) disappears.

It remains now to let (at the same time) n go to infinity and μ_i to zero; before doing this, let us remark that it is always possible to arrange things in such a way that $n\mu_i \to \overline{\mu}$ (total volume of the space of states, taken to be finite for simplicity); it is in fact enough to take $\mu_i = \overline{\mu}/n$. Then the discrete distribution $(P_1, P_2, \ldots, P_n)$ tends to a continuous distribution P and eqn (A6.7) becomes:

$$V(P)) = -\overline{\log(P\overline{\mu})} = -\int P \log(P\overline{\mu})\,\mathrm{d}\mu = -\int P \log P\,\mathrm{d}\mu - \log \overline{\mu} \int P\,\mathrm{d}\mu$$

$$= -\int P \log P\,\mathrm{d}\mu - \log \overline{\mu}. \tag{A6.8}$$

This formula coincides with (A6.1) apart from an inessential constant (which is exactly zero if the measure is normalized in such a way that the total measure $\overline{\mu}$ equals 1).

Let us underline the difficulty which we remarked upon in the main text and which did not escape Boltzmann's attention, that if we change the variable through a transformation with non-constant Jacobian, $V(P)$ in the new variables is not equal to $V(P)$ in the old ones. Hence there is a class of privileged (canonical) variables, singled out by the fact that the volume element in the state space is invariant during the time evolution of the system (thanks to Liouville's theorem); to choose these variables means, from a physical viewpoint, that the sets of equal volume (according to the choice that we have made) are equiprobable in the state space.

Some applications of this expression for the likelihood are given in Appendix 7.1.

Appendix 7.1
The canonical distribution for equilibrium states

The basic problem of determining the distribution of microscopic states in a macroscopic equilibrium state can evidently be solved only if we restrict the systems under consideration. If one makes the assumption that no region of phase space is privileged, i.e. if the trajectories of the point representing the mechanical state of the system pass the same number of times (in a very long time interval) in each region having a given volume, then it is reasonable to think that the N-particle distribution function P (see Chapter 4) is constant. Of course, the existence of constants of the motion modifies this conclusion; if one makes the assumption (called *ergodic* in modern terminology) that there is only one important integral of the motion, the total energy, then the distribution function will be constant only on each hypersurface of constant energy. Gibbs considers systems in which this is more or less explicitly assumed. The distribution that is obtained is the microcanonical distribution, which correctly describes an isolated system with an assigned energy. In principle one can develop thermodynamics starting from this distribution, but the treatment is made complicated by the necessity of passing to the limit when $N \to \infty$ in each expression. It is then more convenient to pass to the so-called canonical distribution, which may be obtained for the subsystems with a finite number of degrees of freedom, when the same number is allowed to go to infinity for the entire system.

The canonical system can be easily obtained from the microcanonical one by imagining the system we are interested in as weakly coupled with a thermostat which may be assumed to coincide with a perfect gas. Weak coupling means that the interaction is sufficient to reach thermal equilibrium, but the corresponding interaction energy is negligible with respect to the thermal one. The circumstance that the system we are interested in, S, is coupled to a thermostat T means that the latter has a number of degrees of freedom much larger than the s degrees of freedom possessed by S, in such a way that the number of degrees of freedom of T can be assumed to tend to infinity.

In fact, let H_S and H_T be the Hamiltonian functions of the system and of the thermostat, and E be the total energy (system + thermostat); then the distribution function P for the enlarged system, including the thermostat, is given by:

$$P = [A(E)]^{-1}, \tag{A7.1}$$

where, if $V(E)$ is the volume of the region in phase space defined by $H_S + H_T \leq E$, we have let $A(E) = \mathrm{d}V/\mathrm{d}E$ (an invariant measure of the area of the hypersurface bounding the region under consideration). We can also compute the areas $A_S(\epsilon)$ and $A_T(E - \epsilon)$ corresponding to the boundaries of the two regions $H_S \leq \epsilon$, $H_T \leq E - \epsilon$: it is clear that there must be a simple relation between these two areas and $A(E)$. To find it, we remark that the volumes of the regions $H_S \leq E$ and $H_T \leq E_T$ (where E and E_T are arbitrary) and the region which is their Cartesian product are obviously related by

$$V(E, E_T) = V_S(E)V_T(E_T) = \int_0^E \int_0^{E_T} (\mathrm{d}V_S/\mathrm{d}\epsilon)(\mathrm{d}V_T/\mathrm{d}\epsilon_T) \, \mathrm{d}\epsilon \, \mathrm{d}\epsilon_T. \tag{A7.2}$$

If instead we take just the volume of the region where $H_S + H_T \leq E$, we must choose different limits of integration to obtain

$$V(E) = \int_0^E \int_0^{E-\epsilon} (\mathrm{d}V_S/\mathrm{d}\epsilon)(\mathrm{d}V_T/\mathrm{d}\epsilon_T) \, \mathrm{d}\epsilon \, \mathrm{d}\epsilon_T = \int_0^E A_S(\epsilon)V_T(E - \epsilon) \, \mathrm{d}\epsilon. \tag{A7.3}$$

If we now differentiate this formula with respect to E, we obtain

$$A(E) = \int_0^E A_S(\epsilon)A_T(E - \epsilon) \, \mathrm{d}\epsilon, \tag{A7.4}$$

since $V(0) = 0$.

We can now obtain P_S, the distribution function of the system by simply integrating (A7.1) with respect to the variables describing the thermostat, to obtain:

$$P_S = [A(E)]^{-1} A_T(E - H_S) \tag{A7.5}$$

or, using (A7.4) to express $A(E)$:

$$P_S = \left[\int_0^E A_S(\epsilon) A_T(E - \epsilon) \, d\epsilon \right]^{-1} A_T(E - H_S). \tag{A7.6}$$

Let us now remark that if the thermostat is a perfect gas made up of N molecules with mass m, A_T is simply the product of V^N (where V is the volume occupied by the gas) by the area of the surface of a hypersphere in the $3N$-dimensional velocity space, because the equation $H_T = E$ may be written in the form $\sum_{k=1}^{N} m|\xi_k|^2 = 2E$, and hence $A_T = V^N \omega_{3N} (2E/m)^{(3N-1)/2}$, where ω_{3N} is the area of the unit sphere in $3N$-dimensional space. Then:

$$[A_T(E)]^{-1} A_T(E - H_S) = (1 - H_S/E)^{(3N-1)/2} \tag{A7.7}$$

and assuming that the energy per particle $E/N = 3/(2\beta)$ in the thermostat is independent of N (i.e. of the size of the thermostat), as is actually the case for a perfect gas, one obtains:

$$\lim_{N \to \infty} [A_T(E)]^{-1} A_T(E - H_S) = \lim_{N \to \infty} (1 - 2\beta H_S/(3N))^{(3N-1)/2} = e^{-\beta H_S}. \tag{A7.8}$$

Let us multiply and divide by $A_T(E)$ in (A7.6) and pass to the limit when $N \to \infty$ with the help of (A7.8) (used twice). We finally obtain the canonical distribution:

$$P_S = \left[\int_0^E A_S(\epsilon) e^{-\beta \epsilon} \, d\epsilon \right]^{-1} e^{-\beta H_S}. \tag{A7.9}$$

Note that if we have several systems coupled with a single thermostat, we obtain a canonical distribution with the same value of β for all of them; then this parameter has the properties required from a universal function of temperature. Henceforth we shall no longer need the thermostat and shall therefore suppress the index S in (A7.9). We rewrite it in the following form:

$$P = Z^{-1} e^{-\beta H}, \tag{A7.10}$$

where Z, the partition function, is none other than the integral of $e^{-\beta H}$ over the entire phase space of the system. The average energy of the system is given by

$$\overline{E} = Z^{-1} \int H e^{-\beta H} \prod d\xi_i \, dx_i = -\frac{\partial}{\partial \beta} \log Z. \tag{A7.11}$$

If we then give a small variation to the parameter β and to one of the other parameters a_k (such as volume), we obtain:

$$\delta \overline{E} = - \left(\frac{\partial^2}{\partial \beta^2} \log Z \right) \delta \beta - \left(\frac{\partial^2}{\partial \beta \partial a_k} \log Z \right) \delta a_k. \tag{A7.12}$$

We must now introduce the work done on the system by the external forces; the simplest case is obtained by varying a parameter a_k in the external forces (and, at least formally, one can always consider this case). Then the work $\delta^* L$ done on the gas is given by

$$\delta^* L = Z^{-1} \int \frac{\partial H}{\partial a_k} e^{-\beta H} \prod d\xi_i \, dx_i \, \delta a_k = -\frac{1}{\beta} Z^{-1} \frac{\partial}{\partial a_k} \int e^{-\beta H} \prod d\xi_i \, dx_i \, \delta a_k$$

$$= \frac{1}{\beta} \frac{\partial}{\partial a_k} (\log Z) \delta a_k. \tag{A7.13}$$

The difference $\delta \overline{E} - \delta^* L$ is none other than the heat $\delta^* Q$ supplied by the system, and hence

$$
\begin{aligned}
\beta \delta^* Q &= \beta(\delta \overline{E} - \delta^* L) \\
&= -\beta \left(\frac{\partial^2}{\partial \beta^2} \log Z \right) \delta\beta - \beta \left(\frac{\partial^2}{\partial \beta \partial a_k} \log Z \right) \delta a_k + \left(\frac{\partial}{\partial a_k} \log Z \right) \delta a_k. \quad \text{(A7.14)}
\end{aligned}
$$

It is now clear that the last expression equals $\delta[\log Z - \beta \frac{\partial}{\partial \beta} \log Z]$ and is therefore an exact differential. Thus β can be identified with $(kT)^{-1}$, where k is a constant (the so-called Boltzmann constant) and T the absolute temperature, whereas $\log Z - \beta \frac{\partial}{\partial \beta} \log Z$ is to be identified (apart from an additive constant) with S/k, where S is the entropy of the system. By a simple calculation we obtain the formula:

$$
S = \frac{\partial}{\partial T}(kT \log Z). \quad \text{(A7.15)}
$$

From this identification all the formulae of thermodynamics follow. As an example, we consider a perfect gas, for which Z equals $V^N (2\pi kT)^{3N/2}$ and the gas pressure is given by

$$
p = \frac{\partial}{\partial V}(kT \log Z) = NkT/V, \quad \text{(A7.16)}
$$

i.e. the equation of state of perfect gases with the identification of the relation between the constant k and the gas constant R.

Alternatively, the canonical distribution can be characterized as the one that minimizes the average value of $\log P$ under suitable constraints. Gibbs's considerations on the extremizing properties start from the elementary inequality:

$$
x \log x - x + 1 \geq 0, \quad \text{(A7.17)}
$$

holding for any real positive x (x may even be zero, provided that we define, by continuity, $x \log x = 0$ for $x = 0$). The equality sign holds if and only if $x = 1$. As we remarked in Chapter 6, this reduces, on changing x to $1/x$, to the simpler inequality $\log x \leq x - 1$, stating that the logarithmic curve is always below its tangent at $(1, 0)$.

Let us put $x = P/G$ in (A7.17), where P and G are two probability densities in phase space. We obtain the so-called *Gibbs inequality*:

$$
\begin{aligned}
&\int P \log P \prod d\boldsymbol{\xi}_i \, d\mathbf{x}_i \geq \int P \log G \prod d\boldsymbol{\xi}_i \, d\mathbf{x}_i \\
&\left(\int P \prod d\boldsymbol{\xi}_i \, d\mathbf{x}_i = \int G \prod d\boldsymbol{\xi}_i \, d\mathbf{x}_i = 1 \right)
\end{aligned} \quad \text{(A7.18)}
$$

and the equality sign holds if and only if P and G coincide almost everywhere.

This inequality can be used to prove the minimum properties of the microcanonical and canonical distributions. Note that if we impose the constraint of given total energy and we do not impose any additional constraint besides normalization, then, rather than integrating over the entire phase space, we integrate only over the hypersurface of constant energy and we can let G be a constant in the equation corresponding to (A7.18) and obtain that this constant distribution minimizes the average value of $\log P$ (microcanonical distribution). If on the contrary we just assign the average value of energy, letting $G = Z^{-1} e^{-\beta H}$ in (A7.18), we obtain

$$
\begin{aligned}
&\int P \log P \prod d\boldsymbol{\xi}_i \, d\mathbf{x}_i \geq -\beta \int PH \prod d\boldsymbol{\xi}_i \, d\mathbf{x}_i - \log Z = -\beta \overline{E} - \log Z \\
&\left(\int P \prod d\boldsymbol{\xi}_i \, d\mathbf{x}_i = \int G \prod d\boldsymbol{\xi}_i \, d\mathbf{x}_i = 1; \right. \\
&\left. \int PH \prod d\boldsymbol{\xi}_i \, d\mathbf{x}_i = \int GH \prod d\boldsymbol{\xi}_i \, d\mathbf{x}_i = \overline{E} \right).
\end{aligned} \quad \text{(A7.19)}
$$

Since

$$\int G \log G \prod \mathrm{d}\boldsymbol{\xi}_i \, \mathrm{d}\boldsymbol{x}_i = -\beta \int G H \prod \mathrm{d}\boldsymbol{\xi}_i \, \mathrm{d}\boldsymbol{x}_i - \log Z = -\beta \overline{E} - \log Z, \qquad \text{(A7.20)}$$

eqn (A7.19) shows that the canonical distribution minimizes the average value of $\log P$ under the constraints of normalization and given average energy.

The Gibbs inequality has several further applications. If for example we assign the one particle distribution function, $P^{(1)}$, and we let $G = \prod_{i=1}^{N} P^{(1)}(\boldsymbol{x}_i, \boldsymbol{\xi}, t)$ in (A7.18), we find that the factorized distribution minimizes the mean value of $\log P$ and, as an additional result, that the mean value of the logarithm of the one-particle distribution function is not greater than $1/N$ times the mean value of $\log P$. Another application is provided by the procedure of smoothing P by averaging over small cells, as indicated in the main text of this chapter. In fact, if G is the distribution which is constant over each cell, obtained by averaging P, then it will satisfy (A7.18). If at time $t = 0$, P is constant over each cell, it will no longer be so at later times, but thanks to the properties of the Liouville equation the average of $\log P$ will equal its initial value and hence the initial value of the average of the logarithm of G: the latter average will be at any time no greater than its initial value.

Appendix 8.1
The H-theorem for classical polyatomic molecules

In this appendix, following ref. [12], we shall be concerned with the Boltzmann equation with particular reference to the H-theorem, which provoked Lorentz's objections that we mentioned in the main text. First of all, let us remark that there are no formal difficulties in writing a Boltzmann equation for molecules with $n > 3$ degrees of freedom. Together with the position vector of the centre of mass x and the corresponding velocity ξ, we shall also need other variables; we shall denote by p a vector in a $(2n - 3)$-dimensional space which includes all the variables except the coordinates of the centre of mass (while the components of ξ are not excluded). Then the equations of the motion between two subsequent collisions are

$$\dot{x} = \xi, \qquad \dot{p} = P(x, p), \tag{A8.1}$$

where P is a vector in $2n - 3$ dimensions that describes the partial derivatives (with the appropriate sign) of the Hamiltonian with respect to the coordinates and momenta different from ξ (but x will now be included). The Boltzmann equation for the distribution function $f(x, p, t)$ can be written as follows:

$$\frac{\partial f}{\partial t} + \xi \cdot \frac{\partial f}{\partial x} + P \cdot \frac{\partial f}{\partial p}$$
$$= \int_{\Re^{6n-9}} [f' f'_* W(p', p'_* \to p, p_*) - f f_* W(p, p_* \to p', p'_*)] \, dp_* \, dp' \, dp'_*, \tag{A8.2}$$

where f', f'_* and f_* denote, as usual, that the function f has (besides x and t) arguments p', p'_*, and p_*, respectively, whereas $W(p', p'_* \to p, p_*)$ (which will include some factors given by the Dirac delta function, in order to ensure conservation of momentum and energy) is essentially the differential scattering cross-section multiplied by the relative speed and in fact we have

$$\int \int W(p', p'_* \to p, p_*) \, dp' \, dp'_* = |\xi - \xi_*| \Sigma_t, \tag{A8.3}$$

where Σ_t is the total scattering cross-section (assumed to be finite). The latter may of course depend on p and p_*. However, we can get rid of this dependence by a trick that makes the proof much simpler. Since the total cross-section is assumed to be finite, there will be a maximum distance r_0 beyond which the molecules do not mutually interact. We can then let $\Sigma_t = \pi r_0^2$, provided that we introduce false collisions in which no change occurs for those values of p and p_* for which possibly there is no interaction, although $r < r_0$. The case in which the collision cross-section is infinite can be obtained by first cutting off the interactions occurring for $r > r_0$ and then letting $r_0 \to \infty$ in the final inequality that we are going to prove.

The microscopic motion equations (A8.1) are of course assumed to be time-reversible; this means that there exists a transformation $(x, p, t) \to (x, p^-, -t)$ (where, typically, a component of p^- is to be equal to the corresponding one of p if it has the meaning of a coordinate, and opposed to the corresponding one of p if it has the meaning of a momentum canonically conjugated to a coordinate). Incidentally, we remark that one frequently treats the changes in the coordinates as if they could be ignored in the average and just considers the changes in the momenta; this aspect however will not enter into what follows.

It is important to notice that the transformation of variables from the p's to the p^-'s, with fixed x, preserves the volume in the space described by the variables p.

The time reversibility of the microscopic equations even during a collision implies that the following relation, called *reciprocity*, holds:

$$W(p', p'_* \to p, p_*) = W(p^-, p^-_* \to p^{-'}, p^{-'}_*). \tag{A8.4}$$

If the interaction possesses spherical symmetry, as is the case for mass points or perfectly smooth hard spheres, then the following stronger property holds:

$$W(p', p'_* \to p, p_*) = W(p, p_* \to p', p'_*), \tag{A8.5}$$

which is called *detailed balance*. This property is more or less explicitly used in the proof of the H-theorem for monatomic gases. Whenever it holds, the proof given in Chapter 5 can be transferred without change to the case of polyatomic gases. Boltzmann's mistake, pointed out by Lorentz, lies precisely in the implicit assumption that eqn (A8.5) holds for a generic polyatomic gas. If this property fails, one can introduce Boltzmann's argument on the "closed cycle of collisions", which, although not so convincing, contains the basic idea of the proof that we shall presently give, following ref. [12], i.e. the fact that one must consider not single collisions but subsets of collisions.

The key point is eqn (A8.3), whose right-hand side is clearly invariant under time reversal, and the reciprocity property, expressed by eqn (A8.4). Let us integrate both sides of the latter relation with respect to p and p_*. We obtain

$$\int \int W(p', p'_* \to p, p_*) \, dp \, dp_* = \int \int W(p^-, p_*^- \to p^{-\prime}, p_*^{-\prime}) \, dp \, dp_*$$

$$= \int \int W(p, p_* \to p^{-\prime}, p_*^{-\prime}) \, dp \, dp_*, \tag{A8.6}$$

where, in the last step, we have changed the integration variables from p, p_* to p^-, p_*^- (using the aforementioned invariance of the volume element) and subsequently abolished the superscript "$-$", which is no longer required. But the last integral is that which appears on the left-hand side of (A8.3), apart from the presence of the superscript "$-$" in the second pair of variables; but because of the invariance of the right-hand side of the same equation with respect to the transformation from variables with superscript "$-$" to variables without the same superscript, we can suppress the latter in the last integral of (A8.6) and get

$$\int \int W(p', p'_* \to p, p_*) \, dp \, dp_* = \int \int W(p, p_* \to p', p'_*) \, dp \, dp_*. \tag{A8.7}$$

This is the new relation that we shall use to prove the H-theorem for polyatomic molecules. The importance of a relation of this kind was underlined by Waldmann [9], who, guided by an analogy with the quantum case, wrote it without proof in the particular case of dumb-bell shaped classical molecules, remarking that "one must get the (purely mechanical) normalization property" expressed by (A8.7). As a possible proof, the same author [9] seems to hint at a complete calculation with the simplifying assumption of "averaging over all possible phase angles [...] before and after collision". This average, albeit useful in some cases as indicated above, to simplify the relations, is not required in the proof which we have just discussed of ref. [12], according to which (A8.7) is a general property which follows from the time reversibility of the microscopic equations of the molecular motion.

Having shown that (A8.7) holds, it is now a relatively simple matter to prove the H-theorem, or, it is better to say, the Boltzmann lemma (whence the H-theorem follows). According to this lemma, if we let

$$Q(f, f) = \int_{\Re^{6n-9}} [f' f'_* W(p', p'_* \to p, p_*) - f f_* W(p, p_* \to p', p'_*)] \, dp_* \, dp' \, dp'_*, \tag{A8.8}$$

we get

$$\int \log f \, Q(f, f) \, \mathrm{d}\boldsymbol{p} \leq 0. \tag{A8.9}$$

To prove this result, let us multiply (A8.8) by $\log f$ and integrate with respect to $\boldsymbol{p}$, thus obtaining

$$\int \log f \, Q(f, f) \, \mathrm{d}\boldsymbol{p} =$$
$$\frac{1}{2} \int_{\mathfrak{R}^{8n-12}} f f_* \log\left(\frac{f' f'_*}{f f_*}\right) W(\boldsymbol{p}, \boldsymbol{p}_* \to \boldsymbol{p}', \boldsymbol{p}'_*) \, \mathrm{d}\boldsymbol{p} \, \mathrm{d}\boldsymbol{p}_* \, \mathrm{d}\boldsymbol{p}' \, \mathrm{d}\boldsymbol{p}'_*. \tag{A8.10}$$

This relation can be easily obtained by the same manipulations as discussed in the Appendices to Chapter 5, i.e. with suitable changes of variables and indices, without using any property of $W(\boldsymbol{p}, \boldsymbol{p}_* \to \boldsymbol{p}', \boldsymbol{p}'_*)$. Equation (A8.10) however does not permit the argument used in Chapter 5 to be used; to this end, in fact, we should sum (A8.10) and the relation that can be obtained from it by exchanging the primed and unprimed variables and use the property of detailed balance, (A8.5), which however does not generally hold. At this point we must make use of a trick, apparently first used by Pauli [10] in quantum statistical mechanics. Together with the identity expressed by (A8.10), let us also consider the following one:

$$\int_{\mathfrak{R}^{8n-12}} f' f'_* W(\boldsymbol{p}', \boldsymbol{p}'_* \to \boldsymbol{p}, \boldsymbol{p}_*) \, \mathrm{d}\boldsymbol{p} \, \mathrm{d}\boldsymbol{p}_* \, \mathrm{d}\boldsymbol{p}' \, \mathrm{d}\boldsymbol{p}'_*$$
$$= \int_{\mathfrak{R}^{8n-12}} f f_* W(\boldsymbol{p}, \boldsymbol{p}_* \to \boldsymbol{p}', \boldsymbol{p}'_*) \, \mathrm{d}\boldsymbol{p} \, \mathrm{d}\boldsymbol{p}_* \, \mathrm{d}\boldsymbol{p}' \, \mathrm{d}\boldsymbol{p}'_*. \tag{A8.11}$$

This can be obtained by an exchange of name of the variables and expresses the conservation of the number of particles in a collision. We can now make use of (A8.7) on the left-hand side of (A8.11) and rewrite it as follows:

$$\int_{\mathfrak{R}^{8n-12}} f' f'_* W(\boldsymbol{p}, \boldsymbol{p}_* \to \boldsymbol{p}', \boldsymbol{p}'_*) \, \mathrm{d}\boldsymbol{p} \, \mathrm{d}\boldsymbol{p}_* \, \mathrm{d}\boldsymbol{p}' \, \mathrm{d}\boldsymbol{p}'_*$$
$$= \int_{\mathfrak{R}^{8n-12}} f f_* W(\boldsymbol{p}, \boldsymbol{p}_* \to \boldsymbol{p}', \boldsymbol{p}'_*) \, \mathrm{d}\boldsymbol{p} \, \mathrm{d}\boldsymbol{p}_* \, \mathrm{d}\boldsymbol{p}' \, \mathrm{d}\boldsymbol{p}'_* \tag{A8.12}$$

or, equivalently:

$$\frac{1}{2} \int_{\mathfrak{R}^{8n-12}} f f_* \left(\frac{f' f'_*}{f f_*} - 1\right) W(\boldsymbol{p}, \boldsymbol{p}_* \to \boldsymbol{p}', \boldsymbol{p}'_*) \, \mathrm{d}\boldsymbol{p} \, \mathrm{d}\boldsymbol{p}_* \, \mathrm{d}\boldsymbol{p}' \, \mathrm{d}\boldsymbol{p}'_* = 0. \tag{A8.13}$$

We can then subtract the integral appearing on the left-hand side of (A8.13) from the right-hand side of (A8.10), without changing anything. We then have:

$$\int \log f \, Q(f, f) \, \mathrm{d}\boldsymbol{p}$$
$$= \frac{1}{2} \int_{\mathfrak{R}^{8n-12}} f f_* \left[\log \frac{f' f'_*}{f f_*} - \left(\frac{f' f'_*}{f f_*} - 1\right)\right] W(\boldsymbol{p}, \boldsymbol{p}_* \to \boldsymbol{p}', \boldsymbol{p}'_*) \, \mathrm{d}\boldsymbol{p} \, \mathrm{d}\boldsymbol{p}_* \, \mathrm{d}\boldsymbol{p}' \, \mathrm{d}\boldsymbol{p}'_*. \tag{A8.14}$$

Let us now make use of the fact that both f and W are non-negative and that the following (previously used) elementary inequality holds:

$$\log x - (x - 1) \leq 0 \qquad (= 0 \quad \text{iff } x = 1). \tag{A8.15}$$

We can then conclude that (A8.9) has been proved and that the equality sign holds if and only if (almost everywhere)

$$f' f'_* = f f_*,$$ (A8.16)

i.e. if the distribution function describes an equilibrium state.

Appendix 8.2
The equipartition problem

Let us consider a Hamiltonian system with n degrees of freedom, i.e. described by the following system of $2n$ ordinary differential equations:

$$\dot{q}_k = \frac{\partial H}{\partial p_k},$$
$$\dot{p}_k = -\frac{\partial H}{\partial q_k}, \tag{A8.17}$$

where q_k, p_k are the n coordinates and the n conjugated momenta, and $H = H(p_j, q_j)$ is the Hamiltonian of the system.

It is convenient to introduce [28] the Poisson bracket

$$\{F, G\} = -\sum_{k=1}^{n} \left(\frac{\partial F}{\partial p_k} \frac{\partial G}{\partial q_k} - \frac{\partial F}{\partial q_k} \frac{\partial G}{\partial p_k} \right). \tag{A8.18}$$

As is well known, Poisson brackets possess many properties which make them useful when studying the analytical and geometrical properties (we are talking here about the so-called symplectic properties) of the Hamiltonian systems in phase space. In particular, the Poisson theorem on the constants of the motion holds: if F and G are two constants of the motion, then $\{F, G\}$ too is a constant of the motion. If $\{F, G\} = 0$, F and G are said to be *in involution* and the theorem turns out to be trivial.

A very important case occurs when one finds n constants of the motion (functionally independent and single-valued) F_j $(j = 1, 2, \ldots, n)$, which are pairwise in involution, i.e.

$$\{F_j, F_k\} = 0 \quad (j, k = 1, 2, \ldots, n). \tag{A8.19}$$

In this case a theorem due to Liouville [28] holds, according to which the equations of motion can be solved by quadratures (i.e. by means of integrations) and there exists in addition a canonical transformation (i.e. a transformation that does not alter the Hamiltonian structure of the equations) from the variables p_k, q_k to new variables I_k and θ_k (called *action–angle variables*), such that the action variables I_k (which are functions of the constants of the motion F_j) are constant in time and the angle variables θ_k increase linearly with time. When the energy hypersurface $H = E$ is (closed and) bounded, as is commonly the case, then the angle variables are actually angles and the motion is multiperiodic with constant (angular) frequency $\omega_k(I_j) = \partial H / \partial I_k$. Geometrically the angle variables describe an n-dimensional torus T_n and each motion takes place on the surface of such a torus, fixed by the initial conditions. Typical examples of integrable systems are the conservative systems with just one degree of freedom, the two-body problem, the inertial motion of a solid point with a fixed point. A rather particular example is provided by the system of N point masses interacting through ideal springs, equivalent to the system of n harmonic oscillators; in this case the Hamiltonian when expressed in terms of action–angle variables is linear in the I_k's and consequently the frequencies do not depend upon the action variables (isochronous oscillations).

The most interesting developments of the Hamiltonian dynamics after Boltzmann's times refer to quasi-integrable systems, i.e. the systems obtained by perturbing a Hamiltonian $h(I_j)$, which has been already expressed in terms of its own action variables, in such a way as to obtain a Hamiltonian of the following form:

$$H(I_j, \theta_j; \epsilon) = h(I_j) + \epsilon f(I_j, \theta_j). \tag{A8.20}$$

Given a system of this kind, the first question that one may ask is: what is the time scale over which the perturbation is felt and on which then the variables I_k significantly deviate from their own initial value? An obvious consideration leads to the conservative estimate $t \cong (\epsilon)^{-1}$.

One can of course imagine more favourable cases, e.g. the existence of a canonical transformation differing from identity by terms of order ϵ and leading to new action–angle variables (I_k', θ_k'), in terms of which $H(I_j, \theta_j; \epsilon) = h'(I_j')$; in this case the I_k's would stay close to their initial values for any real t. It is easy to give counterexamples to the statement that this is always possible and, on the contrary, a theorem by Poincaré [33] leads one to suspect that the impossibility of performing this transformation is the generic case. As a matter of fact, this is the case if one wants the stability statement to hold true for generic initial data. The KAM theorem [26, 28, 29, 30], which has been hinted at in the main text, has altered the situation. A possible formulation of this theorem reads as follows.

Theorem (KAM): Consider a system with a Hamiltonian of the form (A8.20). Let us assume that H is analytic in a multistrip of the complex planes of the coordinates on $B \times T^n$, where B is a suitable domain in $\mathfrak{R}^n$, the determinant of the Hessian matrix, with elements $\partial^2 H / \partial I_k \partial I_j$ being strictly positive in B; then there is an $\epsilon_0 > 0$ (depending on the parameters in the Hamiltonian) such that, if $\epsilon \leq \epsilon_0$, then there exist a canonical transformation, an integrable Hamiltonian $h'(I_j')$, and a set $B_\epsilon \subset B$ (of relative Lebesgue measure tending to 1 for $\epsilon \to 0$), such that, by the canonical transformation under consideration, the old Hamiltonian transforms into the new one in the subset B_ϵ.

This theorem shows that for randomly chosen initial data there will be a high probability that the actions I_k will stay close to their initial values for any t; however, there are initial data, improbable but physically indistinguishable from the previous ones, for which the theorem does not apply. This implies, for $n > 2$ (when the bad set is also connected), that $I(t)$ may arbitrarily depart (on $H = E$) from its initial value. This phenomenon is the so-called Arnold diffusion, which implies among other things that an isolated system of mass points kept together by gravitational forces may tend to assume looser and looser configurations (stars and planets may increase indefinitely their mutual distances; explicit examples have been constructed).

However, if we look at this matter from a physical rather than a mathematical viewpoint, the thesis of the KAM theorem, the assumptions of which contain a condition which may not be physically ascertained, can even be too precise. Eventually, we are not interested in every t; values of t reaching a few billion years, a non-negligible slice of eternity, might suffice.

In order to attack this kind of question, one may try to make use of classical perturbation theory. Then of course one meets with the celebrated theorem by Poincaré [33], which has been already recalled. Then, in a sense, the perturbation method does not help us. But, as taught by the KAM theorem, we can obtain important results by slightly modifying the statement of what we want to prove.

The first attempts to avoid the stumbling block were due to the American mathematician George D. Birkhoff (1884–1944), who studied [34] the neighbourhood of a point of stable equilibrium and hence the perturbation of a system of harmonic oscillators, assuming that the corresponding frequencies are non-resonant, i.e. that there is no n-dimensional vector with integer components $\boldsymbol{k}$ orthogonal to the vector $\boldsymbol{\omega}$ of the frequencies, and that at each order in ϵ the coefficient of ϵ^s is an s-degree polynomial in $\cos \theta$ and $\sin \theta$. It is then possible [33] to construct, in an explicit fashion, a finite sequence of canonical transformations such that the part of the Hamiltonian depending on the angle variables becomes (in the new action–angle variables) of order ϵ^{r+1}, where $r \geq 1$ is an arbitrary integer. If we suppress the assumption that the dependence on ϕ is an s-degree polynomial in $\cos \theta$ and $\sin \theta$ at each order s in ϵ, then we must strengthen the assumption on the non-resonance condition by introducing a Diophantine condition according to which $|\boldsymbol{k} \cdot \boldsymbol{\omega}| \geq$

$\gamma |\boldsymbol{k}|^{-n}$ with $|\boldsymbol{k}| = |k_1| + |k_2| + \ldots + |k_n|$, $\gamma > 0$ and make use of an analyticity assumption (the removal of this assumption was Moser's contribution [30], which we cannot dwell on here). Of course, using Birkhoff's method leads us to improve the estimate of the time during which the system does not significantly depart from its initial state, from ϵ^{-1} to $C_r \epsilon^{-(r+1)}$. We must of course show that the constant C_r does not decrease in a catastrophic manner. An apparently poor estimate, i.e. $C_r > AB^{-r} r^{-rm}$ $(A, B, r > 0)$, leads us to a non-trivial result, the celebrated Nekhoroshev theorem [20, 21], when we show that one can take r as depending on ϵ in the form, roughly speaking, $r(\epsilon) \cong (\epsilon_0/\epsilon)^{1/m}$ (we said "roughly speaking" because r must be an integer). As a matter of fact, the case of the harmonic oscillators was firstly studied by Gallavotti [35], whereas Nekhoroshev had considered the case when the assumptions of the KAM theorem hold.

To see how this theorem, supplying us with an exponential estimate for the stability time, can be applied to the problem of the internal degrees of freedom, let us consider the problem of the collision in one dimension of a diatomic molecule with a wall (simulated by a potential that decreases exponentially with distance). Then the time required to perform a significant exchange of energy increases exponentially with a power $1/m$ of the frequency, where m is of the order of the number of degrees of freedom. The problem which shows up with the estimates provided by Nekhoroshev is that when the number of degrees of freedom increases (as is the case when we consider one or more diatomic molecules in the presence of N monatomic molecules), we lose the advantage provided by the exponential factor, because of the power $1/m$ that we have just mentioned. This inconvenience has been removed in ref. [32], where the authors show that the exchanges of energy are significant during an interaction, but their average is practically zero over the entire interaction. In the case when there are several diatomic molecules, let us say two identical molecules, then the interaction vibrational energy of each molecule will not be conserved, because of resonance phenomena (which make the estimates extremely complicated) but the sum of the two energies will admit an exponentially long time estimate. This circumstance is associated with the well-known fact that in classical perturbation theory, every resonance decreases the number of conserved quantities by one.

In order to arrive at results having a practical meaning, one must obtain more realistic estimates of the constants than those terribly pessimistic supplied by a theoretical analysis. This has been done for a one-dimensional model [25], as indicated in the main text.

Appendix 9.1
The Stefan–Boltzmann law

As indicated in the main text, Boltzmann's proof of the law suggested by Stefan on the basis of rather inaccurate experimental data rests on the concept of radiation pressure. Let us imagine an enclosure closed by a (slowly) movable piston with a reflecting surface. Electromagnetic waves exert a pressure on the piston which, as indicated in the text, is $p = e/3$. This pressure is due to the momentum which the electromagnetic field carries with it, according to Maxwell's equations. Since this momentum density has a magnitude $g = e/c$, where c is the speed of light, the pressure can be computed as was done in Chapter 3 for a gas; the only difference is that the speed of the waves is constant and equal to c. The fact that we obtain $p = e/3$, rather than $p = 2e/3$ as in the kinetic theory of gases, is due to the fact that $e = gc$ rather than $e = gc/2$ (light cannot be treated as a non-relativistic particle).

We can then write, if V is the volume of the enclosure and S its entropy:

$$T\,dS = d(eV) + p\,dV = e\,dV + V\frac{de}{dT}dT + \frac{1}{3}e\,dV = V\frac{de}{dT}dT + \frac{4}{3}e\,dV.$$

Since we have computed the differential of entropy, we can deduce its partial derivatives with respect to T and V:

$$\frac{\partial S}{\partial T} = \frac{V}{T}\frac{de}{dT}, \qquad \frac{\partial S}{\partial V} = \frac{4}{3}\frac{e}{T},$$

and hence, using Schwarz's theorem on the mixed second derivatives, equate the derivative of $\partial S/\partial V$ with respect to T to the derivative of $\partial S/\partial T$ with respect to V, thus obtaining

$$\frac{4}{3}\left(\frac{1}{T}\frac{de}{dT} - \frac{e}{T^2}\right) = \frac{1}{T}\frac{de}{dT},$$

or, simplifying:

$$\frac{de}{dT} = 4\frac{e}{T}$$

and, integrating by separation of the variables:

$$e = \sigma T^4,$$

where σ is an integration constant. An easy argument also gives $S = (4/3)\sigma V T^3$.

Appendix 9.2
Wien's law

Wien's argument rests, as indicated in the main text, on a clever use of the Doppler effect, according to which the frequency v of a wave emitted by a moving source appears to be altered to an observer at rest. The frequency shift Δv equals $v(v/c)\cos\theta$, where v is the speed of the source, c the speed of the wave, and θ the angle between the line of observation and the direction along which the source is moving. This formula is correct up to terms of order $(v/c)^2$, which is sufficient because we need to consider a speed v tending to zero. Now if a mirror moves with velocity v and a wave is reflected on it, we may think of the reflected wave as coming from an imaginary source behind the mirror; this image will move with velocity $2v$ in the direction of the normal to the mirror. Hence for normal reflection the wave will undergo a change of frequency $2vv/c$, and for oblique incidence at an angle θ the frequency after reflection v' will be

$$v' = v\left(1 + \frac{2v}{c}\cos\theta\right).$$

The energy of the radiation in the range $(v, v + dv)$, e_v, will then be changed by an amount

$$\frac{\partial e_v}{\partial v}(v' - v) = \frac{\partial e_v}{\partial v}\left(\frac{2v}{c}\cos\theta\right).$$

This change due to the radiation falling from the solid angle $d\mathbf{n}$ on the area of the mirror A in the time interval dt will be $\partial e_v/\partial v (2v/c\cos\theta)c/4\pi\cos\theta\,d\mathbf{n}\,A\,dt$ and, integrating over the solid angle, we obtain a change $(1/3)v\partial e_v/\partial v\,dV$, where $dV = Av\,dt$ is the volume change, if we imagine that the mirror is the surface of a piston acting on the radiation.

The last expression of course equals $d(e_v V)$. We remark that we have oversimplified this calculation by not taking into account the changes associated with the change in radiation intensity and in the length of the frequency range which occur during the reflection; it can be shown however that these two changes exactly compensate each other. We have thus arrived at the following result:

$$(1/3)v\frac{\partial e_v}{\partial v}dV = d(e_v V).$$

This gives the following partial differential equation for e_v:

$$V\frac{\partial e_v}{\partial V} = (1/3)v\frac{\partial e_v}{\partial v} - e_v.$$

Let us introduce the following change of variables: $\phi(x, z) = e_v v^{-3}$, $y = v^3 V$, $z = v$; then the previous equation simplifies to

$$\frac{\partial\phi}{\partial z} = 0.$$

In other words, ϕ does not depend on z but is an arbitrary function of y, $\phi = \phi(y)$. Going back to the original variables, this gives:

$$e_v = v^3\phi(v^3 V),$$

where ϕ is an arbitrary function.

Now since we can assume that not only the piston but the entire cavity is specularly reflecting, there is no heat exchange at the wall. In addition, the process can be performed as slowly as we like, i.e. the above formulae hold for a reversible process and the lack of heat exchange implies constant entropy. By the formula established at the end of Appendix 9.1, then V is proportional to

T^{-3} and the arbitrary function ϕ can be thought of as depending on just $(\nu/T)^3$. Thus we conclude that

$$e_\nu = \nu^3 F\left(\frac{\nu}{T}\right),$$

which is the so-called Wien displacement law mentioned in the main text.

REFERENCES

The references are listed according to the chapter in which they are cited. The references quoted in the appendices are listed together with those of the relevant chapter. Several references to the journal Annalen der Physik und Chemie up to 1899 *(from 1900, entitled just* Annalen der Physik*) are given under its informal title* Poggendorff's Annalen *(1824–77) and* Wiedemann's Annalen *(1877–99).*

Chapter 1

[1] I.M. Fasol-Boltzmann, Ed. (1990). *Ludwig Boltzmann: Principien der Naturalfilosofi.* Springer-Verlag, Berlin.

[2] D. Flamm, Ed. (1985). *Hochgeehrter Herr Professor! Innig geliebter Louis! Ludwig Boltzmann, Henriette von Aigentler, Briefwechsel.* Böhlau Verlag, Vienna.

[3] D. Flamm (1973). Life and personality of Ludwig Boltzmann, in *The Boltzmann equation: theory and application*, ed. E.G.D. Cohen and W. Thirring, pp. 3–16. Springer-Verlag, Vienna.

[4] W. Höflechner (1993). A version of the life of Ludwig Boltzmann, in *Proceedings of the International Symposium on Ludwig Boltzmann*, ed. G. Battimelli, M.G. Ianniello, and O. Kresten. Verlag der Österreichischen Akademie der Wissenschaften, Vienna.

[5] W. Höflechner, ed. (1994). *Ludwig Boltzmann: Leben und Briefe*, Publikationen aus dem Archiv der Universität Graz, Band 30. Akademische Druck- und Verlagsanstalt, Graz.

[6] L. Boltzmann (1905). *Populäre Schriften*. Leipzig. [Most of the content of this book, but not all, is available in a English translation under the title *Theoretical physics and philosophical problems*, ed. B. McGuinness, Reidel, Dordrecht. (1974). We shall stick to this translation whenever possible.]

[7] W. Höflechner and A. Honester (1985). *Ludwig Boltzmann, 1844–1906, eine Dokumentation*. Graz.

[8] P. Coveney and R. Highfield (1991). *The arrow of time*. W. H. Allen, New York.

[9] G.H. Bryan (1906). Prof. Ludwig Boltzmann. *Nature*, **74**, 569–70.

[10] Anon. (1906). Der Tod des Hofrates Boltzmann. *Die Zeit* (Vienna), No. 1421, Morgenblatt, 8 Sept., p.5.

[11] L. Flamm (1944). Die Persönlichkeit Boltzmanns. *Wiener Chemiker Zeitung*, **47**, 30.

[12] A. Höfler (1906). Ludwig Boltzmann als Mensch und als Philosoph. *Süddeutsche Monats-hefte*, October 3, unpaginated.

[13] E. Mach (1906). Ludwig Boltzmann. *Die Zeit* (Vienna), no. 1420, Abendblatt, 7 Sept., p.1.

[14] Anon. (1906). Weitere Nachrichten die letzen Tage. *Die Zeit* (Vienna), no. 1420, Abendblatt, 7 Sept., p.1.

[15] Private communication from Frau Dr. Lili Hahn, quoted in: J.T. Blackmore (1972). *Ernst Mach: his work, life and influence*. University of California Press, Berkeley.

[16] W. Moore (1989). *Schrödinger: life and thought*. Cambridge University Press.

[17] G.H. von Wright (1955). Ludwig Wittgenstein, a biographical sketch. *Philosophical Review*, **64**, 527–44.

[18] A. Janik and S. Toulmin (1973). *Wittgenstein's Vienna*. Simon and Schuster, New York.

[19] R.M. Rilke (1939). *Duino elegies*, transl. J.B. Leishman and S. Spender. Norton, New York.

[20] A. J. May (1968). *The Habsburg monarchy 1867–1914*. Norton Library, New York.

[21] R. Musil (1953–60). *The man without qualities*, transl. E. Wilkins and E. Kaiser, 3 vols. Secker and Warburg.

[22] E. Broda (1973). Philosophical biography of Ludwig Boltzmann, in *The Boltzmann equation: theory and application*, ed. E.G.D. Cohen and W. Thirring, pp. 17–52. Springer-Verlag, Vienna.

[23] P. Schick (1965). *Karl Kraus in Selbstzeugnissen und Bilddokumenten*. Rowohlt, Reinbeck bei Hamburg.

[24] L. Boltzmann (1897–1904). *Vorlesungen über die Principe der Mechanik*, 2 vols. Barth, Leipzig.

[25] V.I. Ul'yanov (V.I. Lenin) (1909). *Materialism and empirio-criticism* [in Russian]. Izdanie "Zveno" Moscow.

[26] L. Boltzmann (1974). On the principles of mechanics, in *Theoretical physics and philosophical problems* [English translation from *Populäre Schriften*, Leipzig (1905)], ed. B. McGuinness. Reidel, Dordrecht.

[27] H. Schnitzler (1954). Gay Vienna—Myth and Reality. *Journal of the History of Ideas*, **15**, 94–118.

[28] F. Rohrlich (1992) A poem by Ludwig Boltzmann *American Journal of Physics*, **60**, 972–3.

Chapter 2

[1] L. Boltzmann (1974). On the development of the methods of theoretical physics in recent times, in *Theoretical physics and philosophical problems* [English translation from *Populäre Schriften*, Leipzig (1905)], ed. B. McGuinness. Reidel, Dordrecht.

[2] I. Newton (1704). *Opticks*. London.

[3] R. Boscovich (1763). *Theoria philosophiae naturalis*, Venice. [English translation: *A theory of natural philosophy*, Open Court, Chicago (1922); reprinted by MIT Press, Cambridge, MA (1966)].

[4] M. Born (1964). *Natural philosophy of cause and chance*. Dover, New York.

[5] L. Boltzmann (1895). On certain questions of the kinetic theory of gases. *Nature*, **51**, 413–5.

[6] P.S. de Laplace (1814). *Essai philosophique sur les probabilités*. Paris. [reprinted by Gauthiers-Villars, Paris (1921)].

[7] G.W. Leibniz (1686). Brevis demonstratio erroris memorabilis Cartesii et aliorum circa legem naturalem, secundum qua volunt a Deo eandem semper quantitatem motus conservari, qua et in re mechanica abutuntur. *Acta eruditorum (Leipzig)* [English translation in:

R.B. Lindsay, *Energy, historical development of the concept*, Dowden, Hutchinson and Ross, Stroudberg, PA (1975)].

[8] G.W. Leibniz (1840). *Opera philosophica*, p. 775. Erdmann, Berlin.

[9] J. Hutton (1795). *Theory of the Earth*. Edinburgh.

[10] J.R. Mayer (1842). Bemerkungen über die Kräfte der unbelebten Natur. *Liebig's Annalen der Chemie und Pharmazie*, **42**, 239.

[11] J.R. Mayer (1845). *Die organische Bewegung in ihrem Zusammenhang mit dem Stoffwechsel. Ein Beitrag zur Naturkunde*. Verlag der C. Drechslerschen Buchhandlung, Heilbronn.

[12] J.R. Mayer (1848). *Beiträge zur Dynamik des Himmels in populärer Darstellung*. Verlag von Johann Ulrich Landherr, Heilbronn.

[13] S. Carnot (1824). *Reflexions sur la puissance motrice du feu et sur les machines propre à développer cette puissance*. Bachelier, Paris [English translation by R.H. Thurston in: S. Carnot, *Reflections on the motive power of fire and other papers on the second law of thermodynamics*, ed. E. Mendoza, Dover, New York (1960)].

[14] É. Clapeyron (1834). Mémoir sur la puissance motrice de la chaleur. *Journal de l'École Polytechnique*, **14**, 153–90 [English translation by E. Mendoza in: S. Carnot, *Reflections on the motive power of fire and other papers on the second law of thermodynamics*, ed. E. Mendoza, Dover, New York (1960)].

[15] M. Klein (1974). Closing the Carnot cycle, in *Sadi Carnot et l'essor de la thermodynamique*. CNRS, Paris.

[16] J.C. Maxwell (1871). *Theory of heat*. Longmans, Green, London.

[17] W. Thomson (Lord Kelvin) (1852). On a universal tendency in nature to the dissipation of mechanical energy, *Philosophical Magazine*, Ser. 4, **4**, 304.

[18] H. von Helmholtz (1847). *Über die Erhaltung der Kraft*. Berlin.

[19] H. von Helmholtz (1871). Über die Wechselwirkung der Naturkräfte. Ein populärwissenschaftlicher Vortrag, gehalten am 7. Februar 1854 in Könisberg in Preußen, in *Populäre Vorträge*, 2. Heft. Vieweg, Braunschweig.

[20] J.C. Maxwell (1881). *Treatise on electricity and magnetism*, 2nd edn. Clarendon Press, Oxford.

[21] G. Gamow (1961). *Biography of physics*. Harper Torchbooks, New York.

[22] L. Boltzmann (1905). G. Kirchhoff, in *Populäre Schriften*. Barth, Leipzig.

[23] L. Boltzmann (1974). On the indispensability of atomism in natural science, in *Theoretical physics and philosophical problems* [English translation from *Populäre Schriften*, Leipzig (1905)], ed. B. McGuinness, Reidel, Dordrecht.

[24] L. Euler (1914). *Opera omnia*, vol. XIII, 9. Teubner, Lepzig.

Chapter 3

[1] D. Bernoulli (1738). *Hydrodynamica*. Argentorati, Strassburg.

[2] C. Cercignani (1990). *Mathematical methods in kinetic theory*, rev. edn. Plenum Press, New York [1st edn 1969].

[3] C. Cercignani (1988). *The Boltzmann equation and its applications*. Springer-Verlag, New York.

[4] C. Cercignani, R. Illner, and M. Pulvirenti (1994). *The mathematical theory of dilute gases*. Springer-Verlag, New York.

[5] R. Clausius (1858). Über die mittlere Länge der Wege, welche bei der Molekularbewegung gasförmiger Körper von den einzelnen Molekülen zurückgelegt werden; nebst einigen

anderen Bemerkungen über die mechanische Wärmetheorie. *Poggendorff's Annalen*, **105**, 239–58.

[6] J.C. Maxwell (1860). Illustration of the dynamical theory of gases. Part I: On the motions and collisions of perfectly elastic spheres. Part II: On the process of diffusion of two or more kinds of moving particles among one another. Part III: On the collisions of perfectly elastic bodies of any form. *Philosophical Magazine*, Ser. 4, **19**, 48–62, 63–76, 76–80.

[7] J.C. Maxwell (1867). On the dynamical theory of gases. *Philosophical Transactions of the Royal Society*, **157**, 49–88.

[8] L. Boltzmann (1872). Weitere Studien über das Wärmegleichgewicht unter Gasmolekülen. *Sitzungsberichte der Akademie Wissenschaften, Wien*, II, **66**, 275–370.

[9] S. Carnot (1824). *Reflexions sur la puissance motrice du feu et sur les machines propre à développer cette puissance*. Bachelier, Paris.

[10] É. Clapeyron (1834). Mémoir sur la puissance motrice de la chaleur. *Journal de l'École Polytechnique*, **14**, 153–90 [English translation by E. Mendoza in: S. Carnot, *Reflections on the motive power of fire and other papers on the second law of thermodynamics*, ed. E. Mendoza, Dover, New York (1960)].

[11] W. Thomson (Lord Kelvin) (1849). An account of Carnot's theory of the motive power of heat; with numerical results deduced from Regnault's experiments on steam. *Transactions of the Royal Society of Edinburgh*, **16**, 113–55.

[12] M. Klein (1974). Closing the Carnot cycle, in *Sadi Carnot et l'essor de la thermodynamique*. CNRS, Paris.

[13] W. Thomson (Lord Kelvin) (1852). On a universal tendency in nature to the dissipation of mechanical energy. *Philosophical Magazine*, Ser. 4, **4**, 304.

[14] H. von Helmholtz (1847). *Über die Erhaltung der Kraft*. Berlin.

[15] R. Clausius (1850). Über die bewegende Kraft der Wärme, und die Gesestze, welche sich daraus für die Wärmelehre selbst ableiten lassen. *Poggendorff's Annalen*, **79**, 368–500 [English translation by W.F. Magie in: S. Carnot, *Reflections on the motive power of fire and other papers on the second law of thermodynamics*, ed. E. Mendoza, Dover, New York (1960)].

[16] R. Clausius (1856). On a modified form of the second fundamental theorem in the mechanical theory of heat. *Philosophical Magazine*, **12**, 86.

[17] R. Clausius (1865). Über verschiedene für die Anwendung bequeme Formen der Hauptgleichungen der mechanischen Wärmetheorie. *Poggendorff's Annalen*, **125**, 353.

[18] C.G. Knott (1911). *Life and scientific work of Peter Guthrie Tait*. Cambridge University Press, London.

[19] J.C. Maxwell (1871). *Theory of heat*, Chapter XXII. Longmans, London.

[20] M. Klein (1970). Maxwell, his demon, and the second law of thermodynamics. *American Scientist*, **58**, 84–97.

[21] P.G. Tait (1879–80). Clerk-Maxwell's scientific work. *Nature*, **21**, 321.

[22] W.J.M. Rankine (1865). On the second law of the thermodynamics. *Philosophical Magazine*, Ser. 4, **30**, 241–45.

[23] W.J.M. Rankine (1867). Sur la nécessité de vulgariser la seconde loi de la thermodynamique. *Annales de Chimie et de Physique*, Ser. 4, **12**, 258–66.

[24] L. Boltzmann (1870). Theorie der Wärme. *Fortschritte der Physik*, **26**, 441–504.

[25] L. Boltzmann (1866). Über die mechanische Bedeutung des zweiten Hauptsatzes der Wärmetheorie. *Wiener Berichte*, **53**, 195–220.

[26] R. Clausius (1871). Über Zurückführung des zweiten Hauptsatzes der mechanischen Wärmetheorie auf allgemeine mechanische Prinzipien. *Poggendorff's Annalen*, **142**, 433–61.

[27] L. Boltzmann (1871). Zur Priorität der Auffindung der Beziehung zwischen dem zweiten Hauptsatze der mechanischen Wärmetheorie und dem Prinzip der kleinsten Wirkung. *Poggendorff's Annalen*, **143**, 211–30.

[28] R. Clausius (1871). Bemerkungen zu der Prioritätsreclamation des Hrn. Boltzmann. *Poggendorff's Annalen*, **144**, 265–80.

[29] J.W. Gibbs (1889). Rudolf Julius Emmanuel Clausius. *Proceedings of the American Academy*, **16**, 458–65.

[30] A. Einstein (1904). Allgemeine molekulare Theorie der Wärme. *Annalen der Physik*, **14**, 354–62.

Chapter 4

[1] L. Boltzmann (1872). Weitere Studien über das Wärmegleichgewicht unter Gasmolekülen. Sitzungsberichte der Akademie der Wissenschaften, Wien, II, **66**, 275 [English translation in: S.G. Brush, *Kinetic theory*, Vol. 2, *Irreversible processes*, pp. 88–175, Pergamon Press, Oxford (1966)].

[2] J.C. Maxwell (1867). On the dynamical theory of gases. *Philosophical Transactions of the Royal Society*, **157**, 49–88.

[3] L. Boltzmann (1868). Studien über das Gleichgewicht der lebendigen Kraft zwischen bewegten materiellen Punkten. *Sitzungsberichte der Akademie der Wissenschaften, Wien*, **58**, 517.

[4] L. Boltzmann (1871). Über das Wärmegleichgewicht zwischen mehratomigen Gasmolekülen. *Sitzungsberichte der Akademie der Wissenschaften, Wien*, **63**, 397.

[5] L. Boltzmann (1871). Einige allgemeine Sätze über Wärmegleichgewicht. *Sitzungsberichte der Akademie der Wissenschaften, Wien*, **63**, 679.

[6] M.J.Klein (1973). The development of Boltzmann's statistical ideas, in *The Boltzmann equation: theory and applications*, ed. E.G.D. Cohen and W. Thirring. Springer-Verlag, Vienna.

[7] P. and T. Ehrenfest (1911). Begriffliche Grundlagen der statistischen Auffassung in der Mechanik. in *Enzyklopädie der mathematischen wissenschaften*, Vol. IV, Part 32. Leipzig.

[8] L. Boltzmann (1871). Analytischer Beweis des zweiten Hauptsatzes der mechanischen Wärmetheorie aus den Sätzen über das Gleichgewicht der lebendigen Kraft. *Sitzungsberichte der Akademie der Wissenschaften, Wien*, **63**, 712.

[9] M. Planck (1948). *Wissenschaftliche Selbstbiographie*. Leipzig.

[10] C. Cercignani (1988). *The Boltzmann equation and its applications*. Springer-Verlag, New York.

[11] C. Cercignani, R. Illner, and M. Pulvirenti (1994). *The mathematical theory of dilute gases*. Springer-Verlag, New York

[12] R.K. Alexander (1975). The infinite hard sphere system. Ph D thesis, University of California at Berkeley.

[13] C. Cercignani (1972). On the Boltzmann equation for rigid spheres. *Transport Theory and Statistical Physics*, **2**, 211.

[14] H. Spohn (1984). Boltzmann hierarchy and Boltzmann equation, in *Kinetic theories and the Boltzmann equation*, ed. C. Cercignani, LNM 1048. Springer-Verlag, Berlin.

[15] D. Enskog (1922). Kinetische Theorie der Wärmeleitung, Reibung und Selbstdiffusion in gewissen verdichteten Gasen und Flüssigkeiten. Kungliga Svenska Vetenskapsakademiens Handlingar, **63**, 3–44.

Chapter 5

[1] L. Boltzmann (1872). Weitere Studien über das Wärmegleichgewicht unter Gasmolekülen. *Sitzungsberichte der Akademie der Wissenschaften, Wien*, II, **66**, 275–370 [English translation in: S.G. Brush, *Kinetic theory*, Vol. 2, *Irreversible processes*, pp. 88–175, Pergamon Press, Oxford (1966)].

[2] T. Carleman (1933). Sur la théorie de l'équation intégro-differentielle de Boltzmann. *Acta Mathematica*, **60**, 91.

[3] T. Carleman 1957. *Problèmes mathématiques dans la théorie cinétique des gaz*. Almqvist & Wiksell, Uppsala.

[4] H.A. Lorentz (1887). Über das Gleichgewicht der lebendingen Kraft unter Gasmolekülen. *Sitzungsberichte der Akademie der Wissenschaften, Wien*, **95**, 115–52.

[5] L. Boltzmann (1887). Neuer Beweis zweier Sätze über das Wärmegleichgewicht unter mehratomigen Gasmolekülen. *Sitzungsberichte der Akademie der Wissenschaften, Wien*, **95**, 153–64.

[6] L. Boltzmann (1895–8). *Vorlesungen über Gastheorie*, 2 vols. Barth, Leipzig [English translation: *Lectures on gas theory*, transl. S.G. Brush, University of California Press (1964)].

[7] R.C. Tolman (1938). *The principles of statistical mechanics*. Oxford.

[8] E.C.G. Stueckelberg (1952). Théorème *H* et unitarité de *S*. *Helvetica Physica Acta*, **25**, 577–80.

[9] L. Waldmann (1958). Transporterscheinungen in Gasen von mittlerem Druck, in *Handbuch der Physik*, ed. S. Flügge, Vol. XII. Springer-Verlag.

[10] C. Cercignani and M. Lampis (1981). On the *H*-theorem for polyatomic gases. *Journal of Statistical Physics*, **26**, 795.

[11] A. Kox (1990). H.A. Lorentz contributions to kinetic gas theory. *Annals of Science*, **47**, 591–606.

[12] P. Hein (1966). *Grooks*. MIT Press.

[13] W. Thomson (Lord Kelvin) (1874). The kinetic theory of the dissipation of energy. *Proceeding of the Royal Society of Edinburgh*, **8**, 325–34.

[14] J. Loschmidt (1876). Über den Zustand des Wärmegleichgewichtes eines Systems von Körpern mit Rucksicht auf die Schwerkraft. *Wiener Berichte*, **73**, 139.

[15] L. Boltzmann (1877). Über die Beziehung eines allgemeine mechanischen Satzes zum zweiten Hauptsatze der Wärmetheorie. *Sitzungsberichte der Akademie der Wissenschaften, Wien*, II, **75**, 67–73 [English translation in: S.G. Brush, *Kinetic theory*, Vol. 2, *Irreversible processes*, pp. 188–193, Pergamon Press, Oxford (1966)].

[16] F. Nietzsche (1926). *Der Wille zur Macht*, in *Gesammelte Werke*, Vol. 19, Book 4, Part 3. Musarion Verlag, Munich.

[17] H. Poincaré (1893). Le mécanisme et l'expérience. *Revue de Metaphysique et de Morale*, **4**, 534 [English translation in: S.G. Brush, *Kinetic theory*, Vol. 2, *Irreversible processes*, pp. 203–7, Pergamon Press, Oxford (1966)].

[18] H. Poincaré (1890). Sur les problème des trois corps et les équations de la dynamique. *Acta Mathematica*, **13**, 1–270 [Extracts translated into English in: S.G. Brush, *Kinetic theory*, Vol. 2, *Irreversible processes*, pp. 194–202, Pergamon Press, Oxford (1966)].

[19] E. Zermelo (1886). Über einen Satz der Dynamik und die mechanische Wärmetheorie. *Wiedemann's Annalen*, **57**, 485–94 [English translation in: S.G. Brush, *Kinetic theory*, Vol. 2, *Irreversible processes*, pp. 208–17, Pergamon Press, Oxford (1966)].

[20] L. Boltzmann (1896). Entgegnung auf die wärmetheoretischen Betrachtungen des Hrn. E. Zermelo. *Wiedemann's Annalen*, **57**, 773–84 [English translation in: S.G. Brush, *Kinetic theory*, Vol. 2, *Irreversible processes*, pp. 218–28, Pergamon Press, Oxford (1966)].

[21] E. Zermelo (1896). Über mechanische Erklarungen irreversibler Vorgange. *Wiedemann's Annalen*, **59**, 793–801 [English translation in: S.G. Brush, *Kinetic theory*, Vol. 2, *Irreversible processes*, pp. 229–237, Pergamon Press, Oxford (1966)].

[22] L. Boltzmann (1897). Zu Hrn. Zermelo Abhandlung über die mechanische Erklarungen irreversibler Vorgange. *Wiedemann's Annalen*, **60**, 392–8 [English translation in: S.G. Brush, *Kinetic theory*, Vol. 2, *Irreversible processes*, pp. 238–45, Pergamon Press, Oxford (1966)].

[23] H. Grad (1949). On the kinetic theory of rarified gases. *Communications on Pure and Applied Mathematics*, **2**, 325.

[24] H. Grad (1958). Principles of the kinetic theory of gases, in *Handbuch der Physik*, ed. S. Flügge, Vol. XII, Springer-Verlag.

[25] H. Grad (1961). The many faces of entropy. *Communications on Pure and Applied Mathematics*, **14**, 323.

[26] E.L. Hahn (1950). Spin echoes. *Physical Review*, **80**, 580.

[27] B.J. Alder and T.E. Wainright (1960). Studies in molecular dynamics II. Behavior of a small number of elastic spheres. *Journal of Chemical Physics*, **33**, 1439.

[28] A. Bellemans and J. Orban (1967). Velocity inversion and irreversibility in a dilute gas of hard disks. *Physics Letters*, **24A**, 620.

[29] D. Levesque and L. Verlet (1993). Molecular dynamics and time reversibility. *Journal of Statistical Physics*, **72**, 519–37.

[30] C. Cercignani (1972). On the Boltzmann equation for rigid spheres. *Transport Theory and Statistical Physics*, **2**, 211.

[31] O.E. Lanford III (1975). Time evolution of large classical systems, in *Dynamical systems, theory and applications*, ed. J. Moser, LNP 38, 1. Springer-Verlag, Berlin.

[32] D. Morgenstern (1954). General existence and uniqueness proof for spatially homogeneous solutions of the Maxwell–Boltzmann equation in the case of Maxwellian molecules. *Proceedings of the National Academy of Sciences (U.S.A.)*, **40**, 719–21.

[33] L. Arkeryd (1972). On the Boltzmann equation. Part I: Existence. *Archives for Rational Mechanical Analysis*, **45**, 1–16.

[34] T. Nishida and K. Imai (1977). Global solutions to the initial value problem for the nonlinear Boltzmann equation. *Publications of the Research Institute for Mathematical Sciences, Kyoto University*, **12**, 229–39.

[35] Y. Shizuta and K. Asano (1974). Global solutions of the Boltzmann equation in a bounded convex domain. *Proceedings of the Japan Academy*, **53**, 3–5.

[36] S. Ukai (1974). On the existence of global solutions of mixed problem for non-linear Boltzmann equation. *Proceedings of the Japan Academy*, **50**, 179–84.

[37] C. Cercignani 1977. *The Boltzmann equation and its applications*. Springer-Verlag, New York.

[38] C. Cercignani, R. Illner, and M. Pulvirenti (1994). *The mathematical theory of dilute gases*. Springer-Verlag, New York.

[39] R. Illner and M. Shinbrot (1984). The Boltzmann equation: global existence for a rare gas in an infinite vacuum. *Communications in Mathematical Physics*, **95**, 217–26.

[40] R. Illner and M. Pulvirenti (1986). Global validity of the Boltzmann equation for a two-dimensional rare gas in a vacuum. *Communications in Mathematical Physics*, **105**, 189–203.

[41] R. Illner and M. Pulvirenti (1989). Global validity of the Boltzmann equation for two- and three-dimensional rare gas in vacuum: erratum and improved result. *Communications in Mathematical Physics*, **121**, 143–46.

[42] R. DiPerna and P.-L. Lions (1989). On the Cauchy problem for Boltzmann equations. *Annals of Mathematics*, **130**, 321–66.

[43] L. Boltzmann (1974). On the development of the methods of theoretical physics in recent times, in *Theoretical physics and philosophical problems* [English translation from *Populäre Schriften*, Leipzig (1905)], ed. B. McGuinness. Reidel, Dordrecht.

[44] R. Penrose (1989). *The emperor's new mind*. Oxford University Press.

[45] A. Eddington 1959. *New pathways in science*. University of Michigan Press, Ann Arbor.

[46] D.W. Sciama (1971). *Modern cosmology*. Cambridge University Press, London.

[47] T. Gold (1962). The arrow of time, in *Recent developments in general relativity*, pp. 225–34. Pergamon Press, New York.

[48] J.W. Cronin (1981). CP symmetry violation—the search for its origin. *Reviews of Modern Physics*, **53**, 373.

[49] C. Cercignani (1989). Le radici fisiche e matematiche dell'irreversibilità temporale [with translation into English]. *Alma Mater Studiorum*, **2**, 37–52.

[50] R. Illner and H. Neunzert (1987). The concept of irreversibility in the kinetic theory of gases. *Transport Theory and Statistical Physics*, **16**, (1), 89–112.

[51] M. Ageno (1992). *Le origini della irreversibilità*. Bollati Boringhieri, Turin.

[52] K. Gödel (1949). A remark about the relationship between relativity theory and idealistic philosophy, in *Albert Einstein, philosopher-scientist*, ed. P.A. Schilpp, pp. 557–62. Open Court, La Salle, IL.

[53] H. Reichenbach (1956). *The direction of time*. University of California Press, Berkeley.

[54] J.L. Mackie (1974). *Causation: the cement of the universe*. Oxford University Press.

[55] J. Earman (1974). An attempt to add a little direction to 'The problem of the direction of time'. *Philosophy of Science*, **41**, 15–47.

[56] G.H. von Wright (1971). *Explanation and understanding*. Cornell University Press, Ithaca, NY.

[57] W.C. Salmon (1984). *Scientific explanation and the causal structure of the world*. Princeton University Press.

[58] P. Horwich (1987). *Asymmetries in time. Problems in the philosophy of science*. MIT Press, Cambridge, MA.

[59] I. Prigogine and I. Stengers (1979). *La nouvelle alliance. Metamorphoses de la science*. Gallimard, Paris.

[60] I. Prigogine and I. Stengers (1988). *Entre le temps et l'éternité*. Fayard, Paris.

[61] H. Bergson (1907). *L'evolution créatrice*. F. Alcan, Paris.

[62] W. Heisenberg (1958). *The physicist's conception of nature*. Hutchinson, London.

[63] M. Born (1949). *Natural philosophy of cause and chance*. Clarendon Press, Oxford.

[64] K. Popper (1956). *The open universe. An argument for indeterminism*. Rowman & Littlefield, Totowa, NJ.

[65] K. Popper (1956). *Quantum theory and the schism in physics*. Rowman & Littlefield, Totowa, NJ.

[66] K. Popper (1974). *Intellectual autobiography*, in *The philosophy of Karl Popper*, ed. P.A. Schilpp, Book I, pp. 3–181. Open Court, La Salle, IL.

[67] K. Popper (1958). Irreversibility; or, entropy since 1905. *British Journal for the Philosophy of Science*, **8**, 151–63.

[68] J.C. Maxwell (1878). Tait's 'Thermodynamics'. *Nature*, **17**, 257–9, 278–80.

[69] P.K. Feyerabend (1975). *Against the method*. New Left Books, London.

[70] J. Bricmont (1995). Science of chaos or chaos in science?. *Physicalia Magazine*, **17**, 159–208.

[71] I. Prigogine (1995). Science of chaos or chaos in science: a rearguard battle. *Physicalia Magazine*, **17**, 213–18.

[72] J. Bricmont (1995). The last word from the rearguard. *Physicalia Magazine*, **17**, 219–21.

[73] J.L. Lebowitz (1993). Boltzmann's entropy and time's arrow. *Physics Today*, **46**, (9), 32–38.

[74] H. Price (1996). *Time's arrow and Archimedes' point*. Oxford University Press, New York.

[75] P. Coveney and R. Highfield (1990). *The arrow of time—a voyage through science to solve time's greatest mystery*. W.H. Allen, London.

[76] C. Cercignani (1990). *Mathematical methods in kinetic theory*, 2nd edn. Plenum Press, New York [1st edn 1969].

[77] L. Boltzmann (1875). Über das Wärmegleichgewicht von Gasen, auf welche äussere Kräfte wirken. *Sitzungsberichte der Akademie der Wissenschaften, Wien*, **72**, 427–57.

[78] L. Boltzmann (1876). Über die Aufstellung und Integration der Gleichungen, welche die Molekularbewegungen in Gasen bestimmen. *Sitzungsberichte der Akademie der Wissenschaften, Wien*, **74**, 503–52.

[79] T.H. Gronwall (1915). A functional equation in the kinetic theory of gases. *Annals of Mathematics* (2), **17**, 1–4.

[80] T.H. Gronwall (1916). Sur une équation fonctionelle dans la théorie cinétique des gaz. *Comptes Rendus de l'Académie des Sciences (Paris)*, **162**, 415–8.

[81] C. Truesdell and R.G. Muncaster (1980). *Fundamentals of Maxwell's kinetic theory of a simple monatomic gas*. Academic Press, New York.

[82] C. Cercignani (1990). Are there more than five linearly independent collision invariants for the Boltzmann equation?. *Journal of Statistical Physics*, **58**, 817–23.

[83] L. Arkeryd (1972). On the Boltzmann equation. Part II: The full initial value problem. *Archives for Rational Mechanical Analysis*, **45**, 17–34.

[84] L. Arkeryd and C. Cercignani (1990). On a functional equation arising in the kinetic theory of gases. *Rendiconti Matematiche, Accademia dei Lincei*, s.9, **1**, 139–49.

[85] B. Wennberg (1992). On an entropy dissipation inequality for the Boltzmann equation. *Comptes Rendus de l'Académie des Sciences (Paris)*, I, **315**, 1441–6.

[86] B. Touscheck and G. Rossi (1970). *Meccanica statistica*. Boringhieri, Turin.

Chapter 6

[1] L. Boltzmann (1872). Weitere Studien über das Wärmegleichgewicht unter Gasmolekülen. *Sitzungsberichte der Akademie der Wissenschaften, Wien*, II, **66**, 275–370 [English translation in: S.G. Brush, *Kinetic theory*, Vol. 2, *Irreversible processes*, pp. 88–175, Pergamon Press, Oxford (1966)].

[2] M. J. Klein (1973). The development of Boltzmann's statistical ideas. in *The Boltzmann equation: theory and applications*, ed. E.G.D. Cohen and W. Thirring. Springer-Verlag, Vienna.

[3] L. Boltzmann (1877). Über die Beziehung eines allgemeine mechanischen Satzes zum zweiten Hauptsatze der Wärmetheorie. *Sitzungsberichte der Akademie der Wissenschaften, Wien*, II, **75**, 67–73 [English translation in: S.G. Brush, *Kinetic theory*, Vol. 2, *Irreversible processes*, pp. 188–202, Pergamon Press, Oxford (1966)].

[4] L. Boltzmann (1871). Zur Priorität der Auffindung der Beziehung zwischen dem zweiten Hauptsatze der mechanischen Wärmetheorie und dem Prinzip der kleinsten Wirkung. *Poggendorff's Annalen*, **143**, 211–30.

[5] C.G. Knott (1911). *Life and scientific work of Peter Guthrie Tait*. Cambridge University Press, London.

[6] J.C. Maxwell (1871). *Theory of heat*, Chapter XXII. Longmans, London.

[7] J.C. Maxwell to J.W. Strutt, 6 December 1870. [Reprinted in R.J. Strutt, *Life of John William Strutt, Third Baron Rayleigh*, p. 47, Madison (1856)].

[8] L. Boltzmann (1877). Über die Beziehung zwischen dem zweiten Hauptsatze der mechanischen Wärmetheorie und der Wahrscheinlichkeitsrechnung respektive den Sätzen über das Wärmegleichgewicht. *Wiener Berichte*, **76**, 373–435.

[9] W. Ebeling (1993). Entropy and information in processes of self-organization: uncertainty and unpredictability. *Physica A*, **194**, 563–75.

[10] A. Einstein (1904). Allgemeine molekulare Theorie der Wärme. *Annalen der Physik*, **14**, 354–62.

[11] A. Einstein (1905). Über einen die Erzeugung und Verwandlung des Lichtes betreffenden heuristischen Gesichtspunkt. *Annalen der Physik*, **17**, 132–48.

[12] A. Einstein (1909). Zur gegenwärtigen Stand der Strahlungsproblems. *Physikalische Zeitschrift*, **10**, 185–93.

[13] A. Einstein (1910). Theorie der Opaleszenz von homogenen Flüssigkeiten und Flüssigkeitsgemischen in der Nähe des kritischen Zustand. *Annalen der Physik*, **33**, 1275–1298.

[14] A. Einstein (1903). Eine Theorie der Grundlagen der Thermodynamik. *Annalen der Physik*, **11**, 170–87.

[15] L. Boltzmann (1895–8). *Vorlesungen über Gastheorie*, 2 vols. Barth, Leipzig [English translation: *Lectures on gas theory*, transl. by S.G. Brush, University of California Press (1964)].

[16] A. Einstein (1911). Bemerkungen zu den P. Hertzschen Arbeiten: Mechanische Grundlagen der Thermodynamik. *Annalen der Physik*, **34**, 175–76.

[17] A. Einstein (1915). Theoretische Atomistik, in *Kultur der Gegenwart: Die Physik*, ed. E. Lecher, pp. 251–63. Teubner, Leipzig.

[18] E.P. Culverwell (1890). Note on Boltzmann's kinetic theory of gases, and on Sir W. Thomson's address to Section A, British Association, 1884. *Philosophical Magazine*, **30**, 95.

[19] E.P. Culverwell (1894). Dr. Watson's proof of Boltzmann's theorem on permanence of distributions. *Nature*, **50**, 617.

[20] S.H. Burbury (1894). Boltzmann's minimum function. *Nature*, **51**, 78.

[21] L. Boltzmann (1895). On certain questions of the theory of gases. *Nature*, **51**, 413–15.

[22] L. Boltzmann (1974). The Second Law of thermodynamics, in *Theoretical physics and philosophical problems* [English translation from *Populäre Schriften*, Leipzig (1905)]. Reidel, Dordrecht.

[23] L. Boltzmann (1898). Über die sogenannte *H*-curve. *Mathematische Annalen*, **50**, 325–32.

[24] P. Ehrenfest and T. Ehrenfest (1911). Begriffliche Grundlagen der statistischen Auffassung in der Mechanik, in *Enzyklopädie der mathematischen wissenschaften*, Vol. IV, Part 32. Leipzig [English translation: P. Ehrenfest and T. Ehrenfest, *The conceptual foundations of the statistical approach to mechanics*, Dover, New York (1990)].

[25] M. Kaç (1959). *Probability and related topics in physical sciences*. Interscience, London.

[26] L. Boltzmann (1897). Zu Hrn. Zermelo Abhandlung über die mechanische Erklärungen irreversibler Vorgange. *Wiedemann's Annalen*, **60**, 392–8 [English translation in: S.G. Brush, *Kinetic theory*, Vol. 2, *Irreversible processes*, pp. 238–45, Pergamon Press, Oxford (1966)].

[27] C. Cercignani (1988). *The Boltzmann equation and its applications*. Springer-Verlag, New York.

[28] E. Artin (1964). *The gamma function*. Holt, Rinehart and Winston, New York.

Chapter 7

[1] L. Boltzmann (1884). Über die Möglichkeit der Begründung einer kinetischen Gastheorie auf anziehende Kräfte allein. *Wiener Berichte*, **89**, 714.

[2] J.W. Gibbs (1902). *Elementary principles in statistical mechanics, developed with special reference to the rational foundations of thermodynamics.* Yale University Press.

[3] M.J. Klein (1973). The development of Boltzmann's statistical ideas, in *The Boltzmann equation: theory and application*, ed. E.G.D. Cohen and W. Thirring, pp. 53–106. Springer-Verlag, Vienna.

[4] M.J. Klein (1983). The scientific style of Josiah Willard Gibbs, in *Springs of scientific creativity: essays on founders of modern science*, ed. R. Aris, H.T. Davis and R.H. Stuewer, pp. 142–62. University of Minnesota Press, Minneapolis.

[5] M.J. Klein (1990). The physics of J. Willard Gibbs in his time, in *Proceedings of the Gibbs Symposium*, pp. 1–21. American Mathematical Society, Providence, NJ.

[6] J.W. Gibbs (1906). Graphical methods in the thermodynamics of fluids, in *The scientific papers of J. Willard Gibbs*, ed. H.A. Bumstead and R.G. Van Name, Vol. I, pp. 1–32. Longmans, Green, New York.

[7] T. Andrews (1869). On the continuity of the gaseous and liquid states of matter. *Philosophical Transactions of the Royal Society*, **159**, 575–90.

[8] J.C. Maxwell (1871). *Theory of heat.* Longmans, Green, London.

[9] J.W. Gibbs (1906). A method of geometrical representation of the thermodynamic properties of substances by means of surfaces, in *The scientific papers of J. Willard Gibbs*, ed. H.A. Bumstead and R.G. Van Name, Vol. I, pp. 33–54. Longmans, Green, New York.

[10] J. Thomson (1871). Considerations on the abrupt changes of boiling or condensing in reference to the continuity of the fluid state of matter. Proceedings of the Royal Society, **20**, 1–8.

[11] J.D. van der Waals (1873). *Over de continuiteit van den gas en vloeistoftoestand.* A.W. Sijthoff, Leiden.

[12] L. P. Wheeler (1952). *Josiah Willard Gibbs. The history of a great mind*, 2nd edn. Yale University Press, New Haven, CT.

[13] J.C. Maxwell (1875). *Theory of heat*, 4th edn. Longmans, Green, London.

[14] J.C. Maxwell (1875). On the dynamical evidence of the molecular constitution of bodies. *Nature*, **11**, 357–9, 374–7.

[15] J.W. Gibbs (1906). On the equilibrium of heterogeneous substances, in *The scientific papers of J. Willard Gibbs*, ed. H.A. Bumstead and R.G. Van Name, Vol. I, pp. 55–353. Longmans, Green, New York.

[16] J.W. Gibbs (1906). Abstract of the equilibrium of heterogeneous substances, in *The scientific papers of J. Willard Gibbs*, ed. H.A. Bumstead and R.G. Van Name, Vol. I, pp. 354–71. Longmans, Green, New York.

[17] P. Duhem (1908). *Josiah Willard Gibbs à propos de la publication de ses mémoires scientifiques.* A. Hermann, Paris.

[18] E.B. Wilson (1931). Josiah Willard Gibbs, in *Dictionary of American Biography*, Vol. VII, pp. 248–51.

[19] J.W. Gibbs (1906). List of titles. In *The scientific papers of J. Willard Gibbs*, ed. H.A. Bumstead and R.G. Van Name, Vol. I, pp. 418–34. Longmans, Green, New York.

[20] G. Helm (1898). *Die Energetik nach ihrer geschichtlichen Entwicklung.* Veit, Leipzig.

[21] H.A. Bumstead (1906). Josiah Willard Gibbs. In *The scientific papers of J. Willard Gibbs*, vol. I, pp. XIII–XXVII. Longmans, Green, New York.

[22] J.C. Maxwell to P.G. Tait, August 1873 [Quoted in a footnote, p. 114, of: C.G. Knott, *Life and scientific work of Peter Guthrie Tait*, Cambridge (1911)].

[23] G. Gallavotti (1995). Ergodicity, ensembles, irreversibility in Boltzmann and beyond. *Journal of Statistical Physics*, **78**, 1571–90.

[24] P. Ehrenfest and T. Ehrenfest (1911). Begriffliche Grundlagen der statistischen Auffassung in der Mechanik, in *Enzyklopädie der mathematischen wissenschaften*, Vol. IV, Part 32. Leipzig [English translation: P. Ehrenfest and T. Ehrenfest, *The conceptual foundations of the statistical approach to mechanics*, Dover, New York (1990)].

[25] C. Seelig (1960). *Albert Einstein*, p. 176. Europa Verlag, Zurich.

[26] N. Bohr (1972). *Collected works*, Vol. 6, p. 320. North-Holland, Amsterdam.

[27] L. Boltzmann (1895–8). *Vorlesungen über Gastheorie*, 2 vols. Barth, Leipzig.

[28] J. Hadamard (1906). La mécanique statistique. *Bulletin of the American Mathematical Society*, **12**, 194–210.

[29] H. Poincaré (1906). Allocution de la séance publique annuelle du 17 décembre 1906. *Comptes Rendus de l'Académie des Sciences (Paris)*, **143**, 997.

[30] H. Poincaré (1906). Réflexions sur la théorie cinétique des gaz. *Journal de Physique*, Ser. 4, **5**, 369–403.

[31] S. Brush (1976). *The kind of motion we call heat*. Elsevier, Amsterdam.

[32] J.H. Jeans (1902). On the conditions necessary for equipartition of energy. *Philosophical Magazine*, Ser. 6, **4**, 585–96.

[33] L. Boltzmann (1884). Über die Eigenschaften monocyclischer und anderer damit verwandter Systeme. *Wiener Berichte*, **90**, 231.

[34] J.C. Maxwell (1879). On Boltzmann's theorem on the average distribution of energy in a system of material points. *Transactions of the Cambridge Philosophical Society*, **12**, 547.

[35] E. Borel (1915). In *Encyclopédie des Sciences Mathématiques*, Vol. 4, (1.1), p. 188.

[36] G. Birkhoff (1939). The mean ergodic theorem. *Duke Mathematical Journal*, **5**, 635–46.

[37] J. von Neumann (1932). Zur Operatorenmethode in der klassischen Mechanik. *Annals of Mathematics*, **33**, 587–648.

[38] M. Planck (1926). *Treatise on thermodynamics*. Dover, New York.

[39] M. Planck (1904). Über die mechanische Bedeutung der Temperatur und der Entropie, in *Festschrift Ludwig Boltzmann*, pp. 113–22. Barth, Leipzig.

Chapter 8

[1] L. Boltzmann (1974). On the development of the methods of theoretical physics in recent times, in *Theoretical physics and philosophical problems* [English translation from *Populäre Schriften*, Leipzig (1905)], ed. B. McGuinness. Reidel, Dordrecht.

[2] L. Boltzmann (1876). Über die Aufstellung und Integration der Gleichungen, welche die Molecularbewegung in Gasen bestimmen. *Wiener Berichte*, **74**, 553–60.

[3] L. Boltzmann (1872). Weitere Studien über das Wärmegleichgewicht unter Gasmolekülen. *Sitzungsberichte der Akademie der Wissenschaften, Wien*, II, **66**, 275–370.

[4] H.A. Lorentz (1887). Über das Gleichgewicht der lebendingen Kraft unter Gasmolekülen. *Sitzungsberichte der Akademie der Wissenschaften, Wien*, **95**, 115–52.

[5] L. Boltzmann (1887). Neuer Beweis zweier Sätze über das Wärmegleichgewicht unter mehratomigen Gasmolekülen. *Sitzungsberichte der Akademie der Wissenschaften, Wien*, **95**, 153–64.

[6] L. Boltzmann (1895–8). *Vorlesungen über Gastheorie*, 2 vols. Barth, Leipzig [English translation: *Lectures on gas theory*, transl. by S.G. Brush, University of California Press (1964)].

[7] R.C. Tolman (1938). *The principles of statistical mechanics*. Oxford.

[8] G.E. Uhlenbeck (1973). The validity and the limitations of the Boltzmann equation, in *The Boltzmann equation: theory and application*, ed. E.G.D. Cohen and W. Thirring, pp. 107–19. Springer-Verlag, Vienna.

[9] L. Waldmann (1973). On kinetic equations for particles with internal degrees of freedom, in *The Boltzmann equation: theory and application*, pp. 223–46, ed. E.G.D. Cohen and W. Thirring. Springer-Verlag, Vienna.

[10] E.C.G. Stueckelberg (1952). Théorème *H* et unitarité de *S*. *Helvetica Physica Acta*, **25**, 577–80.

[11] L. Waldmann (1958). Transporterscheinungen in Gasen von mittlerem Druck, in *Handbuch der Physik*, ed. S. Flügge, Vol. XII. Springer-Verlag.

[12] C. Cercignani and M. Lampis (1981). On the *H*-theorem for polyatomic gases. *Journal of Statistical Physics*, **26**, 795.

[13] A. Kox (1990). H.A. Lorentz contributions to kinetic gas theory. *Annals of Science*, **47**, 591–606.

[14] L. Galgani (1993). Boltzmann and the problem of equipartition of energy, in *Proceedings of the International Symposium on Ludwig Boltzmann*, ed. G. Battimelli, M.G. Ianniellom, and O. Kresten, pp. 193–202. Verlag der Österreichischen Akademie der Wissenschaften, Vienna.

[15] L. Boltzmann (1895). On certain questions of the theory of gases. *Nature*, **51**, 413–15.

[16] J.H. Jeans (1905). A comparison between two theories of radiation. *Nature*, **72**, 293.

[17] J.H. Jeans (1905). On the partition of energy between matter and ether. *Philosophical Magazine*, **10**, 91.

[18] J.H. Jeans (1903). On the vibrations set up in molecules by collisions. *Philosophica Magazine*, **6**, 279.

[19] L. Landau and E. Teller (1965). On the theory of sound dispersion. In *Collected papers of L.D. Landau*, ed. D. ter Haar, p. 147. Pergamon Press, Oxford [from *Physikalische Zeitschrift der Sowietunion*, **10**, 34 (1936)].

[20] N.N. Nekhoroshev (1977). Exponential estimate of the stability time of near-integrable Hamiltonian systems. *Uspekhi Matematicheskikh Nauk*, **32**, 1.

[21] N.N. Nekhoroshev (1979). Exponential estimate of the stability time of near-integrable Hamiltonian systems. *Trudy Sem. Petrowski*, **5**, 5.

[22] G. Benettin, L. Galgani, and A. Giorgilli (1985). A proof of Nekhoroshev's theorem for the stability time of near-integrable Hamiltonian systems. *Celestial Mechanics*, **37**, 1–25.

[23] G. Benettin, L. Galgani, and A. Giorgilli (1984). Boltzmann's ultraviolet cutoff and Nekhoroshev's theorem on Arnold diffusion. *Nature*, **311**, 444–5.

[24] G. Benettin and G. Gallavotti (1986). Stability of motions near resonances in quasi-integrable Hamiltonian systems. *Journal of Statistical Physics*, **44**, 293–338.

[25] G. Benettin, L. Galgani, and A. Giorgilli (1987). Exponential law for the equipartition times among translational and vibrational degrees of freedom. *Physics Letters A*, **120**, 23–7.

[26] A. N. Kolmogorov (1954). On the conservation of quasi-periodic motions for a small change in the Hamiltonian function. *Doklady Akademii Nauk SSSR*, **98**, 527–30 [English translation in: G. Casati and G. Ford (Eds), *Lecture Notes on Physics*, No. 93. Springer-Verlag, Berlin (1979)].

[27] E. Fermi, J. Pasta, and S. Ulam (1955). Studies of non linear problems. *Los Alamos Reports*, no. LA-1940.

[28] V.I. Arnold (1978). *Mathematical methods of classical mechanics* [translated from Russian]. Springer-Verlag, New York.

[29] V.I. Arnol'd (1963). Small denominators and problems of stability of motion in classical and celestial mechanics. *Russian Mathematical Surveys*, **18**, 85–193.

[30] J.E. Moser (1962). On invariant curves of area-preserving mappings of an annulus. *Nachrichten Akademie der Wissenschaft Göttingen*, No. 1.

[31] G. Benettin, L. Galgani, and A. Giorgilli (1987). Realization of holonomic constraints and freezing of high frequency degrees of freedom in the light of classical perturbation theory. Part I. *Communications in Mathematics Physics*, **113**, 87–103.

[32] G. Benettin, L. Galgani, and A. Giorgilli (1989). Realization of holonomic constraints and freezing of high frequency degrees of freedom in the light of classical perturbation theory. Part II. *Communications in Mathematics Physics*, **121**, 557–601.

[33] H. Poincaré (1957). *Les méthodes nouvelles de la mécanique céleste*. Dover, New York.

[34] G. Birkhoff (1927). *Dynamical systems*. New York.

[35] G. Gallavotti (1976). Quasi-integrable mechanical systems, in *Critical phenomena, random systems, gauge theories*, ed. K. Osterwalder and R. Stora. North-Holland, Amsterdam.

Chapter 9

[1] L. Boltzmann (1865). Über die Bewegung der Elektrizität in kummen Flächen. *Wiener Berichte*, **52**, 214–21.

[2] L. Boltzmann (1874). Über die Verschiedenheit der Dielectricitätsconstante des kristallisirten Schwefels nach verschiedenen Richtungen. *Wiener Sitzberichte*, 8 Oct.

[3] L. Boltzmann (1874). Zur Theorie der elastischen Nachwirkung. *Wiener Berichte*, **70**, 275–306.

[4] W. Weber (1835). Über die Elasticität der Seidenfäden. *Poggendorf's Annalen*, **34**, 247–57.

[5] F. Kohlrausch (1864). Über die elastische Nachwirkung bei der Torsion. *Poggendorf's Annalen*, **119**, 337–68.

[6] F. Kohlrausch (1866). Beiträge zur Kenntniss der elastischen Nachwirkung. *Poggendorf's Annalen*, **128**, 1–20, 207–27, 399–419.

[7] O.E. Meyer (1874). Theorie de elastischen Nachwirkung. *Poggendorf's Annalen*, **151**, 108–19.

[8] E. Picard (1907). La mécanique classique et ses approximations successives. *Scientia*, **1**, 4–15.

[9] V. Volterra (1909). Sulle equazioni integro-differenziali della teoria dell'elasticità. *Rendiconti Academie dei Lincei*, Ser. 5^a, **18**, 295–301.

[10] V. Volterra (1909). Equazioni integro-differenziali della elasticità nel caso della isotropia, *Rendiconti Academia dei Lincei*, Ser. 5^a, **18**, 577–86.

[11] M.G. Ianniello and G. Israel (1993). Boltzmann's "Nachwirkung" and hereditary mechanics, in *Proceedings of the International Symposium on Ludwig Boltzmann*, ed. G. Battimelli *et al.*, pp. 113–33, Verlag der Österreichischen Akademie der Wissenschaften, Vienna.

[12] H. Markovitz (1977). Boltzmann and the beginnings of linear viscoelasticity. *Transactions of the Society of Rheology*, **21**, 381–98.

[13] L. Boltzmann (1890). Über die Hertz'schen Versuche. *Wiedemann's Annalen*, **40**, 399–400.

[14] L. Boltzmann (1891–3). *Vorlesungen uber Maxwells Theorie der Elektrizität und des Lichtes*, 2 vols. Barth, Leipzig.

[15] L. Boltzmann (1897). Über irreversible Strahlungsvorgänge. I. *Berliner Berichte* [*Berichte der Deutschen Chemischen Gesellschaft*], 660–2.

[16] J. Stefan (1879). Über die Beziehung zwischen der Wärmestrahlung und der Temperatur. *Wiener Berichte*, **79**, 391–428.

[17] L. Boltzmann (1884). Ableitung des Stefan'schen Gesetzes, betreffend die Abhängigkeit der Wärmestrahlung von der Temperatur aus der electromagnetischen Lichttheorie. *Wiedemann's Annalen*, **22**, 291–4.

[18] H.A. Lorentz (1907). Ludwig Boltzmann. *Verhandlungen der Deutschen Physikalischen Gesellschaft*, **9**, 206–38.

[19] M. Planck (1931). *J.C. Maxwell*. Macmillan, New York.

[20] W. Wien (1893). Eine neue Beziehung der Strahlung schwarzer Körper zum zweiten Haupsatz der Wärmetheorie. *Sitzungsberichte der Akademie der Wissenschaften*. Berlin, 9 Feb., p. 55.

[21] L. Boltzmann (1974). More on atomism, in *Theoretical physics and philosophical problems* [English translation from *Populäre Schriften*, Leipzig (1905)], ed. B. McGuinness. Reidel, Dordrecht.

[22] L. Boltzmann (1868). Über die Integrale linearer Differentialgleichungen mit periodischen Koeffizienten. *Wiener Berichte*, **58**, 54–9.

[23] I.M. Fasol-Boltzmann (Ed.) (1990). *Ludwig Boltzmann: Principien der Naturalfilosofi*. Springer-Verlag, Berlin.

[24] H. Hertz (1894). *Die Prinzipien der Mechanik in neuen Zusammenhange dargestellt*. Barth, Leipzig.

[25] L. Boltzmann (1897–1904). *Vorlesungen über die Principe der Mechanik*, 2 vols. Barth, Leipzig.

[26] L. Boltzmann (1974). *Theoretical physics and philosophical problems*, ed. B. McGuinness. Reidel, Dordrecht.

[27] H. Motz (1982). Did the germ of general relativity come from Boltzmann, in *Ludwig Boltzmann Gesamtausgabe: 8. Ausgewählte Abhandlungen der Internationalen Tagung Wien 1881*, ed. R. Sexl and J. Blackmore, pp. 355–61. Akademische Druck- und Verlagsanstalt, Graz.

[28] S. Wagner (1982). Ludwig Boltzmann and the special theory of relativity, in *Ludwig Boltzmann Gesammtausgabe: 8. Ausgewählte Abhandlungen der Internationalen Tagung Wien 1881*, ed. R. Sexl and J. Blackmore, pp. 341–54. Akademische Druck- und Verlagsanstalt, Graz.

[29] A. Einstein (1905). Zur Elektrodynamik bewegter Körper. *Annalen der Physik*, **17**, 891–921 [English translation in: H.A. Lorentz, A. Einstein, H. Minkowski, and H. Weyl *The principle of relativity*, Methuen, London (1923)].

[30] P. Frank (1979). *Einstein, sein Leben und seine Zeit*. Vieweg, Braunschweig [English translation: P. Frank, *Einstein, his life and time*, transl. G. Rosen, Knopf, New York (1953)].

Chapter 10

[1] L. Boltzmann (1905). *Populäre Schriften*. Barth, Leipzig.

[2] L. Boltzmann (1974). *Theoretical physics and philosophical problems* [partial English translation of *Populäre Schriften*, Leipzig (1905)], ed. B. McGuinness. Reidel, Dordrecht.

[3] I. M. Fasol-Boltzmann (Ed.) (1990). *Ludwig Boltzmann: Principien der Naturalfilosofi*. Springer-Verlag, Berlin.

[4] I. Kant (1969). Prolegomena to any future metaphysics, in *Ten great works of philosophy*, ed. R.D. Wolff. London.

[5] E. Mach (1914). *The analysis of sensations*. Open Court, La Salle, IL.

[6] H. Hertz (1894). *Die Prinzipien der Mechanik in neuen Zusammenhange dargestellt*. Barth, Leipzig.

[7] T. Kuhn (1970). *The structure of scientific revolutions*. University of Chicago Press.

[8] T. Kuhn (1977). *The essential tension*. University of Chicago Press.

[9] I.B. Cohen (1977). *Revolution in science*. Belknap Press, Cambridge, MA.

[10] E. Scheibe (1988). The physicists' conception of progress. *Studies in History and Philosophy of Science*, **19**, 141–59.

[11] K. Popper (1974). Intellectual autobiography, in *The philosophy of Karl Popper*, Book I, ed. P.A. Schilpp. Open Court, La Salle, IL.

[12] E. Broda (1973). Philosophical biography of Ludwig Boltzmann, in *The Boltzmann equation: theory and application*, ed. E.G.D. Cohen and W. Thirring, pp. 7–52. Springer-Verlag, Vienna.

[13] A.D. Wilson (1993). Boltzmann's philosophical education and its bearing on his mature scientific epistemology, in *Proceedings of the International Symposium on Ludwig Boltzmann*, ed. G. Battimelli *et al.*, pp. 57–69. Verlag der Österreichischen Akademie der Wissenschaften, Vienna.

[14] R. Zimmermann (1852). *Philosophische Propädeutik*. Vienna.

[15] S.G. Brush (1976). *The kind of motion we call heat*. North-Holland, Amsterdam.

[16] P. Clark (1976). Atomism versus thermodynamics, in *Method and appraisal in the physical sciences*, ed. C. Howson. Cambridge University Press.

[17] M.R. Gardner (1979). Realism and instrumentalism in 19th century atomism. *Philosophy of Science*, **46**, 1–34.

[18] J. Nyhof (1988). Philosophical objections to the kinetic theory. *British Journal for the Philosophy of Science*, **19**, 81–109.

[19] E. Mach (1883). *Die Mechanik in ihrer Entwickelung historisch-kritisch dargestellt*. Barth, Leipzig [English translation: *The science of mechanics*, Open Court, La Salle, IL (1942)].

[20] L. Boltzmann (1895–8). *Vorlesungen über Gastheorie*, 2 vols. Barth, Leipzig [English translation: *Lectures on gas theory*, transl. by S.G. Brush, University of California Press (1964)].

[21] A. Einstein (1949). Autobiographical notes, in *Albert Einstein, philosopher-scientist*, ed. P.A. Schilpp. Open Court, La Salle, IL.

[22] H.W. De Regt (1996). Philosophy and the kinetic theory of gases. *British Journal for the Philosophy of Science*, **47**, 31–62.

[23] Y. Elkana (1974). Boltzmann's scientific research program and its alternatives, in *The interaction between science and philosophy*, ed. Y. Elkana, pp. 243–79. Humanities Press, Atlantic Highlands.

[24] E.N. Hiebert (1978). Boltzmann's conception of theory construction: the promotion of pluralism, provisionalism, and pragmatic realism, in *Proceedings of the 1978 Pisa Conference on the History and Philosophy of Science*, ed. J. Hintikka *et al.*, pp. 175–98. Reidel, Dordrecht.

[25] L. Boltzmann (1895). On certain questions of the theory of gases. *Nature*, **51**, 413–15.

[26] J.T. Blackmore (1982). Boltzmann's concessions to Mach's philosophy of science, in *Ludwig Boltzmann Gesammtausgabe: 8. Ausgewählte Abhandlungen der Internationalen Tagung Wien 1881*, ed. R. Sexl and J. Blackmore, pp.155–90. Akademische Druck- und Verlagsanstalt, Graz.

[27] G.H. Bryan (1894). Prof. Boltzmann and the kinetic theory of gases. *Nature*, **51**, 176.

Chapter 11

[1] J.C. Maxwell (1860). Illustration of the dynamical theory of gases. Part I: On the motions and collisions of perfectly elastic spheres. Part II: On the process of diffusion of two or more kinds of moving particles among one another. Part III: On the collisions of perfectly elastic bodies of any form. *Philosophical Magazine*, Ser. 4, **19**, 48–62, 63–76, 76–80.

[2] R. Clausius (1858). Über die mittlere Länge der Wege, welche bei der Molekularbewegung gasförmiger Körper von den einzelnen Molekülen zurückgelegt werden; nebst einigen anderen Bemerkungen über die mechanische Wärmetheorie. *Poggendorff's Annalen*, **105**, 239–58.

[3] J.C. Maxwell (1867). On the dynamical theory of gases. *Philosophical Transactions of the Royal Society*, **157**, 49–88.

[4] L. Boltzmann (1872). Weitere Studien über das Wärmegleichgewicht unter Gasmolekülen. *Sitzungsberichte der Akademie der Wissenschaften, Wien*, II, **66**, 275–370.

[5] J.C. Maxwell (1866). On the viscosity or internal friction of air and other gases. *Philosophical Transactions of the Royal Society*, **156**, 249–68.

[6] L. Boltzmann (1868). Studien über das Gleichgewicht der lebendingen Kraft zwischen bewegten materiellen Punkten. *Sitzungsberichte der Akademie der Wissenschaften, Wien*, **58**, 517.

[7] L. Boltzmann (1871). Über das Wärmegleichgewicht zwischen mehratomigen Gasmolekülen. *Sitzungsberichte der Akademie der Wissenschaften, Wien*, **63**, 397.

[8] J.C. Maxwell (1873). On the final state of a system of molecules in motion subject to forces of any kind. *Nature*, **8**, 537–8.

[9] J.C. Maxwell (1879). On stresses in rarified gases arising from inequalities of temperature. *Philosophical Transactions of the Royal Society*, **170**, 231–56.

[10] C. Cercignani (1990). *Mathematical methods in kinetic theory*, rev. edn. Plenum Press, New York [1st edn 1969].

[11] C. Cercignani (1988). *The Boltzmann equation and its applications*. Springer-Verlag, New York.

[12] C. Cercignani, R. Illner, and M. Pulvirenti (1994). *The mathematical theory of dilute gases*. Springer-Verlag, New York.

[13] J.C. Maxwell to P.G. Tait, August 1873. [Quoted in a footnote, p. 114, of: C.G. Knott, *Life and scientific work of Peter Guthrie Tait*, Cambridge (1911)].

[14] J.C. Maxwell (1879). On Boltzmann's theorem on the average distribution of energy in a system of material points. *Transactions of the Cambridge Philosophical Society*, **12**, 90–3.

[15] H.K. Kuiken (1996). H.A. Lorentz: sketches of his work on slow viscous flow and some other areas in fluid mechanics and the background against which it arose. *Journal of Engineering Mathematics*, **30**, 1–18.

[16] G.L. de Haas-Lorentz (Ed.) (1957). *H.A. Lorentz: impressions of his life and work*. Amsterdam.

[17] A. Einstein (1954). *Ideas and opinions*. Bonanza Books, New York.

[18] H.A. Lorentz (Committee President) (1926). *Verslag Staatscommissie Zuiderzee 1918–1926*. Algemeene Landsdrukkerij, The Hague.

[19] A. Knox (1900). H.A. Lorentz contributions to kinetic gas theory. *Annals of Science*, **47**, 591–606.

[20] H.A. Lorentz (1891). Zur Molekulartheorie verdünnter Lösungen. *Zeitschrift für Physikalische Chemie*, **7**, 36–54.

[21] L. Boltzmann (1890). Die Hypothese van't Hoffs über den osmotischen Druck vom Standpunkte der kinetischen Gastheorie. *Zeitschrift für Physikalische Chemie*, **6**, 474–80.

[22] L. Boltzmann (1891). Nachtrag zur Betrachtung der Hypothese van't Hoffs vom Stand-
      punkte der kinetischen Gastheorie. *Zeitschrift für Physikalische Chemie*, **7**, 88–90.

[23] H.A. Lorentz (1907). Ludwig Boltzmann. *Verhandlungen der Deutschen Physikalischen
      Gesellschaft*, **9**, 206–38.

[24] W. Ostwald (1924). *L'Energie*, French translation by E. Philippi. Alcan, Paris.

[25] J. Dalton (1808, 1810, 1827). *New system of chemical philosophy*, Vol. 1, Parts 1 and 2,
      Vol. 2. Bickerstaff.

[26] D. Mendeleev (1891). *The principles of chemistry*, Vol. 1, translated from the 5th Russian
      edn by G. Kamenski. Greenaway, London.

[27] L. Boltzmann (1905). *Populäre Schriften*. Barth, Leipzig.

[28] G. Helm (1890). Ueber die analytische Verwendung des Energieprincip in der Mechanik.
      Zeitschrift für Mathematik und Physik, **35**, 307–220.

[29] J.C. Maxwell (1856). On Faraday's lines of force. *Transactions of the Cambridge Philosoph-
      ical Society*, **10**, 27–83.

[30] H. Hertz (1892). *Untersuchungen über die Ausbreitung der elektrischen Kraft*. Barth,
      Leipzig.

[31] Lord Salisbury (1894). Inaugural address at the British Association. *Nature*, **50**, 339–43.

[32] L. Boltzmann (1895–8). *Vorlesungen über Gastheorie*, 2 vols. Barth, Leipzig.

[33] G. Helm (1896). Zur Energetik. *Wiedemann's Annalen*, **57**, 646–59.

[34] L. Boltzmann (1974). In *Theoretical physics and philosophical problems* [English translation
      from *Populäre Schriften*, Leipzig (1905)], ed. B. McGuinness. Reidel, Dordrecht.

[35] S. Brush (1976). *The kind of motion we call heat*. Elsevier, Amsterdam.

[36] M. Planck (1909). Zur Mach's Theorie der physikalischen Erkenntniss. Eine Erwiderung.
      Physikalische Zeitschrift, **11**, 1186–90.

[37] M. Planck (1904). Über die mechanische Bedeutung der Temperatur und der Entropie, in
      Festschrift Ludwig Boltzmann, pp. 113–22. Barth, Leipzig.

[38] M. Planck (1909). Die Einheit des physikalischen Weltbildes. *Physikalische Zeitschrift*, **10**,
      62–75.

[39] M. Planck (1933). *Ursprung und Auswirkung wissenschaftlicher Ideen*. Berlin.

[40] W. Nernst (1893). *Theoretische Chemie*. Enke, Stuttgart [English translation: W. Nerst,
      Theoretical chemistry, London (1911)].

[41] W. Nernst (1922). Zum Gültigkeitsbereich der Naturgesetze. *Die Naturwissenschaften*, **10**,
      489–95.

[42] L. Meitner (1964). Looking back. *Bulletin of the Atomic Scientists*, **20**, (Nov.), 3.

[43] L. Flamm (1956). Zum 50. Todestag von Ludwig Boltzmann. *Physikalische Blätter*, **12**, 408–
      11.

[44] G. Rabel (1920). Mach und die Realität der Aussenwelt. *Physikalische Zeitschrift*, **21**, 433–7.

[45] Anon. (1906). Der Lebenslauf Boltzmanns. *Die Zeit* (Vienna), No. 1420, 7 Sept., p. 2.

[46] S. Brush (1968). Mach and atomism. *Synthèse*, **18**, 207.

[47] J.T. Blackmore (1972). *Ernst Mach: his work, life and influence*. University of California
      Press, Berkeley.

Chapter 12

[1] L. Boltzmann (1895–98). *Vorlesungen über Gastheorie*, 2 vols. Barth, Leipzig.

[2] L. Boltzmann (1896). Entgegnung auf die wärmetheoretischen Betrachtungen des Hrn.
     E. Zermelo. *Annalen der Physik*, **57**, 773–84 [English translation in: S.G. Brush, *Kinetic
     theory*, Vol. 2, *Irreversible processes*, pp. 218–28, Pergamon Press, Oxford (1966)].

[3] G. Cantoni (1868). Su alcune condizioni fisiche dell'affinità, e sul moto browniano. *Rend. R. Ist. Lomb. Science Lett.*, **1**, 56–67.

[4] J.H. Poincaré (1905). The present crisis in physics, in *Congress of Arts and Science, Universal Exposition, St. Louis, 1904*, I, p. 604. Houghton, Mifflin, Boston and New York [Reprinted in *The value of science*, transl. by G.B. Halsted, pp. 96–105, Dover, New York (1958)].

[5] A. Sommerfeld (1917). Zum Andenken an Marian von Smoluchowski. *Physikalische Zeitschrift*, **18**, 533–39.

[6] A. Pais (1982). *'Subtle is the Lord…': the science and the life of Albert Einstein*. Oxford University Press.

[7] A. Einstein (1949). Autobiographical notes. In *Albert Einstein, philosopher-scientist*, ed. P.A. Schilpp. Open Court, La Salle, IL.

[8] A. Beck and P. Havas (Eds.) (1987). *The collected papers of Albert Einstein*. Vol. I, *The early years*, English translation by A. Beck. Princeton University Press.

[9] M. Einstein (1924). *Albert Einstein, Beitrag für sein Lebensbild*. Manuscript, Einstein Archives, Princeton.

[10] C. Seelig (1960). *Albert Einstein*. Europa Verlag, Zurich.

[11] A. Einstein (1920). *On the special and the general relativity theory: a popular exposition*. Methuen, London [translation of *Über die spezielle und allgemeine Relativitätstheorie (gemeinverständlich)*, Vieweg, Braunschweig (1917)].

[12] M. Planck (1897). Über irreversible Strahlungsvorgänge. Erste Mitteilung. *Berliner Berichte* [*Berichte der Deutschen Chemischen Gesellschaft*], 57–68.

[13] M. Planck (1897). Über irreversible Strahlungsvorgänge. Zweite Mitteilung. *Berliner Berichte*, 715–7.

[14] M. Planck (1897). Über irreversible Strahlungsvorgänge. Dritte Mitteilung. *Berliner Berichte*, 1122–45.

[15] M. Planck (1898). Über irreversible Strahlungsvorgänge. Vierte Mitteilung. *Berliner Berichte*, 446–76.

[16] M. Planck (1899). Über irreversible Strahlungsvorgänge. Fünfte Mitteilung. *Berliner Berichte*, 440–80.

[17] M. Planck (1900). Über irreversible Strahlungsvorgänge. *Annalen der Physik*, **1**, 69–122.

[18] T.S. Kuhn (1978). *Black-body theory and the quantum discontinuity, 1894–1912*. Clarendon Press, Oxford.

[19] L. Boltzmann (1897). Über irreversible Strahlungsvorgänge. *Berliner Berichte*, 615–7.

[20] M. Planck (1901). Über das Gesetz der Energieverteilung im Normalspektrum. *Annalen der Physik*, **4**, 553–63.

[21] M. Planck (1920). Die Entstehung und bisherige Entwicklung der Quantentheorie. Nobel-Vortrag, Stockholm.

[22] M. Planck (1900). Über eine Verbesserung der Wien'schen Spektralgleichung. *Verhandlungen der Deutschen Physikalischen Gesellschaft*, **2**, 202–4.

[23] L. Boltzmann (1877). Über die Beziehung zwischen dem zweiten Hauptsatze der mechanischen Wärmetheorie und der Wahrscheinlichkeitsrechnung respektive den Sätzen über das Wärmegleichgewicht. *Wiener Berichte*, **76**, 373–435.

[24] M. Planck (1900). Zur Theorie des Gesetzen der Energieverteilung im Normalspektrum. *Verhandlungen der Deutschen Physikalischen Gesellschaft*, **2**, 237–45.

[25] A. Einstein (1905). Über einen die Erzeugung und Verwandlung des Lichtes betreffenden heuristischen Gesichtpunkt. *Annalen der Physik*, **17**, 132–48.

[26] A. Einstein (1906). Zur Theorie der Lichterzeugung und Lichtabsorption. *Annalen der Physik*, **19**, 289–306.

[27] A. Einstein (1903). Eine Theorie der Grundlagen der Thermodynamik. *Annalen der Physik*, **11**, 170–87.

[28] A. Einstein (1905). Über die von der molekular kinetische Theorie der Wärme geforderte Bewegung von in ruhenden Flüssigkeiten suspendierten Teilchen. *Annalen der Physik*, **17**, 549–60.

[29] A. Einstein (1909). Zur gegenwärtigen Stand des Strahlungsproblems. *Physikalische Zeitschrift*, **10**, 185–93.

[30] A. Einstein (1907). Die Plancksche Theorie der Strahlung und die Theorie der spezifischen Wärme. *Annalen der Physik*, **22**, 180–90.

[31] L. Boltzmann (1877). Über die Natur der Gasmoleküle. *Wiener Berichte*, **74**, 553–60.

[32] A. Avogadro (1833). Mémoire sur les chaleurs spécifiques des corps solides et liquides. *Annales Chimie et de Physique*, **55**, 80–111.

[33] H.F. Weber (1872). Die spezifische Wärme des Kohlenstoffs. *Poggendorf's Annalen*, **147**, 311–9.

[34] H.F. Weber (1875). Die spezifische Wärme der Elemente Kohlenstoff, Bor und Silicium. *Poggendorf's Annalen*, **147**, 367–423.

[35] H.F. Weber (1875). Die spezifische Wärme der Elemente Kohlenstoff, Bor und Silicium. *Poggendorf's Annalen*, **147**, 533–82.

[36] J. Dewar (1872). On the specific heat of carbon at high temperatures. *Philosophical Magazine*, **44**, 461–67.

[37] J. Dewar (1905). Studies with the liquid hydrogen and air-calorimeters. *Proceedings of the Royal Society*, **76**, 325–40.

[38] L. Boltzmann (1871). Analytischer Beweis des zweiten Hauptsatzes der mechanischen Wärmetheorie aus den Sätzen über das Gleichgewicht der lebendingen Kraft. *Sitzungsberichte der Akademie der Wissenschaften, Wien*, **63**, 712–32.

[39] W. Nernst (1911). Untersuchungen über die spezifischen Wärme bei tiefen Temperaturen. *Sitzungsberichte, Preussische Akademie der Wissenschaften*, 306–15.

[40] P. Debye (1912). Zur Theorie der spezifischen Wärmen. *Annalen der Physik*, **39**, 789–839.

[41] M. Born and T. Kármán (1912). Über Schwingungen in Raumgittern. *Physikalische Zeitschrift*, **13**, 297–311.

[42] M. Born and T. Kármán (1913). Zur Theorie der spezifischen Wärmen. *Physikalische Zeitschrift*, **14**, 15–19.

[43] A. Pais (1991). *Niels Bohr's times, in physics, philosophy and polity*. Clarendon Press, Oxford.

[44] D. Hilbert (1916–17). Begründung der kinetischen Gastheorie. *Mathematische Annalen*, **72**, 562–77.

[45] S. Chapman (1916–17). The kinetic theory of simple and composite gases: viscosity, thermal conduction and diffusion. *Proceedings of the Royal Society*, **A93**, 1–20.

[46] D. Enskog (1917). *Kinetische Theorie der Vorgänge in mässig verdünnten Gasen, I. Allgemeiner Teil*. Almqvist & Wiksell, Uppsala.

[47] H.A. Lorentz (1909). *The theory of electrons*. Dover, New York.

[48] T. Carleman (1933). Sur la théorie de l'équation intégro-differentielle de Boltzmann. *Acta Mathematica*, **60**, 91–146.

[49] T. Carleman (1957). *Problèmes mathématiques dans la théorie cinétique des gaz*. Almqvist & Wiksell, Uppsala.

[50] H. Grad (1949). On the kinetic theory of rarified gases. *Communications on Pure and Applied Mathematics*, **2**, 331–407.

[51] C. Truesdell (1956). On the pressures and the flux of energy in a gas according to Maxwell's kinetic theory, II. *Journal of Rational Mechanics and Analysis*, **5**, 55–128.

[52] V.S. Galkin On a solution of a kinetic equation. *Prikladnaya Matematika i Mekhanika*, **20**, 445–6.

[53] V.S. Galkin (1958). On a class of solutions of Grad's moment equations. *PMM Journal of Applied Mathematics and Mechanics*, **22**, 532–6.

[54] D. Morgenstern (1954). General existence and uniqueness proof for spatially homogeneous solutions of the Maxwell–Boltzmann equation in the case of Maxwellian molecules. *Proceedings of the National Academy of Sciences (U.S.A.)*, **40**, 719–21.

[55] L. Arkeryd (1972). On the Boltzmann equation. Part I: Existence. *Archives for Rational Mechanical Analysis*, **45**, 1–16.

[56] L. Arkeryd (1972). On the Boltzmann equation. Part II: The full initial value problem. *Archives for Rational Mechanical Analysis*, **45**, 17–34.

[57] C. Cercignani (1988). *The Boltzmann equation and its applications*. Springer-Verlag, New York.

[58] C. Cercignani (1990). *Mathematical methods in kinetic theory*, rev. edn. Plenum Press, New York [1st edn 1969].

[59] C. Cercignani, R. Illner, and M. Pulvirenti (1994). *The mathematical theory of dilute gases*. Springer-Verlag, New York.

[60] L. Boltzmann (1905). *Populäre Schriften*. Barth, Leipzig.

[61] L. Boltzmann (1974). *Theoretical physics and philosophical problems* [English translation from *Populäre Schriften*, Leipzig (1905)], ed. B. McGuinness. Reidel, Dordrecht.

[62] B. D'Espagnat (1995). *Veiled reality. An analysis of present-day quantum mechanical concepts*. Addison-Wesley, Reading, MA.

Epilogue

[1] D. Flamm (1983). Ludwig Boltzmann and his influence on science. *Studies in History and Philosophy of Science*, **14**, 255–78.

[2] A. Einstein (1905). Ist die Trägheit eines Körpers von seinem Energieinhalt abhängig?. *Annalen der Physik*, **18**, 639–41.

[3] A. Einstein (1905). Über die von der molekular kinetischen Theorie der Wärme geforderte Bewegung von in ruhenden Flüssigkeiten suspendierten Teilchen. *Annalen der Physik*, **17**, 549–60.

Index